The Economics of Commodity Promotion Programs

PETER LANG
New York • Washington, D.C./Baltimore • Bern
Frankfurt am Main • Berlin • Brussels • Vienna • Oxford

The Economics of Commodity Promotion Programs

Lessons from California

Edited by
Harry M. Kaiser, Julian M. Alston,
John M. Crespi, and Richard J. Sexton

PETER LANG
New York • Washington, D.C./Baltimore • Bern
Frankfurt am Main • Berlin • Brussels • Vienna • Oxford

Library of Congress Cataloging-in-Publication Data

Commodity promotion programs in California:
evaluation and legal issues / edited by Harry M. Kaiser ... [*et al.*].
p. cm.
Includes bibliographical references and index.
1. Agricultural subsidies—California. 2. Produce trade—California.
3. Commercial products—California. I. Kaiser, Harry Mason.
HD1775.C2C66 338.1'8794—dc22 2004027474
ISBN 0-8204-7271-9

Bibliographic information published by Die Deutsche Bibliothek.
Die Deutsche Bibliothek *lists this publication in the* "Deutsche
Nationalbibliografie"; *detailed bibliographic data is available
on the* Internet *at http://dnb.ddb.de/.*

Cover design by Joni Holst

275 Seventh Avenue, 28th Floor, New York, NY 10001
www.peterlangusa.com

Table of Contents

Illustrations

Tables

Preface

Many agricultural commodities in the United States have mandated marketing programs designed to improve producer incomes through the provision of industry collective goods such as generic promotion and advertising, applied research, and quality assurance. Each of these programs is financed by an assessment collected from all producers of the relevant commodity. Participation is mandatory once a program has been voted in by a defined majority of the producers of the commodity.

Mandated marketing programs are economically important, especially advertising and promotion programs—the *Dancing Raisins*; the *Incredible Edible Egg*; *Pork, the Other White Meat*; *Beef, It's What's for Dinner*; and *Got Milk?*, for example—on which the lion's share of assessment revenue has been spent. In recent years, more than a billion dollars has been spent annually on generic promotion programs.

Besides being economically important, mandated marketing programs are also the subject of much dispute and litigation. California is home to the largest number of these programs and much of the controversy and litigation has arisen here. The bulk of the legal challenges have been on the grounds that mandatory programs violate individual producers' free-speech rights under the First Amendment. Some cases have gone as far as the U.S. Supreme Court, and various programs have been ruled unconstitutional. The fate of these programs is currently at a watershed and there is much consternation within the industry as to how the litigation will play out.

This book takes a comprehensive look at the economic consequences and resulting legal implications of mandated commodity marketing programs in California. It presents the history of how these programs evolved in California, a synopsis of the coverage of programs that currently operate in the state, and an analysis of the litigation over mandatory promotion programs and how the programs and litigation arrived at their current state. The bulk of the book draws from a number of economic benefit-cost studies that have been performed on California programs in recent years. While not all commodities are covered, we have selected a wide range of products, types of organization, and types of organizational activities to be representative of California agriculture. The book emphasizes commodity promotion programs but includes several case studies investigating the economics of various nonpromotion activities by some of the mandatory programs.

Our overall aim is to help inform the ongoing debate by bringing together and making sense of the findings from the California case studies. Our goal has been to make the findings accessible to a broad audience of noneconomists and economists alike. The book provides a critique of economists' methodology for examining the benefits and costs of mandatory programs and lays a foundation for improving future economic analyses. We believe the book will be of benefit to anyone interested in the economic consequences of mandatory commodity programs, whether in California or elsewhere. It should also serve as a valuable tool for parties involved in the many court cases that are ongoing in California and other states.

A debt of thanks is owed to a number of individuals and organizations who helped to bring this book to fruition. First, the authors of the individual chapters have done an excellent job and it has been our pleasure to work with them. Special thanks go to Natalie Karst, who was the technical and copy editor of the book. Natalie did a wonderful job of editing the book and did so in a timely fashion under rigid time constraints. We are also grateful to Humei Wang for help in constructing the index to this book. Thanks are extended also to the various academic journals that granted us permission to republish several previous articles as chapters in the book. These journals include *Agribusiness: An International Journal*, *Journal of Agricultural and Resource Economics*, *Review of Agricultural Economics*, and *Journal of Dairy Science*. Finally, this book would not have been possible without the financial support of the Gianinni Foundation of Agricultural Economics and the National Institute of Commodity Promotion and Research Evaluation. We very much appreciate the support of these two organizations.

Julian M. Alston
John M. Crespi
Harry M. Kaiser
Richard J. Sexton

Part I

Setting the Scene

1

Introduction to the Economics of Commodity Promotion Programs

Julian M. Alston, John M. Crespi, Harry M. Kaiser & Richard J. Sexton

Mandated commodity promotion programs are important and controversial. In recent years in the United States, such programs have spent on the order of $1 billion per year in funds raised using commodity "check-offs," a type of tax that is mandatory if approved by a sufficient majority of producers. These assessments have been the subject of much dispute and litigation, and several cases have gone to the U.S. Supreme Court, where a number of the programs have been ruled unconstitutional. Much of both the promotion activities and the legal disputes has taken place in California, where many such programs operate. The results of a number of benefit-cost studies evaluating individual commodity promotion programs in California have been published in various academic journals, monographs, and research reports. As these programs continue to be debated in the courts and elsewhere, this book aims to help inform the debate by bringing together and making sense of those findings and making them accessible to a broad audience of both economists and noneconomists who are interested in the issues.

Overview

Agricultural commodity check-offs are used to finance *collective* goods, such as commodity-specific promotion, market and nutrition research, industry public relations, agricultural research and development, and development, implementation, and enforcement of uniform grade and packaging standards and minimum quality regulations.[1] The check-off programs are made possible by government policy through application of the government's coercive taxing powers to collect the funds, exemption of check-off programs from some types of antitrust regulations, use of government resources to develop and implement the programs, and, in some cases, provision of government funds to support them. Proponents argue that, because of free-rider problems, producers acting individually do not find it profitable to fund these activities in amounts sufficient to maximize net benefits for the industry (or the nation) as a whole, and that hence there is a case for the government to intervene. The programs are voluntary in the sense that support from a defined majority of relevant producers is required to implement them, but participation is mandatory once a program has been authorized.

In California, mandated commodity programs are authorized under both federal and state law, and they include a variety of federal marketing orders and state marketing orders, commissions, and councils. In 2003, the sixty-one active programs in California covered more than 50 percent of California's agricultural production value and spent more than $200 million on commodity promotion, research, and other activities in 2003/04 (see Chapter 2). An additional sixteen national programs are important for some California commodities.

This book compiles existing evidence on the economic impacts of mandatory commodity promotion programs in California. The first part of the book, including the present chapter and the next two, sets the scene for the case studies that make up the bulk of the book. Chapter 2 documents the institutional arrangements for the California programs and Chapter 3 summarizes the history of recent litigation in which various programs have been challenged under the U.S. Constitution. Chapters 4 through 11 report eight case studies that cover domestic or export commodity promotion programs. These cases are presented in chronological order based upon the date when the research findings were first published: table grapes (1997), eggs (1997), prunes or dried plums (1998), avocados (1998), almonds (2000), walnuts (2001), raisins (2003), and strawberries (2004). These case studies represent significant, though not comprehensive, coverage of commodity promotion programs specific to California. Chapter 12 presents a summary of findings from benefit-cost studies of the national dairy, beef, pork, and cotton commodity promotion programs in which California agriculture also participates. Check-off funds are also used to enhance demand for farm products in other ways. Chapters 13 through 16 document the results of four case studies: nutrition education for school children funded by a dairy check-off, public-relations investments to counter a food scare, research to develop new uses for whey products, and quality testing and certification for California pistachios. Chapter 17 summarizes and synthesizes the findings and concludes the book.

As noted, the presentation here is meant to make this work accessible to a wide audience of people with an interest in the operation and consequences of California's mandated commodity marketing and promotion programs. Some of the subject matter is inherently technical, but the chapter authors and the editors have sought to minimize that aspect. Most of the work reported here is based on longer, more detailed research reports or academic publications, and those documents contain detailed technical support for the findings, which are summarized here.

The benefit-cost analysis in each case study reported in this book rests on a common general conception of a commodity market and how that market is affected both by the collection of a check-off and by the use of the funds to enhance demand. This common conception is often expressed in terms of a model of supply and demand, and the remainder of this chapter is devoted to presenting key elements of this basic supply-and-demand model.

A Model for Analyzing the Benefits and Costs of Check-Off-Funded Promotion

A supply-and-demand model of a commodity market is useful for illustrating some key points about the benefits and costs of promotion, the check-off used to finance it, and methods used by economists to measure these benefits and costs. In Figure 1.1, S_0 represents the market supply of a commodity (quantity supplied as a function of price, holding constant other factors that might affect supply) and D_0 represents market demand (quantity demanded as a function of price, holding constant other factors that might affect demand). Supply slopes up, indicating that higher prices result in additional quantities supplied; demand slopes down, indicating that lower prices result in increases in quantities consumed. Market equilibrium is indicated by the intersection of supply and demand at point *b*, and P_0 represents the unique (i.e., equilibrium) price at which the quantity supplied is equal to the quantity demanded, Q_0.

Applying some additional assumptions that are standard in much of applied economics, the same supply and demand curves can be used to measure the total costs and benefits from consumption. Specifically, these assumptions (known as Harberger's three postulates of applied welfare economics) state that (1) the area beneath the demand curve represents total consumer benefits from consuming the commodity, (2) the area beneath the supply curve represents total variable costs of production, and (3) we can add up benefits

Figure 1.1. A Commodity Market Model of Check-Off-Funded Promotion

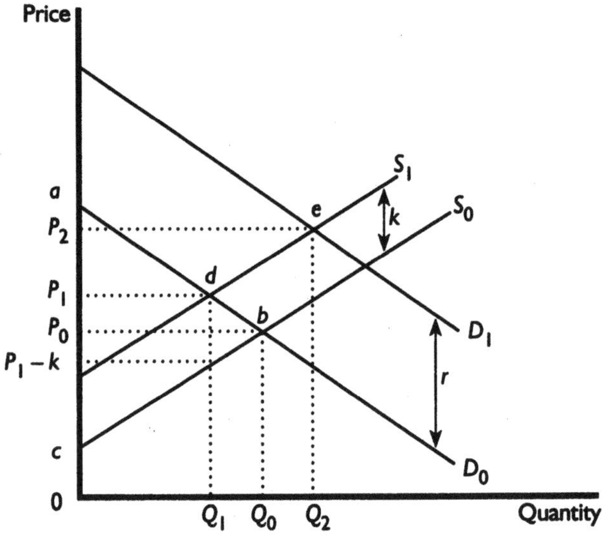

and costs across producers and consumers. Hence, for the initial quantity of Q_0, total benefits from consumption are equal to the trapezoidal area $OabQ_0$ and, given consumer expenditure of P_0Q_0 = area OP_0bQ_0, consumers enjoy a net benefit called "consumer surplus" that is equal to the triangle area abP_0. This area represents the difference between the amount that consumers would be willing to pay for consumption of Q_0 and what they actually pay. Similarly, for the initial quantity of Q_0, total variable costs of production are equal to the trapezoidal area $OcbQ_0$ and, given total revenue of P_0Q_0 = area OaP_0Q_0, producers enjoy a net benefit called "producer surplus" that is equal to the triangle area P_0bc. (Note that fixed costs, if any, in this setting are irrelevant because they are incurred regardless of the price and output in the market.) This corresponds to total profit on the production of Q_0. The total net benefit in this market is equal to the sum of producer surplus and consumer surplus, the area *abc*.

Now suppose a check-off of *k* per unit is collected from producers of the commodity, which we can model as a parallel shift in supply from S_0 to S_1 (i.e., an increase in the supply price by the amount of *k* per unit for any quantity). As a result, the industry equilibrium shifts from point *b* to point *d*, the market price (inclusive of the check-off) increases to P_1, and the quantity demanded falls to Q_1, which is also the quantity supplied at a producer net price of $P_1 - k$.

Figure 1.2. Benefits and Costs of Check-Off-Funded Promotion

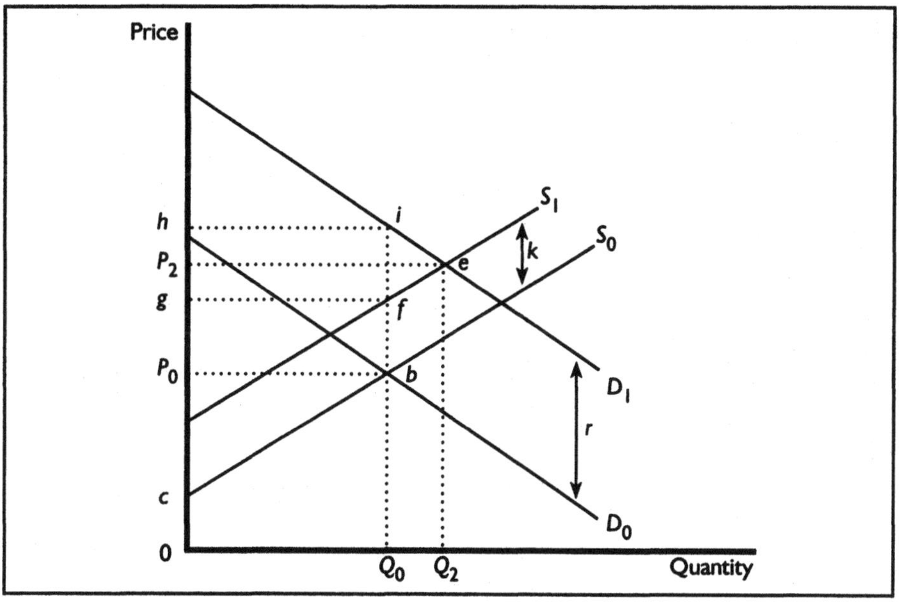

Both producers and consumers are worse off at this point, but this is not the final outcome. The check-off also generates revenue equal to $R = kQ_1$, which is used to finance some activity (promotion, for example) designed to enhance demand. Suppose the expenditure of R results in a shift in demand from D_0 to D_1, reflecting an increase in consumers' willingness-to-pay for the commodity by r per unit for any quantity. As a result, the industry equilibrium shifts from point d to point e, the market price (inclusive of the check-off) increases to P_2, and the quantity demanded increases to Q_2, which is also the quantity supplied at a producer net price of $P_2 - k$. This represents the final equilibrium, reflecting both the promotion-induced increase in willingness-to-pay of r per unit and the check-off of k per unit that financed the promotion.

The implications for consumer, producer, and national welfare can be represented in terms of changes in the respective areas in Figure 1.1, but for clarity only the key elements of Figure 1.1 that are required for the welfare economics analysis are replicated in Figure 1.2. First, the change in producer surplus between the initial equilibrium at point b (P_0, Q_0) and the final equilibrium reflecting the check-off and the induced demand shift, point e (P_2, Q_2), must be measured. This producer net benefit can be measured as the area of additional producer surplus associated with the increase in production from Q_0 to Q_1, measured along the supply curve that includes the check-off, S_1–i.e., the trapezoidal area P_2efg (it also could be measured as the change in producer surplus associated with the same increase in production along S_0).

All benefit-cost studies of returns to generic commodity promotion take some measure of producer benefits that corresponds to this producer surplus measure. Many commodity promotion studies measure only producer benefits because the legislation authorizing the programs often specifies that the program objective is to enhance producer welfare. However, some studies also seek to measure benefits to consumers and the nation. These broader analyses have been hampered by a lack of definitive results in the literature on the appropriate way to measure consumer welfare impacts. Some studies have assumed that the consumer benefits from promotion, as from processing research, can be reasonably approximated using changes in the standard measure of consumer surplus (e.g., see Alston, Freebairn, and James 2003). Using this assumption, we can measure consumer net benefits as the area of additional consumer surplus associated with the increase in consumption from Q_0 to Q_1, measured along the final demand curve, D_1–i.e., the trapezoidal area $hieP_2$ (it also could be measured, equivalently, as the change in consumer surplus associated with the same increase in consumption along D_0). Finally, combining the producer and consumer benefits yields an estimate of net national benefits from check-off-funded promotion programs compared to an absence of those programs, which is equal to the area *efghi* in Figure 1.2. These measures of producer, consumer, and national benefits are all net measures, reflecting the costs of the check-off, the benefits from the promotion-induced demand increase, and the distribution of those costs and benefits between producers and consumers.

Benefit-cost ratios can be computed for producers, consumers, and the nation as a whole. In practice, studies of the returns to check-off-funded promotion programs generally have not measured consumer benefits and thus have not measured national benefits, partly because of uncertainty about the appropriate measure of consumer (and thus national) benefits but also because the authors were interested specifically in benefits to producers. When studies report benefit-cost ratios, generally they refer to the producers' gross benefits from promotion (the gross gain in producer surplus associated with the promotion) divided by the producer incidence of the check-off (the loss of producer surplus associated with the introduction of the check-off, corresponding to area P_1deP_2 in Figure 1.1). Occasionally, however, studies report a benefit-cost ratio computed as producer gross benefits from promotion divided by total expenditures on promotion rather than by the share of that total cost ultimately borne by producers.

The comparison of total benefits and total costs just described yields a measure of an average benefit-cost ratio: comparing the costs and benefits from having the program with a hypothetical alternative of no program. Some studies of benefits from check-off-funded programs compute a marginal benefit-cost ratio. They compare the costs and benefits associated with a small hypothetical change in the size of the program. The average benefit-cost measure indicates whether a program was profitable, while the marginal benefit-cost measure indicates whether it would have been profitable to increase the size of the program (marginal benefit-cost ratio greater than 1) or reduce it (marginal benefit-cost ratio less than 1). Chapter 4, for example, derives both average and marginal benefit-cost ratios, while Chapters 6 and 8 report only marginal ratios.

A further dimension for variation among studies is the dynamic structure. In the model in Figures 1.1 and 1.2, the effects of the check-off and the promotion all occur within a single period and a comparative-static analysis (comparing two market equilibrium outcomes, with and without both the check-off and promotion) is appropriate. In some cases, especially in studies using short time periods for their units of analysis (for instance, monthly rather than annual), the effects of a check-off or promotion in one period may be felt over subsequent periods, so it is necessary to model the dynamics of response and to aggregate the measures of benefits and costs over time. The procedures for this aggregation are well established, but getting the dynamics right is not always simple and the models may be difficult to interpret. In most cases of commodity promotion, the dynamic elements of demand response are a minor part of the story. In some cases, however, the dynamic elements are important enough to warrant a dynamic analysis–in, for instance, the study of the pistachio marketing order discussed in Chapter 16.

In some settings, actual measures have to take into account real-world market characteristics that complicate the models and the analysis. For instance, in some cases international trade alters a market's structure, and it may be important to separate domestic and foreign elements of demand (or supply), not only to represent the consequences of the check-off and promotion for market prices and quantities but also to isolate impacts on domestic producers and consumers. In other cases, it may be appropriate to consider implications not only for the commodity under study but also for closely related commodities. In addition, some markets are influenced by government policies that significantly affect analysis of the impacts of check-off-funded promotion. And in some cases models can be designed to take into account the distortionary effects of environmental externalities or agribusiness firms exercising market power. These complications notwithstanding, the essential story and conceptual apparatus remain as outlined in Figures 1.1 and 1.2.

Empirical implementation of this apparatus to quantify benefits and costs rests on obtaining reasonable measures of the responsiveness of supply and demand to prices (the slopes of the curves in Figures 1.1 and 1.2) and demand response to promotion (the size of the demand shift, r in Figures 1.1 and 1.2, for a given expenditure on promotion, R).[2] Most of the studies of check-off-funded programs have used econometric techniques to estimate the key parameters representing demand response to both price and check-off-funded activities, but some have used other approaches to estimate (or calibrate) at least some of the parameters of supply and demand. The econometric approach is only feasible in an *ex post* analysis in which the consequences of past policies can be observed and measured. In an *ex ante* analysis that measures the future benefits and costs of a policy, noneconometric approaches are used. Most of the case studies in this book are *ex post* and use econometric methods to estimate at least some of the parameters, but two of the case studies are *ex ante* (Chapters 15 and 16). The precision of estimates of benefits and costs depends on both the accuracy of the structure of the model and the precision of the estimates of the parameters, and many of the studies examine the sensitivity of their findings to these elements.

Notes

1 A *collective good* is something for which the benefits are enjoyed by all of the members of the collective (for instance, all of the producers of a particular commodity) even if they do not pay for the good individually. Collective goods have the characteristics of what economists call *public goods*, except that the group of primary beneficiaries is more narrowly defined.

2 These measures of responsiveness of supply and demand to price and promotion are often expressed in terms of *elasticities*. The elasticity of demand (or supply) with respect to price measures the percentage change in the quantity demanded (or supplied) in response to a 1 percent increase in price, holding constant all other factors that might influence demand (or supply). The elasticity of demand with respect to promotion measures the percentage change in the quantity demanded in response to a 1 percent increase in expenditure on promotion.

References

Alston, J.M., J.W. Freebairn, and J.S. James. "Distributional Issues in Check-off Funded Programs." *Agribusiness: An International Journal* 19(3) (2003):277–288.

Harberger, A.C. "Three Basic Postulates for Applied Welfare Economics: An Interpretive Essay." *Journal of Economic Literature* 9(3) (1971):785–797.

2

California's Mandated Commodity Programs

Hoy F. Carman & Julian M. Alston

California agricultural producers have a long tradition of using group action to solve marketing problems. Their initial efforts in the early 1900s focused on the organization of marketing cooperatives. These early cooperatives initially enjoyed some short-run success with voluntary quality control programs and supply-management efforts, but free-rider problems doomed most of their attempts to improve product prices. However, these organized commodity groups were well positioned to embrace both state and federal marketing order programs when marketing order legislation was enacted in the 1930s.

As government-sanctioned marketing programs evolved, California producers and handlers approved a combination of state and federal marketing orders, commodity commissions, and councils. The programs, which operate under broad enabling legislation and individual statutes, covered almost 55 percent of California's agricultural commodities in 2002/03 based on value (Table 2.1). The dairy industry typically ranks number one in value of production among all California commodities and is also number one in terms of marketing program expenditures. Collectively, marketing orders affecting the dairy, beef, and sheep industries apply to almost 79 percent of commercial animal products in California. Marketing programs are also significant for California fruit and nut crops; programs for twenty-one separate commodities cover more than 73 percent of total fruit and nut production and programs for thirteen different vegetable crops account for 43 percent of total vegetable production.

This chapter describes the state marketing programs and federal marketing orders that currently affect California commodities with emphasis on state programs. National research and promotion programs, some of which are important in California, are not included in the discussion.[1] This chapter also

Table 2.1. Production Value for California Commodities Covered by Marketing Programs as a Share of All Commodities in each Crop Category, 2002

Crop Category[1]	Total Value of Production ($1,000)	Value of Production of Commodities under Marketing Programs ($1,000)	Percent of Total Value Covered by Marketing Programs
Field Crops	3,827,859	795,094	20.8
Fruits and Nuts	9,705,335	7,139,711	73.5
Vegetables	6,701,580	2,888,087	43.1
Animal Products	7,090,660	5,586,287	78.8
Nursery and Floral	3,310,099	365,945	11.1
Total	**30,635,533**	**16,775,124**	**54.8**

[1] Fishery and Forestry are excluded.
Source: Data on value of production are from California Agricultural Statistics Service, *Summary of County Agricultural Commissioners' Reports*, 2002.

reviews legislative history, summarizes the current status of marketing programs and the economic issues that surround them, and analyzes their budget allocations, emphasizing research and market promotion activities.

Legislative Background

Federal marketing orders, authorized in the Agricultural Adjustment Acts of 1933 and 1935 (AAA), were developed as part of government efforts to increase and stabilize farm incomes during the 1930s. After the U.S. Supreme Court invalidated part of the AAA in 1935 and 1936, portions of the AAA related to marketing orders and agreements were revised and reenacted as the Agricultural Marketing Agreement Act of 1937 (AMAA). California enacted similar legislation, the California Marketing Act, in 1937 as well. These acts as subsequently amended serve as the enabling legislation for federal and state marketing orders and agreements.[2]

Unlike state orders, federal marketing orders can cover a production region across more than one state. Federal marketing orders tend to focus on quality regulations and sometimes on volume controls; state marketing programs tend to focus more on research and promotion.[3] Existing federal marketing orders apply to milk and specific groups of fruits, vegetables, and other specialty crops, while state marketing programs apply to many other commodities. The legislative evolution of marketing orders, especially of the federal programs, is described in more detail in Chapter 3.

Three types of state marketing programs are in use in California: marketing orders, commissions, and councils. Marketing orders are authorized by the California Marketing Act of 1937. Individual commissions and councils are authorized by specific individual laws. Some of the earliest councils to be established include the Dairy Council of California, organized in 1945, and the Beef Council, formed in 1967. The earliest commission, the California Table Grape Commission, dates back to 1968.

While the stated purpose of the various state marketing programs is to develop more efficient and equitable marketing and aid producers in maintaining their purchasing power, each marketing program operates somewhat differently based on its authorizing legislation.

Federal and state marketing programs are industry-initiated and are sponsored by producers or handlers of a commodity or by both producers and handlers. Marketing programs can institute production and marketing research and development projects, implement advertising and promotion programs, and specify quality standards or standardized packs or containers. Federal programs also may operate to smooth the flow of a commodity to market or permit only a certain portion of the crop to move into specified outlets. Not all programs engage in all of these activities.

Program Establishment

The statutes that authorize each marketing program lay out the rules for program establishment and funding and the activities allowed. While federal and California programs generally have similar provisions, they differ in their details. With this chapter's emphasis on California programs, we first discuss the provisions of the state programs.

California Marketing Programs

Procedures for implementing a California marketing program differ based on the type of program an industry desires. Marketing orders are established under the general enabling authority of the California Marketing Act of 1937 and therefore do not require new legislation. Commissions and councils, on the other hand, are commodity-specific and creation of each one requires special legislation.

Marketing Orders. Marketing orders are initiated by the industry (as are commissions and councils). Since marketing orders are formed under the existing enabling law, the California Marketing Act, and therefore require no new legislation, industry leaders or representatives approach the California Department of Food and Agriculture (CDFA) to request a marketing order. If CDFA agrees that a marketing order could contribute to solving marketing problems facing the industry, CDFA staff members, working with industry representatives, prepare a preliminary draft of the marketing order and the composition of an industry committee. Once the draft order is prepared, a public hearing is held on the proposed provisions to determine whether the order is in the industry's and the public's interest. The last and most important legal requirement before instituting a marketing order is obtaining approval of the proposal by an industry vote.

Approval by the industry requires at least 40 percent of affected producers/handlers to vote. Once that participation threshold is reached, results must show support from at least 65 percent of all producers/handlers by number who produced and/or handled more than 50 percent of the total volume of the commodity produced for a market in a given period, or more than 50 percent of all producers/handlers by number who produced/handled at least 65 percent of total volume produced for market. Lists of producers for voting are usually developed from information submitted by handlers as required by CDFA.

All but minor amendments to a marketing order must be developed in the same manner as the original order. Amendments must be discussed at a public hearing and submitted for an industry vote before they can become effective. All marketing orders are required to submit to a periodic continuation process at least once every five years. Some call for periodic industry votes. For the others, a public hearing must be held a minimum of every five years to ascertain the industry's opinion of the effectiveness of the order and whether it should continue. A public hearing also can be called via a petition signed by at least 25 percent by number and volume of producers/handlers affected. In addition, a marketing order can be terminated by an industry petition by at least 51 percent by number and volume of affected producers/handlers. If a marketing order is inactive for three consecutive seasons, CDFA can terminate the order without a public hearing or vote.

Commissions and Councils. If an industry wishes to establish a state commission or council, it must first get legislative approval since there is no general enabling law for such programs. For either form of organization, industry representatives work with legislators to develop a commission or council law as an assembly or senate bill. After the bill is passed by the state legislature, the commission or council law becomes effective with the governor's signature. Establishing these programs typically requires an industry referendum, and the requirements for the referendum are usually the same as those for implementing a marketing order. Existing commission and council laws can be amended only by new legislation, but they can be terminated by an industry referendum.

All commission programs hold periodic referenda or public hearings to consider their continuation or termination. With a two-thirds vote of commission members, the commission can recommend to the secretary of CDFA that operations be suspended. Producers or handlers may also petition the secretary to suspend operation of a commission.

Each council's provisions for continuation and termination are different. For example, the Beef Council continues to operate unless a producer petition to CDFA triggers a referendum. The Dairy Council has a more conventional continuation procedure–CDFA holds a public hearing every five years to decide whether the council should remain in effect.

Finance. All California marketing programs are self-financed. Each producer/handler directly affected pays an assessment levied on each unit of the commodity marketed to fund expenses incurred in formulating, issuing, operating, and enforcing the program (including state government administrative expenditures).

The California Marketing Act provides for maximum rates of assessment that can be collected under a marketing order. The maximum amount for administrative purposes (rent, insurance, staff salaries) is 2.5 percent of the gross dollar value of sales by producers to handlers. For orders that involve advertising and promotion, an assessment should not exceed 4 percent of the gross dollar value of sales. A maximum of 6.5 percent can be levied on each producer and handler if both are directly affected by the marketing order.

Assessment rates for marketing programs are established for each marketing year. To facilitate collection, handlers and processors are usually required to pay the assessments on behalf of the producers from whom they receive product. Handlers and processors may in turn deduct such assessment payments from any money owed to the producers. For example, the marketing order for dried plums established a rate of assessment from producers of $9.33 per ton delivered to the processor and a further assessment from processors of $4.67 per ton of dried plums accepted for processing during the 2003/04 crop season. For convenience in making collections, each processor was required to remit the combined assessment of $14 per ton on behalf of both producers and the processor to the Dried Plum Advisory Board and to deduct the $9.33 per ton from payments owed to the producers. While the marketing group usually receives the assessments directly, the marketing branch of CDFA handles collection of any delinquent assessments.

Marketing orders and agreements are subject to more government oversight of activities and expenditures while commissions and councils have more program and budget autonomy. The board for each marketing order must obtain approval of its annual budget from CDFA and return quarterly statements of income and expenditures. Commissions are subject to less government oversight. They can develop their own operating plans and budgets with CDFA concurrence and can hire executives and elect commission members without CDFA's prior approval.

Federal Marketing Orders

The AMAA (as amended) provides the enabling legislation for federal marketing orders. A federal marketing order for a commodity is generally initiated by a group of producers who present a formal request for an order to the U.S. Secretary of Agriculture. The Secretary of Agriculture, upon determining that the proposed marketing order meets legislative criteria, schedules a public hearing at which all interested parties can comment on the proposal. The secretary has the authority to determine, based on the outcome of the hearing, whether the proposal serves the public interest. If the proposed marketing order is approved by the secretary, a producer referendum is held on the final version. Passage requires that two-thirds of the producers (three-fourths for California citrus fruit) or producers representing two-thirds of the volume produced in the proposed marketing order area vote to adopt the order. If an order is to be issued with a marketing agreement, handlers who have handled not less than 50 percent of the total volume of the commodity covered by the order must sign the marketing agreement (Neff and Plato 1995).

Like state programs, federal marketing orders are also entirely self-financed. Once a federal marketing order is approved, it is binding on all producers and handlers in the designated production area and all producers/handlers pay mandatory assessments that cover all of the costs of activities authorized by the provisions included in the order.

The Secretary of Agriculture can suspend or terminate an order if it obstructs or does not support the declared policy of the enabling legislation.

An order must be terminated if at least half of the producers or producers representing at least half of the production vote to end it.

Marketing Program Provisions

The enabling legislation for both federal (AMAA) and state (California Marketing Act) marketing order programs spells out the activities that are authorized. California's commission and council laws also include provisions authorizing similar activities. Federal and state marketing programs can include provisions for one or more of the following types of activities: (1) generic advertising and sales promotion; (2) production, processing, and marketing research; (3) quality regulations with inspection; (4) supply management or volume control; (5) standardization of containers or packs; and (6) prohibition of unfair trade practices. Quality standards and supply management are regulatory in nature while advertising and research are nonregulatory, market-support activities. Prohibition of unfair trade practices has been used very infrequently.

Historically, the provisions typically authorized by federal and state programs have varied. In general, federal programs have tended to concentrate more on volume controls and on quality controls in the form of grades and standards. In recent years, state programs have tended to emphasize advertising, promotion, and research activities. Note, however, that there have been federal programs with provisions for advertising, promotion, and research and California programs that included strong controls on the volume of fruit that could be marketed.

Advertising and Promotion

Advertising and promotion programs are designed to increase demand, sales, and prices for relatively homogeneous agricultural commodities. Advertising and promotion include consumer education, public relations, and point-of-purchase programs for retailers. Television, radio, newspapers, mass circulation magazines, and cooking demonstrations are media used for these activities. From the standpoint of expenditures, advertising and promotion are the most important marketing activities under California programs.

Only generic advertising and promotion are permitted by most California programs.[4] A successful promotion usually requires the commodity group to account for a high proportion of the supply of the commodity. Otherwise, producers in other states can "free ride" on the California program. California producers are particularly active in promotion since California is the major or only U.S. producer of such specialty crops as almonds, avocados, table grapes, raisins, dates, prunes, figs, pistachios, walnuts, kiwifruit, olives, and artichokes. Because California producers account for most of the supply of these crops, they are in a position to capture most of the producer benefits from advertising designed to increase demand for them (Carman, Green, and Mandour 1992).

In addition to promotion financed by assessments, supplemental federal funding is available for these commodity groups. The U.S. Department of Agriculture (USDA) grants promotion funds through its Market Access Promotion Program (MAPP, formerly the Targeted Export Assistance program). These funds are specifically targeted for promotion of U.S. farm products in overseas markets.

Research

Research provisions permit the industry to use program funds to conduct farm-related and off-farm (postharvest) research and development projects designed to increase net returns. Projects that lead to increased yields, improved quality, reduced use of inputs, more efficient harvesting, improved transportation and distribution methods, and increased consumer demand contribute to increased net returns. Farm-related research may include, for example, control or eradication of insects and diseases and development of new varieties resistant to a certain disease or with better handling properties.[5] Off-farm research includes efforts such as developing a package that extends shelf life of a product, developing new shipping containers, improving storage technology, and evaluating potential marketing opportunities. Approved projects typically are carried out by University of California researchers.

Quality Standards

Marketing orders often include minimum quality standards and provide mandatory inspection to enforce the standards. Minimum quality standards can require a commodity to meet minimum requirements for size, grade, sugar content, and/or maturity before it can be sold to consumers. Regulations on quality standards are intended to maintain or enhance market demand by keeping inferior products off the market. These standards are also designed to prevent the market failure associated with early season sales of immature fruit that can look good but taste sour.

Volume Control

Supply-management provisions authorize commodity groups to legally regulate the supply of agricultural products marketed as a tool for orderly marketing. Volume controls are generally implemented under federal orders and have not been used by a state marketing order since the 1970s (CDFA 1985). Various volume control methods have been used to divert product to markets with more elastic demand while limiting shipments to markets with the least elastic demand. Reserve pools are used to limit supply to a particular end use or to manage supply within and across crop years (Alston et al. 1995). Flow-to-market restrictions regulate the amount of product shipped to a given market during a given time period. The lemon prorate, for example, limited the quantity of California/Arizona lemons shipped to the domestic fresh market each week during the marketing year (Carman and Pick 1990).

Marketing order volume controls have been controversial because of their potential to increase short-run prices through exercise of monopoly pricing power. There is evidence that use of reserve pools and flow-to-market provisions in marketing orders has enhanced prices through price discrimination. As a result, USDA has discouraged the use of supply control provisions as a price enhancement tool in existing marketing orders and has not approved new programs with supply control provisions. Readers interested in the effect of marketing orders on prices and other market conditions are referred to the review by Sexton and Lavoie (2001).

Standardization of Containers or Packs

Standardization requirements fix the size, capacity, weight, dimensions, or packing of containers that handlers must use in the packaging, sale, shipment, or handling of produce. Standardization of containers and packages can lower marketing costs by eliminating deceptive pricing or lowering handling and transportation costs. However, as with the case of quality and supply controls, these standards have not been without controversy.[6]

Current California Marketing Programs

California currently has sixty-two active marketing programs, including twelve federal marketing orders, twenty-seven state marketing orders and agreements, twenty commissions, and three councils (Table 2.2). The importance of these marketing programs in the state's agricultural industry is indicated by the fact that the sixty-two marketing programs covered almost 55 percent of the value of California's 2002 agricultural production (Table 2.1). Marketing program coverage varies by commodity group. In terms of 2002 value of production, more than 78 percent of animal products, 73 percent of fruit and nut crops, and 43 percent of vegetable crops were covered by state or federal marketing programs. The next section examines how state and federal marketing programs have changed over the years.

California marketing programs continue to change and evolve. While a majority of the programs have been effective for decades, new programs have been added and many past programs have been either terminated or consolidated with others. The number of state marketing programs has been increasing since the early 1980s while the number of federal marketing orders applicable to California commodities has been declining.

During the 1950s and 1960s, there were four separate marketing programs for peaches. Now there is a single state marketing order for cling peaches and a federal marketing order for fresh peaches. There also were individual marketing orders for four types of pears: fresh Bartletts, fresh Bartlett promotion, fresh fall and winter pears, and canning Hardy pears. These programs were consolidated and now only one state marketing order for pears remains. There were four marketing orders for citrus in the early 1990s. Litigation concerning handler violations of prorate provisions of federal marketing orders for California/Arizona citrus (navel oranges, Valencia oranges, and lemons) led the U.S. Secretary of Agriculture to terminate all three orders in August 1994. Now there is one state marketing order for citrus research.

Current California marketing programs with provisions authorized and the year the programs became effective are shown in Table 2.2. There are currently twelve federal marketing orders covering California commodities. All twelve include grade and size provisions and eleven have research provisions, though the research provisions are active for only eight. Only two of the federal orders include active quantity controls (almonds and raisins) and five involve active advertising and promotion programs (Table 2.2). There are twenty-five state marketing orders included in Table 2.2. Of those, twenty-one include active provisions for research and fifteen have active provisions for advertising and promotion. All but one (grape rootstock) of the state commodity commissions and all three of the councils have provisions for advertising and promotion. Similarly, seventeen of the twenty commissions and all three councils have research provisions. Note that seven of the programs authorize only research activities (alfalfa seed, citrus, iceberg lettuce, melons, potatoes, rice, and grape rootstock). Both marketing agreements listed in Table 2.2 include

Table 2.2. Authorized Activities for California Commodity Marketing Programs

	Quality Standards and Inspection	Quantity Controls	Advertising and Promotion	Research	Year Effective
			A = Active I = Inactive		
Federal Marketing Orders					
Almonds	A	A	I	A	1950
Dates	A	I	A	A	1955
Grapes, California Desert	A	I	–	A	1980
Kiwifruit	A	–	–	I	1984
Nectarines	A	–	A	A	1958
Olives	A	–	A	A	1965
Peaches, Fresh	A	–	A	A	1939
Pistachios	A	–	–	–	2004
Potatoes, Oregon/ California	A	–	–	I	1942
Prunes, Dried	A	I	–	I	1949
Raisins	A	A	A	A	1949
Walnuts	A	I	–	A	1948
State Marketing Orders					
Alfalfa Seed Production	–	–	–	A	1973
Artichoke Promotion	–	–	A	I	1960
Dry Beans	I	–	A	A	1979
Cantaloupes	A	–	A	A	1988
Carrots, Fresh	–	–	A	A	1992
Celery	–	–	–	A	1976
Cherries	–	–	A	A	1993
Citrus Research	–	–	–	A	1968
Figs, Dried	A	I	A	A	1944
Garlic and Onion Dehydrators	A	–	–	–	1999
Lettuce Research	–	–	–	A	1973
Melon Research	–	–	–	A	1972
Manufacturing Milk	–	–	A	A	1970

continued on following page

Table 2.2 (continued)

	Quality Standards and Inspection	Quantity Controls	Advertising and Promotion	Research	Year Effective
State Marketing Orders (continued)					
Market Milk	–	–	A	A	1969
Milk, Fluid	–	–	A	–	1969
Peach, Cling	I	–	A	A	1984
Pear	A	–	A	A	1992
Plums	A	–	A	A	1994
Plums, Dried	–	–	A	A	1947
Potato Research	–	–	–	A	1974
Raisins	–	–	A	A	1998
Rice Research	–	–	–	A	1969
Strawberries, Processing	A	–	I	A	1960
Tomatoes, Processing	A	–	–	–	1986
Wild Rice	–	I	A	A	1968
State Commodity Commissions					
Apple Commission	I	–	A	A	1994
Asparagus Commision	–	–	A	A	1990
Avocado Commission	I	–	A	A	1978
Date Commission	–	–	A	–	1996
Cut Flower Commission	–	–	A	A	1990
Forest Products Commission	–	–	A	A	1991
Grape Rootstock Commission	–	–	–	A	1993
Kiwifruit Commission	–	–	A	A	1980
Pepper Commission	–	–	–	A	1988
Pistachio Commission	–	–	A	A	1981
Rice Commission	–	–	A	–	1999
Sea Urchin Commission	A	–	A	A	2004
Sheep Commission	–	–	A	A	1999

continued on following page

Table 2.2 (continued)

	Quality Standards and Inspection	Quantity Controls	Advertising and Promotion	Research	Year Effective
State Commodity Commissions (continued)					
Strawberry Commission, Fresh	-	-	A	A	1955
Table Grape Commission	-	-	A	A	1968
Tomato Commission	-	-	A	A	1996
Walnut Commission	-	-	A	-	1986
Wheat Commission	-	-	A	A	1983
Lake County Winegrape Commission	-	-	A	A	1992
Lodi-Woodbridge Winegrape Commission	-	-	A	A	1991
Councils					
Beef Council	-	-	A	A	1967
Dairy Council	-	-	A	A	1945
Salmon Council	-	-	A	A	1991
Marketing Agreements					
Buy California Marketing Agreement	-	-	A	A	2001
Pistachio Agreement	-	A	-	-	1994

provisions for advertising and promotion and the *Buy California* program also funds research.

The oldest currently effective marketing program for California commodities is the federal marketing order for fresh peaches, which was established in 1939. The federal marketing orders tend to be older than the state programs. There are thirteen programs that were established prior to 1960, including eight of the twelve federal marketing orders. The four most recent federal marketing orders, established after 1960, are for olives (1965), desert grapes (1980), kiwifruit (1984), and pistachios (2004). Of the total, thirty California marketing programs were established prior to 1980 and thirty-two were established after 1980. In terms of number of California marketing programs established by decade, one current marketing order was established in the 1930s, seven were established between 1940 and 1949, five were established between 1950 and 1959, nine were established between 1960 and 1969, eight were established between 1970 and 1979, eleven were established between 1980 and 1989, eighteen were established between 1990 and 1999, the Buy California Marketing Agreement was established in 2001, and the federal marketing order for pistachios and the Sea Urchin Commission were established in 2004 . Note that eleven of the eighteen new programs in the 1990s were California commodity commissions.

While the overall number of California marketing programs has increased over time, the distribution of programs by type has changed. The number of federal marketing orders applicable to California commodities has decreased, but this has been more than offset by development of California marketing programs, especially by those established as commissions. Garoyan and Youde (1975) listed sixteen federal marketing orders applicable to California commodities in 1974. That number decreased to thirteen in 1995 (Lee et al. 1996), fell to eleven in 2003, and then increased to twelve in 2004. There were thirty-seven California marketing programs effective in 1975 (Garoyan and Youde 1975), forty-eight in 1995 (Lee et al. 1996), and fifty in 2004.

California fruit industries have been actively involved in marketing programs since enactment of the AMAA and California legislation in 1937. For instance, of the twelve marketing programs that were established prior to 1950 and continued to exist for at least two decades (since then, some were terminated), seven were for fruit.[7] Tree crops continue to have strong representation in marketing programs with twenty currently effective programs. In addition, nine programs are currently effective for vine crops (table grapes, wine grapes, raisins, and kiwifruit) and there are separate programs for fresh and processing strawberries.

The number of commodities covered under marketing programs has expanded and become more diverse over time. Vegetables have emerged as an important category of marketing programs. Only two vegetable crops (lettuce and Oregon/California potatoes) were included in the twelve commodity marketing programs established prior to 1950, but over time, programs for vegetables have been added steadily. In 1971, there were six programs covering five different vegetables (artichokes, asparagus, Brussels sprouts, lettuce, and potatoes). That number increased to seven programs by the early 1980s, covering artichokes, celery, iceberg lettuce, melons, potatoes, and fresh tomatoes. The asparagus program was terminated during the 1970s and the asparagus commission was initiated in 1990. In the early 1980s, the marketing order for Brussels sprouts was terminated and the marketing order for lettuce was split into two programs by activity–research and promotion.[8] Currently, there are thirteen programs for vegetables covering artichokes, asparagus, cantaloupes, carrots, celery, garlic and onions, lettuce, melons, potatoes (two), fresh tomatoes, processing tomatoes, and peppers.

The evolution of California marketing programs also extends to the types of programs chosen. The popularity of federal marketing orders seems to have declined over time, in part because of the greater amount of time required to implement and amend federal programs relative to state programs. Commodity commissions have gained in popularity because of the greater autonomy provided under commission laws than under the California Marketing Act. A number of the commodities that now have state commissions had previous experience with federal marketing orders. The kiwifruit industry has a commission for advertising and promotion and a federal marketing order for grade and size standards; the date industry conducts advertising and promotion under a federal marketing order and uses the state commission mainly for research; the walnut industry has a federal marketing order for grade and size standards and research plus a state commission for advertising and promotion. Several commodities operating under commission laws once had state marketing orders, including apples, avocados, strawberries, and tomatoes.

California Marketing Program Expenditures

Budgeted expenditures for California marketing programs, estimated at a little more than $208 million for 2003/04, have increased significantly over time. Lee et al., for example, estimated total budgets of $71.35 million in 1985, $124.82 million in 1990, and $112.94 million in 1992. A number of factors have contributed to the observed increase. First, California marketing programs expanded progressively over time to include more agricultural products. One area of expansion was to a greater number of vegetable crops. In addition, increases in the number of programs and of covered commodities were accompanied by increased marketing volumes and higher per-unit commodity values. Finally, the cost of conducting similar activities has increased over the years. Since marketing assessments are based on value or volume, marketing funds also increased. Furthermore, California agriculture has become even more diversified over time as new and previously minor crops such as kiwifruit and pistachios have developed into major products. To develop markets for these products, commodity groups often sought collective action and used government-mandated marketing programs. Supplemental funding from state and federal sources helped commodity groups expand their market promotion activities.[9] To assist producers' efforts to expand markets, the state and federal governments also subsidized export market promotion through their matching-fund programs.

California commodity program budgets for 2003/04 are presented in Table 2.3. Total budgeted expenditures of more than $208 million account for 1.2 percent of the $16.78 billion total value of the crops covered (Table 2.1). While expenditures as a share of total value are relatively small, they have increased significantly over time. Lee et al. traced annual marketing budgets as a proportion of value of production from 1973 to 1992. Budgets increased from less than 0.2 percent of the value of production to approximately 1 percent during the years from 1983 through 1989 and were slightly below 1 percent in 1992. Expenditures by commodity groups demonstrate large differences in total budgets and in their budgets' relationship to the value of the commodities' production. The California dairy industry, the state's number one commodity in terms of value of production (about $3.8 billion), has the greatest total marketing program expenditures. Budgets for manufacturing milk, market milk, fluid milk, and the Dairy Council totaled $68.19 million, which is somewhat more than 1.7 percent of the total 2002/03 value of California milk production. The dairy industry accounts for more than 32 percent of the total budgeted expenditures shown in Table 2.3, and with its large promotion programs, it accounts for almost 42 percent ($59.3 million) of total advertising and promotion.[10] A number of commodity groups, however, spend more relative to value of output than the dairy industry does. The largest budgeted expenditure as a share of 2002/03 value of production was 6.3 percent for raisins (federal and state marketing orders combined). Six other programs budgeted more than 2 percent of their total value of production for their marketing programs: avocados (3.8 percent), figs (3.8 percent), walnuts (federal marketing order and state commission, 3.3 percent), pistachios (2.5 percent), pears (2.3 percent), and cling peaches (2.1 percent). Eleven other programs budgeted between 1 and 2 percent of their total value of production: almonds, nectarines, olives, fresh peaches, cherries, plums, processing strawberries, wild rice, table grapes, kiwifruit, and fresh strawberries. As shown in Table 2.3, the largest nondairy marketing programs, in terms of total budgeted expenditures, were almonds ($20.4 million), table grapes ($14 million), and avocados ($13.6 million).

Table 2.3. California Marketing Program Budgeted Expenditures for Administration, Advertising and Promotion, Inspection, and Research by Type of Program, 2003/04

	Dollars Budgeted				
Commodity Program	Administration	Advertising and Promotion	Inspection	Research	Total
Federal Marketing Orders					
Almonds	3,466,059	13,963,062	1,078,851	1,850,332	20,358,304
Dates	123,710	101,655	0	0	225,365
Grapes, California Desert	80,215	0	0	100,000	180,215
Kiwifruit	67,642	0	0	0	67,642
Nectarines	1,771,448	2,263,061	0	138,929	4,173,438
Olives	347,090	633,500	0	250,000	1,230,590
Peaches, Fresh	1,736,040	2,211,346	0	138,930	4,086,316
Pistachios	0	0	0	0	0
Potatoes, Oregon/ California	0	0	0	0	0
Prunes	324,290	0	0	0	324,290
Raisins	1,986,500	0	2,634,800	0	4,621,300
Walnuts	710,600	1,470,000	0	1,045,000	3,225,600
Subtotal	*10,613,594*	*20,642,624*	*3,713,651*	*3,523,191*	*38,493,060*
State Marketing Orders					
Alfalfa Seed	24,070	0	0	34,095	58,165
Artichoke Promotion	0	0	0	0	0
Cantaloupes	68,420	0	156,084	33,500	258,004
Carrots, Fresh	77,740	75,000	0	370,500	523,240
Celery	63,110	0	0	205,377	268,487
Cherries	223,200	1,452,721	0	165,000	1,840,921
Citrus	602,300	0	0	2,795,967	3,398,267
Dry Beans	186,900	194,000	0	177,691	558,591
Figs, Dried	514,650	256,300	0	27,950	798,900
Garlic and Onion Dehydrators	254,179	0	214,840		469,019

continued on following page

Table 2.3 (continued)

Commodity Program	Dollars Budgeted: Administration	Advertising and Promotion	Inspection	Research	Total
State Marketing Orders (continued)					
Iceberg Lettuce Research	256,358	0	0	784,558	1,040,916
Melon Research	73,500	0	0	201,600	275,100
Manufacturing Milk	94,000	301,000	0		395,000
Market Milk	1,884,000	34,542,000	0	1,900,000	38,326,000
Milk, Fluid	711,675	19,881,153	0	0	20,592,828
Peaches, Cling	151,840	1,825,252	0	355,000	2,332,092
Pears	122,333	1,749,479	0	176,303	2,048,115
Plums	546,736	2,179,903	937,408	99,557	3,763,604
Dried Plums	683,142	5,185,000	0	446,000	6,314,142
Potato Research	50,420	0	0	100,500	150,920
Raisins	1,294,800	4,460,680	0	725,000	6,480,480
Rice Research	194,850	0	0	2,177,338	2,372,188
Strawberries, Processing	449,400	0	482,000	0	931,400
Tomatoes, Processing	200,100	0	3,436,424	37,000	3,673,524
Wild Rice	23,613	40,000		1,000	64,613
Subtotal	*8,751,336*	*72,182,488*	*5,226,756*	*10,813,936*	*96,934,516*
State Commodity Commissions					
Apple Commission	247,135	401,400	0	13,600	662,135
Asparagus Commission	175,650	529,089	0	111,484	816,223
Avocado Commission	2,245,000	9,762,500	0	1,615,600	13,623,100
Date Commission	28,550	0	0	90,000	118,550
Cut Flower Commission	267,285	999,011	0	117,893	1,384,189
Forest Products Commission	196,990	1,334,716	0	30,000	1,561,706
Grape Rootstock Commission	61,406	0	0	251,000	312,406
Kiwifruit Commission	123,200	224,000	0	14,900	362,100
Pepper Commission	54,120	0	0	145,000	199,120

continued on following page

Table 2.3 (continued)

	Dollars Budgeted				
Commodity Program	**Administration**	**Advertising and Promotion**	**Inspection**	**Research**	**Total**
State Commodity Commissions (continued)					
Pistachio Commission	2,295,660	6,138,705	0	563,300	8,997,665
Rice Commission	1,088,563	1,879,166	30,000	26,000	3,023,729
Sea Urchin Commission	0	0	0	0	0
Sheep Commission	78,250	172,492	0	45,237	295,979
Strawberry Commission, Fresh	1,324,625	4,839,383	0	1,576,850	7,740,858
Table Grape Commission	900,000	12,100,000	0	1,000,000	14,000,000
Tomato Commission	499,926	1,095,000	0	318,710	1,913,636
Walnut Commission	711,600	7,115,775	0	625,000	8,452,355
Wheat Commission	367,994	195,029	0	176,606	739,628
Lake County Winegrape Commission	46,112	125,822	0	56,800	228,734
Lodi-Woodbridge Winegrape Commission	197,074	546,705	0	120,087	863,866
Subtotal	*10,909,140*	*47,458,793*	*30,000*	*6,898,066*	*65,295,979*
Councils					
Beef Council	661,325	966,175	0	0	1,627,500
Dairy Council	1,282,110	4,591,372	0	0	5,873,482
Salmon Council	55,650	124,900	0	0	180,550
Subtotal	*1,999,085*	*5,682,447*	*0*	*0*	*7,681,532*
Marketing Agreements					
Buy California Marketing Agreement	0	0	0	0	0
Pistachio Agreement	53,500	40,000	0	0	93,500
Grand Total	**32,326,655**	**145,966,351**	**8,970,407**	**21,235,193**	**208,498,587**

Source: Federal marketing order data provided by Agricultural Marketing Service, U.S. Department of Agriculture. California program data provided by Division of Marketing Services, California Department of Food and Agriculture.

Overall, marketing program budgets for 2003/04 spent 15.5 percent of total funds collected for administration, 70 percent for advertising and promotion, 4.3 percent for inspection, and 10.2 percent for research. The 1992 distribution of budgeted expenditures was 19 percent for administration and inspection, 74 percent for advertising and promotion, and 8 percent for research (Lee et al. 1996). The distribution of budgeted expenditures varies by commodity and commodity group because of provisions included in individual marketing programs. Programs for fruits and nuts and animal products tend to focus on advertising and promotion while programs for field crops and vegetables place more emphasis on research. Administrative expenditures are common to all active federal and state marketing programs. They cover the basic costs of operating a marketing program, including staff salaries, office facilities, equipment, utilities, supplies, committee members' travel, meetings, etc. It is important to note that many programs include the administrative costs for advertising and promotion, inspection, and research as a component of administration rather than allocating those overhead costs to the individual activities. This may tend to exaggerate the administrative component of their budgets.

Fruit and nut programs, which account for 58 percent of the total marketing program expenditures shown in Table 2.3, allocated 65 percent of their budgets to advertising and promotion, 12 percent to research, 4 percent to inspection, and 19 percent to administration. Animal product programs, representing 32 percent of total marketing program expenditures, allocated 90 percent of their budgets to advertising and promotion, 3 percent to research, and 7 percent to administration. Vegetable programs, at almost 5 percent of total budgeted expenditures, allocated 18 percent of their budgets to advertising and promotion, 24 percent to research, 40 percent to inspection (primarily for processing tomatoes), and 18 percent to administration. Field crop programs, the smallest category at almost 4 percent of total budgeted expenditures, allocated 40 percent of their budgets to advertising and promotion, 33 percent to research, and 27 percent to administration and inspection.

Allocation of budgets also varies by type of program. Budgets for federal marketing orders allocated almost 54 percent of total funds to advertising and promotion and 9 percent to research while California marketing orders allocated almost 75 percent of total funds to advertising and promotion and 11 percent to research. California commodity commission allocations were similar to those of the state marketing orders; 73 percent of funds were devoted to advertising and promotion and 11 percent went to research. The council programs allocated 74 percent of their funds to advertising and promotion and did not fund research.[11]

The Buy California Marketing Agreement

Information for the Buy California Marketing Agreement is included in Table 2.3 but budgeted expenditures were not entered. This is because of the voluntary nature of the program and the funds, which came from a government specialty-crop grant in 2001 in addition to industry assessments. The federal government provided a $20 million grant that was initially supplemented with $5 million from California's general fund to augment the Buy California Marketing Agreement. The purpose of the program is to promote consumption of California-grown agricultural products to California consumers, benefiting both public health and the state's economy. This program created a partnership between government (CDFA) and industry (the Buy California Marketing Agreement Advisory Board).

The Buy California Marketing Agreement is a voluntary, multicommodity, generic marketing board that serves as an advisory board to CDFA. Currently, twenty-seven California commodity organizations have signed on to the program. These organizations, listed in Table 2.4, contributed money, time, and support to help make the *California Grown* campaign a reality. Not surprisingly, the signatories to the marketing agreement include many of the marketing

Table 2.4. Buy California Marketing Agreement Advisory Board Members

California Apple Commission	California Nectarine Administrative Committee
California Aquaculture Association	California Peach Commodity Committee
California Asparagus Commission	California Pear Advisory Board
California Avocado Commission	California Plum Marketing Board
California Cherry Marketing Program	California Poultry Federation
California Citrus Mutual	California Raisin Marketing Board
California Cling Peach Board	California Salmon Council
California Cut Flower Commission	California Sheep Commission
California Dried Plum Board	California Strawberry Commission
California Fig Advisory Board	California Table Grape Commission
California Forest Product Commission	California Tomato Commission
California Garlic Producers Association	California Walnut Commission
California Kiwifruit Commission	California Winegrape Growers Association
California Milk Advisory Board	Wine Market Council

Source: CDFA.

programs listed in Table 2.3. Included are eleven commodity commissions, eight state marketing order boards, two federal marketing order boards, and one council. The other five signatories represent commodities that have a history of participating in mandated marketing programs.

The Buy California Marketing Agreement's consumer advertising campaign and public relations effort, which has a stylized blue California license plate with the vanity inscription "CA GROWN" as its logo, was launched in August 2002 at the State Fair in Sacramento. Major retailers are actively involved in the campaign, using the logo in weekly store ads and posting in-store promotion materials. Retailers have been active in planning ways to incorporate the campaign and signatory products in promotions. A series of *California-Grown*-sponsored in-store promotions was launched in April 2003. The promotion featured dozens of products, customized point-of-sale displays, in-store demonstrations, and many other elements. Food retailers participating in the program include Raley's, Safeway, Vons, Save-Mart, Ralphs, Unified, Fleming, and Henry's Market. Lumber and home improvement stores are also beginning to feature the *California Grown* campaign and utilize "CA GROWN" point-of-sale materials. The Buy California Marketing Agreement also partnered with the California Restaurant Association to take the *California Grown* campaign into restaurants. The logo will be placed on menus featuring California cuisine, as well as on promotional materials such as table tents, menu clips, and signage.

In addition to providing funds for California market promotion and development, the *Buy California* Initiative awarded competitive grants for international market promotion. This program, the California International Market Promotion for Agriculture (CIMPA), gave $7.6 million in matching funds to help commodity organizations and private firms conduct activities in export markets to increase sales. Other grants have funded nutrition education that focuses on children ($8.5 million), The Western Institute for Food Safety and Security ($5 million), The California Agricultural Emergency Response Team ($2.7 million), minor-crops research ($1.1 million), research for sustainable agriculture ($1.9 million), the California Institute for the Study of Specialty Crops ($2.9 million), the California Agricultural Research Initiative ($2.7 million) for applied research, quarantine and pest risk analysis ($760,000), and competitive grants to benefit smaller segments of the industry ($6.6 million). The competitive-grants program funded fifty projects in amounts ranging from $13,000 to $280,000 with many of the grantees providing significant matching funds.

Concluding Comments

Marketing programs for California commodities have been in place since 1937, when federal and state legislation became effective. Even though the essential features of the marketing programs have remained unchanged, the commodities they cover and the program activities they emphasize have changed over time. Use of federal marketing orders by California commodity groups has decreased while the formation of state commodity commissions has increased. Recent trends indicate that vegetables have emerged as important program commodities along with tree crops and animal products. In addition, promotion activities have become relatively more important, as measured by outlays for those activities. Promotion expenditures as a share of the value of production have increased over time while research budgets have maintained their small, fairly constant share of production value.

California agricultural industries continue to be strong proponents of self-help marketing programs sponsored by both the state and federal government. New programs are being adopted and periodic votes and public hearings for existing programs typically support program continuation. The growing number of programs with provisions for promotion and the increasing amount of funds devoted to promotion programs indicate the substantial support these programs enjoy among the majority of producers and handlers of each commodity. Several of the following chapters in this book document that producers and handlers' faith in the returns provided by generic commodity promotion programs is well founded.

Notes

1 The sixteen national programs cover blueberries, beef, cotton, dairy, eggs, fluid milk, Hass avocados, honey, lambs, mushrooms, peanuts, popcorn, pork, potatoes, soybeans, and watermelons.

2 Marketing orders and agreements generally are spoken of synonymously. The basic difference is that agreements are voluntary and affect only the producers/handlers who agree to participate, provided that they represent a sufficient share of the industry. A marketing order, on the other hand, is mandatory once implemented (after a public hearing and successful referendum) and affects all producers/handlers in the industry.

3 State marketing programs include commissions, councils, and marketing orders and agreements.

4 The federal marketing order for almonds is an exception. The almond marketing order has assessment credit-back provisions that allow handlers to receive credit toward their assessment for their own advertising expenditures. Briefly, a handler gets $1.00 of assessment credit for each $1.50 of approved advertising and promotion expenditures. The credit-back is limited to one-half of a handler's total assessment.

5 For examples of production research projects funded by the Almond Board of California, see Carman et al. (2004).

6 See Powers (1990) for a more detailed discussion of each of the federal marketing order provisions.

7 The twelve programs included early apples (established in 1948), dried figs (1944), desert grapefruit (1941), lettuce (1942), fresh Bartlett pears (1937), fresh fall and winter pears (1941), poultry and turkey (1947), raisins (1949), wine (1938), processing pears (1938), Oregon/California potatoes (1942), and the Dairy Council (1945).

8 Lettuce research and promotion were administered by the Iceberg Lettuce Advisory Board and the Iceberg Lettuce Commission, respectively. The Iceberg Lettuce Commission was terminated in 1992.

9 For example, in 2003, USDA's Market Access Promotion Program awarded $20.24 million (out of total grants of $110 million) to California trade organizations and marketing programs. Recipients of more than $1 million each included Blue Diamond Growers/Almond Board of California ($1.21 million), California Prune Board ($2.18 million), California Table Grape Commission ($2.25 million), California Tree Fruit Agreement ($1.15 million), California Walnut Commission ($2.81 million), Raisin Administrative Committee ($1.84 million), Sunkist Growers, Inc. ($1.78 million), and Wine Institute ($3.76 million).

10 Note that four of the California programs–the Milk Advisory Board (Market Milk), the Fluid Milk Board ("Got Milk" campaign), the Beef Council, and the Avocado Commission–have close working relationships with national programs for those commodities. While there is some sharing of funds collected from California producers (the Beef Council, for example, collects $1 per head on California cattle sales, sends $0.50 per head to the national program, and keeps $0.50 per head), none of the federal funds are included in budgeted expenditures shown in Table 2.3.

11 Consumer education and information programs for nutrition and food-safety issues account for most council advertising and promotion expenditures.

References

Agricultural Marketing Agreement Act of 1937. U.S. Code. Vol. 7, §601, et seq.

Alston, J.M., H.F. Carman, J.E. Christian, J.H. Dorfman, J.R. Murua, and R.J. Sexton. *Optimal Reserve and Export Policies for the California Almond Industry: Theory, Econometrics and Simulations*. Berkeley CA: University of California Division of Agriculture and Natural Resources, Giannini Foundation Monograph 42, February 1995.

California Agricultural Statistics Service. *Summary of County Agricultural Commissioners' Reports*. Sacramento CA, 2002.

California Department of Food and Agriculture. *The Specialty Crop Block Grant: The Buy California Initiative*. Sacramento CA, 2003. Available online at *www.cdfa.ca.gov/mkt/mkt/pdf/buycal_spltycropblkgrntrpt03.pdf*.

California Department of Food and Agriculture. Data from website available August 2004 at *www.cdfa.ca.gov/mkt/mkt/mktbrds.html*.

California Department of Food and Agriculture, Marketing Branch. *Purpose and Nature of Marketing Order, Commission and Council Programs*. Sacramento CA, 1985.

California Marketing Act of 1937, California Food and Agriculture Code, Division 21.

Carman, H.F., R. Green, and G. Mandour. "Commodity Advertising Pays... or Does It?" *California Agriculture* 46(2) (1992):8–12.

Carman, H.F., K. Klonsky, A. Beaujard, and A.M. Rodriguez. *Marketing Order Impacts on the Organic Sector: Almonds, Kiwifruit and Winter Pears.* Berkeley CA: University of California Division of Agriculture and Natural Resources, Giannini Foundation Research Report No. 346, January 2004. Available online at *http://giannini.ucop.edu/researchreports.htm.*

Carman, H.F., and D.H. Pick. "Orderly Marketing for Lemons: Who Benefits?" *American Journal of Agricultural Economics* 72(2) (1990):346–357.

Ekboir, J., D.A. Sumner, and C.A. Wolf. "A Study of California's Milk Quota System." Working paper, Department of Agricultural Economics, University of California, Davis, October 1995.

Fleishman-Hillard, Inc. *The Economic Impact of Shifts in Consumer Purchasing Patterns to More California Grown Agricultural Commodities.* Sacramento CA, 2003. Available online at *www.californiagrown.org/press_room.asp?id=4&).*

French, B., N. Tamimi, and C. Nuckton. *Marketing Order Program Alternatives: Use and Importance in California, 1949–1975.* Berkeley CA: University of California Division of Agriculture and Natural Resources, Giannini Foundation Information Series 78-2, May 1978.

Garoyan, L., and J.G. Youde. *Marketing Orders in California: A Description.* Berkeley CA: University of California, Division of Agricultural Sciences Leaflet 2719, August 1975.

Lee, H., J.M. Alston, H.F. Carman, and W. Sutton. *Mandated Marketing Programs for California Commodities.* Berkeley CA: University of California Division of Agriculture and Natural Resources, Giannini Foundation Information Series 96-1, August 1996.

Neff, S., and G. Plato. *Federal Marketing Orders and Federal Research and Promotion Programs.* Washington DC: U.S. Department of Agriculture, Economic Research Service Agricultural Economic Report 707, May 1995.

Powers, N.J. *Federal Marketing Orders for Fruits, Vegetables, Nuts, and Specialty Crops.* Washington DC: United States Department of Agriculture, Economic Research Service Agricultural Economic Report 629, March 1990.

Sexton, R.J., and N. Lavoie. "Chapter 15: Food Processing and Distribution: An Industrial Organization Approach." *Handbook of Agricultural Economics.* B.L. Gardner and G.C. Rausser, eds. pp. 863–932. Amsterdam Netherlands: Elsevier, 2001.

Sumner, D.A., and N.L. Wilson. "Creation and Distribution of Economic Rents by Regulation: Development and Evolution of Milk Marketing Orders in California." *Agricultural History* 74 (2000):198–210.

Sumner, D.A., and C.A. Wolf. "Quotas without Supply Control: Effects of Dairy Quota Policy in California." *American Journal of Agricultural Economics* 78 (1996):354–366.

U.S. Department of Agriculture. *Guidelines for Fruit, Vegetable, and Specialty Crop Marketing Orders.* Washington DC, 1982.

3

Generic Advertising's Long History and Uncertain Future

John M. Crespi

After decades of relative calm surrounding commodity programs, the 1980s and 1990s saw a swell of litigation with nearly every commodity promotion program in the country involved in lawsuits over their constitutionality. In 1997, the U.S. Supreme Court appeared to give a definitive ruling in *Glickman v. Wileman Bros. & Elliott, Inc., et al.* (hereafter called *Wileman*) that federally mandated generic advertising for California peaches, plums, and nectarines did not violate the First Amendment. Just prior to the ruling, Mark Houston, president of the California Kiwifruit Administrative Committee, summed up the views of many growers when he said, "This issue needs to be cleared up and not become a full employment act for attorneys....The controversy has pitted farmer against farmer at a time when it is getting tougher and tougher to farm" (Cline 1996). With the *Wileman* ruling, the years of commodity advertising litigation appeared to be coming to an end, having been "cleared up" by the Supreme Court. But such appearances proved to be an illusion. In 2001, the extent of the *Wileman* ruling's impact on generic advertising programs was called into question by a U.S. Supreme Court ruling that a mushroom promotion program violated the First Amendment (*U.S. v. United Foods* 2001). With that ruling, cases that once seemed doomed were suddenly revived. Since then, numerous courts across the country have struck down generic advertising programs; other programs have weathered the assaults and survived and still others believed to be immune from attack following *Wileman* now face new assaults because of the *United Foods* case. As this book was going to press, in fact, the U.S. Supreme Court agreed to hear a new case concerning beef promotion. To understand where the litigation may be heading, it is important to understand how farmers arrived here. This chapter examines the history of these controversial marketing programs and the legal challenges that have dogged them.

This discussion focuses on generic advertising battles that grew out of the 1937 Agricultural Marketing Agreement Act and subsequent state acts that established the federal and state marketing orders. Briefly, a marketing order for promotion and advertising, also called a check-off, compels all growers under the order to contribute funds for joint industry advertising and promotion.[1] The program is administered by a board or commission that is elected by the growers and operates under the auspices of the U.S. Secretary of Agriculture for a federal order or the state secretary for a state order. Forker and Ward in their 1993 book, *Commodity Advertising*, defined generic advertising as "the cooperative effort among producers of a nearly homogeneous product to disseminate information about the underlying attributes of the product to existing and potential consumers for the purpose of strengthening demand for the commodity" (p. 6). The idea is simple. Producers collectively donate funds to pay for the advertising and promotion and reap the benefits of increased demand if revenues subsequently obtained are greater than the costs of the program. The question that often arises in the debate is why producers must be *compelled* to promote their products.

The key is the homogeneity of the products. Peaches are peaches, pork is pork, and milk is milk, or so it is argued. Richards (1995) provides a simple game-theoretic example that helps in understanding. Suppose two growers of the same commodity earn $10 each from sales of their crop. However, if they both contribute money to an advertising program, their profits, net of the advertising expense, may rise, say, to $16 each. Clearly industry profits are higher if both contribute to the advertising. Now, suppose that one contributes to the advertising and the other does not. The net profit for the contributor then is $8 but that for the "free rider" is $18 because the free-rider takes advantage of the fact that consumers do not distinguish between the goods, thus reaping the benefit of the advertising-induced increase in demand without paying for any of the cost. Even in this situation, industry profits ($8 + $18) are lower than if both growers contributed ($16 + $16). Nevertheless, joint contribution will not arise voluntarily. The reason is that *not contributing* is a better strategy than *contributing*: if one grower contributes, the other grower is better off not contributing ($18 > $16) and if one grower does not contribute, the other grower is still better off not contributing ($10 > $8). Because of this, the end result will be that neither grower will contribute to the advertising and both will wind up in the least profitable scenario where each earns only $10. Economists will recognize this as the Nash equilibrium of this two-player game. Obviously, the outcome would change with different choices of numbers, but this game does represent the rationale behind marketing orders: if the cost of advertising makes one worse off in the presence of a free-rider, then the only way profits can increase in the industry is if both growers are compelled to fund the program. The underlying assumption, of course, is that a grower who advertises is worse off in the presence of a free-rider.

Because of the right of freedom of association set forth by the First Amendment, compelling farmers to associate with each other requires more justification than the rationale arising from a simple, two-person game. This chapter attempts to dispel two myths about these programs: (1) that generic advertising is just an anachronistic holdover from a Depression-era act of Congress, and (2) that arguments in the many litigation cases have come about only because of a minority of curmudgeonly free-riders, free-market extremists, or unbridled lawyers. Whether either side of the battle realizes it or not, both the marketing programs and the legal issues surrounding them grew out of a process—a long, evolutionary process of economics, legislation, and litigation—that has been obscured behind the invective.

The Growth of Farmer Associations

It is true that the Agricultural Marketing Agreement Act of 1937, like its precursors, the Agricultural Adjustment Acts of 1933 and 1935, grew out of the suffering of the Great Depression, but the acts are extensions of earlier voluntary attempts by farmers to obtain more control over the production and marketing of their goods. Such control was seen as necessary in agriculture because of what has been popularly referred to as a "farm problem." Whether this problem exists today (see Gardner (1992), for example) and, if so, whether it should be solved through collective action are debatable, but the debate over collectivization is hardly new.

The term "farm problem" has evolved over the years, but its popular, though perhaps inaccurate (again, see Gardner 1992), definition refers to a combination of several presumed idiosyncrasies in agricultural production. Gardner (1992, p. 63) summarized the basic features of the supply-and-demand model of the farm problem as arising from the presumption that in agriculture:

> (1) the demand is very inelastic, (2) the supply is very inelastic, (3) the demand increases slowly over time, and (4) the supply increases more rapidly. An implication is that farm product prices decline over time. Moreover, it requires technological

progress sufficient to generate only a slightly larger rate of increase of supply compared to demand to cause prices to fall a lot, and relatively small transitory output or demand shocks to cause substantial price fluctuations.

The rationale for government support of agriculture has been deduced from this model. However, as Gardner goes on to explain, the conclusion that farmers need government aid is increasingly under attack, not only in the popular press but also by agricultural economists. Regardless of what economists believe about the existence of the problem today, economists of the early part of the twentieth century certainly felt that a problem existed (see Bonnen and Schweikhardt 1998). Even amid today's debate about what constitutes a farm problem, the fact that U.S. Supreme Court Chief Justice Rehnquist, during oral arguments for the *Wileman* case, took this economic model of the agricultural market as a matter of fact shows the extent to which the notion remains popular.

"If there's a good crop, the prices are low, and if the prices are good, there's virtually no crop. It's a totally different situation from most other kinds of marketed goods." (*Wileman* 1997)

Pre-Depression Marketing Associations

Before the introduction of modern technology and farming methods, difficulties in adjusting agricultural production meant that the agricultural sector suffered far more from a roller coaster ride of booms and busts than did the industrial sector, which could more easily control day-to-day production levels. Early attempts to gain some control over agricultural production involved controlling the amount and timing of a good going to market and were seen in voluntary cooperative movements (like those of the Grangers and the Farmers' Alliance in the mid-1800s) that were based on English Rochdale principles of nonprofit collectives. Often referred to as nonstock cooperatives (because earnings were not retained for investment), these early alliances combined the production of their members to place the cooperatives in better bargaining positions with marketing middlemen. The larger the cooperative, the greater the control over production swings that gave rise to volatile prices. To understand why collectivization grew in popularity when it did, it is important to understand that, whether causal or correlative, the burgeoning cooperative movement coincided with an increase in relative U.S.-farm incomes beginning around 1889 (U.S. Department of Agriculture, Economic Research Service

2000). Nevertheless, these early nonstock cooperatives, while politically appealing due to their grassroots, self-help ideals, were constrained in their ability to grow and venture into marketing activities because their capital structure did not allow for the raising of equity. Today this would seem to be an oversight, but the nonstock structure was, in fact, desired by members who sought to distinguish their cooperative enterprises from other corporate forms that were anathema to most farmers and many lawmakers.[2]

Cooperative idealism and political necessities notwithstanding, in order to raise equity to expand and gain greater control over marketing, some cooperatives at the turn of the century began experimenting with different capital structures. Unlike their nonstock counterparts, the capital-stock cooperatives of the early 1900s had to walk a fine line between cooperation and incorporation to avoid antitrust litigation. The earliest successes were California cooperatives for walnuts, almonds, peaches, table grapes, and raisins that, in the words of Woeste (1998, p. 30), "all preached what was becoming the hallmark of California-style cooperation: monopoly control over the crop."[3] These near monopolies developed primarily in California, due in part to a need to counter the bargaining power that buyers had over the state's perishable fruit crops (by comparison, Midwestern cooperatives, which served growers of more storable commodities such as wheat and corn, were typically more numerous and represented a smaller market share than did fruit cooperatives).

While these cooperative farmers may have seen their associations as necessary to gain some control over an inherently unstable marketplace, many Americans (and especially the marketing middlemen) saw them as little different from other trusts. In some ways, the cooperative movement added a new wrinkle to the problem it was attempting to overcome; namely, greater production control led to greater scrutiny. Because a small cooperative is not much more effective in dampening price volatility than an individual farmer, the effectiveness of a cooperative could be strengthened only with greater and greater production control; however, a large cooperative attracted a great deal of political and judicial inquiry.

Nevertheless, the belief of farmers and their representatives that a solution to the farm problem required granting farmers greater control over production and marketing of their goods continued to grow. In his 1913 inaugural address, President Woodrow Wilson took the opportunity to point out that the nation's farm policy, "never yet given the efficiency of great business undertakings," would have a new focus under his administration. The Clayton Act of 1914 (USC 15) exempted nonstock agricultural associations from antitrust laws, but because these associations were constrained in their ability to raise capital, the law had little effect on the new California-style cooperatives (Woeste 1998, pp. 93–107). In the 1920s, as Europe weaned itself from the U.S. imports it had needed following the end of World War I, farm prices in the U.S. fell (Knutson, Penn, and Flinchbaugh 1998, p. 246; Bonnen and Schweikhardt 1998, p. 5). As farm income fell, momentum grew among farm lobbies to seek expanded antitrust protection for stock cooperatives. The Capper-Volstead Act of 1922 (7 USC § 291) extended the Clayton Act's antitrust exemptions to all producer cooperatives but still prevented cooperatives from engaging in predatory market practices to exclude competition.

Cooperatives always face the threat of deviation. If the marketing activities of the cooperative generate very high prices, there is an incentive for an individual producer-member to sell his or her product on the open market rather than through the cooperative. This is the problem facing any cartel, legal or otherwise, unless members are compelled to remain within the cartel. The success of the California Associated Raisin Company (Sun-Maid) between 1910 and 1930, for example, resulted from the cooperative's near monopoly in the California raisin market (and, according to Woeste (1998, pp. 216–223), the often heavy-handed tactics cooperative members used to recruit and retain growers). However, the Sun-Maid cooperative of those years was the exception rather than the rule, and the Capper-Volstead Act's provisions against restraint of trade made such a near monopoly virtually impossible. Thus, while the Capper-Volstead Act might have been initially praised as providing some solution to the farm problem, the limitations of cooperatives in alleviating overproduction or jointly marketing a crop remained as long as cooperatives were prevented from exercising near monopoly control over production or marketing.

Depression-Era Marketing Acts

From 1929 to 1932, the farm-commodities price index fell by 56 percent and net farm income fell by 70 percent (Breimeyer 1983; Ezekial and Bean 1933). In response, some farm groups and cooperatives instituted commodity marketing programs that included promotion and advertising using voluntarily contributed producer funds. Not surprisingly, the voluntary programs failed since many producers chose not to participate and to instead "free ride" on the efforts of the producers that adhered to the voluntary marketing programs (Forker and Ward 1993, pp. 78–80). Federal and state lawmakers responded to growing discontent among farmers with a sequence of acts enabling the development of "marketing orders," which were just one of several pieces of legislation enacted to restore market stability.[4]

The legislation that created federal and state marketing orders is often seen as merely a reaction to the dire economic consequences of the Great Depression. For example, during Supreme Court testimony for the *Wileman* case, Justice Scalia referred to marketing orders for generic advertising as "time-warped" out of that era, as "a remnant of the National Recovery Act." However, as illustrated in the history of the cooperative movement outlined here, the 1937 legislation was not merely an attempt to alleviate a problem created by the Depression. The problem had always existed; it was simply that the Depression made it more obvious. Thus, claims that marketing orders are just remnants of New Deal legislation are not entirely accurate. In fact, the earliest New Deal "solution" to the farm problem, granting of more credit through the Reconstruction Finance Corporation and the Farm Act of 1932, exacerbated the farm problem. An increase in credit led to more production, which in turn led to lower farm prices. As Keller (1990, p. 157) points out, "The Depression made it evident that the great need was not easy credit but higher prices." The Clayton and Capper-Volstead Acts provided some degree of control over marketing, but when the Depression came, their limitations became more pronounced. In the view of the majority of farmers and their legislative allies at the time, "the free-rider problem could not be solved unless the state mandated universal participation" (Woeste 1998, p. 230).

The Depression notwithstanding, the courts' approaches initially remained antitrust and antiregulatory when it came to government price or production programs. Also, while farmers lobbied for government controls that would increase farm prices, handlers and processors naturally resisted any program that would raise the prices they paid farmers. Because of this, any expansion of powers allowing regulation of prices and production was carefully scrutinized (Woeste 1998, p. 230; Iron 1982, p. 112). To provide farmers with greater control over marketing of their goods, the Agricultural Adjustment Act of 1933 gave the U.S. Secretary of Agriculture the power to enter into marketing agreements with individual handlers and to issue licenses granting permission to handle agricultural commodities. Through the act, the secretary was also granted the ability to impose production restraints to reduce commodity surpluses. However, the 1933 act was vaguely worded, granting rather broad legislative and taxing powers to a member of the executive branch (the secretary) without clearly delineating the specifics for these powers. Thus, in 1935, when the U.S. Supreme Court ruled that similar portions of other New

Deal legislation were unconstitutional (*Panama Refining Co. v. Ryan* 1935; *Schecter v. United States* 1935), Congress amended the 1933 act in 1935.[5]

Even with the amendments, the Agricultural Adjustment Act of 1935 had only a short life. Although Congress tried to create an act that would be more amenable to the courts, in 1936 the Supreme Court (*United States v. Butler* 1936) balked at the broad processing-tax provisions of the act, ruling that, since only Congress or the states may tax, giving these powers to the secretary violated the Tenth Amendment. Later that year, a Massachusetts district court (*United States v. David Buthick Co.* 1936) ruled that the tax and marketing order provisions were inseparable and therefore that the validity of the entire act was dubious (Harl 1999). In an attempt to satisfy the courts, Congress passed the 1937 Agricultural Marketing Agreement Act (hereafter referred to as the AMAA). In this act, the vagaries of the previous acts were clarified so that the specific powers of the secretary with regard to the creation, terms and conditions, administration, enforcement, and termination of marketing orders and agreements were sufficiently delineated. Likewise, the tax or assessment provisions were narrowed to provide for costs incurred solely for furtherance of the goals of the act.

Simply, the AMAA provides for four types of actions: (1) regulatory restrictions on the quantity of a commodity that can be sold, either through marketing allotments or reserve pools; (2) regulatory limits on the grade, size, and/or quality of the commodity; (3) regulation of packaging and container sizes; and (4) some limited generic promotion and advertising allowances. This fourth provision of the AMAA is key to today's marketing order challenges. The programs covered under the AMAA were few in number and limited in scope to a few products (milk promotion, for example, was one of the original programs); broader provisions for generic advertising and promotion came with subsequent amendments (as discussed later). Under the AMAA, marketing orders had to be for specific commodities and in as small a region as possible while still furthering the objectives of the order. In addition, a federal marketing order could be set up only by the secretary after a two-thirds majority of affected growers voted in favor of one, and the secretary had to nullify an order upon a simple majority vote by the growers to do so.[6] Nonetheless, with the AMAA, farmers finally gained the means by which to impose a limited form of collectivization to overcome the free-rider problem in marketing their products. Challengers did not wait long.

The First Shot in the Battle: *United States v. Rock Royal CO-OP* (1939)

The first test of the AMAA came on October 27, 1938, when the U.S. Department of Agriculture (USDA) filed a complaint against several milk processors for failing to pay their assessments under a mundanely titled marketing order, Milk Order No. 27. As discussed in Harl (1999, p. 11), the accused processors claimed the order had been inappropriately adopted and that both the milk order and the act were unconstitutional infringements on their due-process rights (Fifth Amendment), on rights reserved only for the states (Tenth Amendment), and on processors' property rights (Fourteenth Amendment). When the district court concurred, the government appealed to the Supreme Court, which heard the case in April of 1939 (*United States v. Rock Royal CO-OP, Inc.* 1939).

Taking their cue from the 1936 *Butler* decision on the 1935 Agricultural Adjustment Act, the milk processors made three main arguments. First, the processors argued that their Fifth Amendment rights had been violated because the broad price-fixing powers given to the secretary violated due-process laws. Second, the processors argued that the Tenth Amendment had been violated since any attempt by the federal government at fixing the price of milk before any interstate commerce had commenced interceded such rights reserved to the states. Third, the processors argued that the delegation of authority granted (1) to the secretary to establish marketing areas, (2) to producers to approve a marketing order without an agreement by the processors, and (3) to cooperatives to cast the votes of producer patrons (so-called "bloc voting") went against the Court's rulings in the *Schecter* case.

The Supreme Court rejected these arguments, upholding the 1937 act and the order by a narrow five-to-four margin. Regarding the first two challenges, the Supreme Court held that the interstate commerce clause granted Congress the authority to regulate economic sectors and that the timing of the regulation did not supersede such authority. Writing for the majority, Justice Reed stated, "Activities conducted within state lines do not by this fact alone escape the sweep of the Commerce Clause. Interstate commerce may be dependant on them." Regarding the third challenge, the Court found that, unlike the previous acts of 1933 and 1935, the provisions of the 1937 act with regard to delegation of authority were sufficiently narrow. Further, since Congress had the power to authorize regulatory powers as it deemed necessary, there was nothing invalid about the cooperative bloc-voting scheme either (*U.S. v. Rock Royal CO-OP, Inc.* 1939, p. 569).

The importance of the *Rock Royal* decision should not be underestimated. By upholding the 1937 act, the Supreme Court established the right of producers in dairy, fruit, and vegetable markets to vote themselves into or out of a regulated industry. True, the years that followed were still occupied with litigation involving marketing orders, but most of those complaints were based on procedural grounds such as the timing or calling of board elections and voting (e.g., *U.S. v. Mills* 1963; *Freeman v. Hygeria Dairy Co.* 1964; *Consolidated-Tomoka Land Co. v. Butz* 1974; *Sequoia Orange Co. v. Yeutter* 1992; *Cecelia Packing v. USDA* 1993) or the secretary's handling of various suspensions (e.g., *Carnation Co. v. Butz* 1974; *Abbots Dairies v. Butz* 1975) and not about the constitutional validity of the programs. The strength of the Supreme Court decision in *Rock Royal* can be seen in the fact that there would be no serious challenge to the constitutionality of the AMAA for nearly fifty years.

Advertising, *Abood*, and *Central Hudson*

In the decades following implementation of the AMAA, marketing orders and agreements became rather commonplace components of U.S. agriculture. The question that naturally arises, then, is why things have changed so much today. It is probably not surprising that few noticed a handful of events that occurred over twenty years, but today's check-off controversy can be traced directly to three seemingly unrelated actions whose effects coalesced to create the biggest challenge to the AMAA since *Rock Royal*.

Event 1: 1954 Advertising Amendment to the AMAA

In 1954, Congress amended the AMAA to authorize the secretary to establish "marketing development projects," including advertising and promotion for a broad range of commodities, to further the goals of the original act. Except for some minor provisions (for milk promotion, for example), generic advertising had been left out of the original 1937 act because USDA was concerned that advertising simply shifted market share from one commodity to another. By the 1950s, however, the government was purchasing a good deal of excess supply to maintain parity prices. Stimulating demand through advertising, it was hoped, would help increase farm prices while relieving pressure for both governmental purchases of excess stocks and expanding governmental support to other commodities. Viewed in this light, generic advertising simply added a demand instrument to the government's toolbox of supply controls. The amendment stipulated that generic advertising under a marketing order must truly be of a generic nature, not a benefit to some growers and not others, and that the assessed money could not be used to promote political or ideological viewpoints.

Perhaps because of the strength of the *Rock Royal* decision, the 1954 amendment was enacted with little notice in the courts. Or perhaps it was that farmers who were unhappy with a marketing order could always vote it away. Besides, given a homogeneous commodity, generic advertisements could not be argued as benefiting one group of producers over another. Thus, one might speculate that growers and handlers were simply content with this new aspect of marketing orders. Whatever the reasons, no challenges to the constitutionality of the new advertising and promotion amendment appeared for nearly twenty-five years, and it would take two Supreme Court decisions, on cases unrelated to agriculture, to change this.

Event 2: *Abood v. Detroit Board of Education* (1977)

The second event had nothing to do with agriculture. According to a provision of a collective bargaining agreement between the Detroit Federation of Teachers (the union) and the Detroit Board of Education, any teacher who did not become a union member within sixty days of hire had to pay a service charge equal to the regular union dues. A Detroit teacher named D. Louis Abood objected to the notion of collective bargaining in the public sector and therefore had ideological reasons for not associating with a union. Further, Abood objected to the service charges because the assessments were being used in part for political endorsements. The case began in a Michigan state court in 1969 and was finally decided by the U.S. Supreme Court in May 1977.

The Supreme Court ruled that if the government deems labor relations to be important in maintaining a healthy economy then compelling payment for collective bargaining is valid even if some members disagree with collective bargaining. However, the union did not have a right to compel speech through union dues or service charges if an individual disagreed with an ideological viewpoint unrelated to collective bargaining, contract administration, or grievance adjustments. The Court ruled that, although Mr. Abood was not being prohibited from expressing his opinion, freedom of speech includes the right not to be compelled to speak. The union could only use a dissenter's money in a way that was germane to the purpose of the compelled association (e.g., collective bargaining). Thus, an *Abood* test in related cases requires that compelled assessments be relevant to the goals of the government interest and cannot be used to fund ideological or political activities.

Event 3: *Central Hudson Gas & Electric v. Public Service Commission of New York* (1980)

In 1973, to temper demand on shrinking fuel stocks, the Public Service Commission of New York ordered electric utilities in the state of New York to suspend all advertising that promoted electrical usage. In 1977, the commission decided to make this ban a permanent fixture in its energy conservation policy in light of the growing energy crisis. Citing a legal history that commercial speech is still protected speech, Central Hudson Gas & Electric Corp. opposed the ban on First Amendment grounds.

The Supreme Court decided the case in June 1980 and set up the so-called "three-prong" test that must be administered in appropriate commercial speech cases. Provided the speech is lawful and not misleading, the three criteria are as follows. First, does the government's program (in this case, the commission's ban on advertising) involve a substantial government interest? Second, does the regulation directly advance that government interest? Third, is the government's program narrowly tailored to minimize adverse impacts on First Amendment rights? Failure of any one of the three prongs of a *Central-Hudson* test means the regulation is unconstitutional.

In the *Central Hudson* case, the Court found that the first two prongs were satisfied: New York State's interest in energy conservation was substantial, and the regulation did directly advance that interest. However, on the third prong, the Court indicated that a more limited promotion ban might have had the same effect and that the commission presented no evidence that a full ban was necessary to achieve its aim. Put simply, denying Central Hudson Gas & Electric its right to promote energy usage went much further than was necessary to foster the state's interest in energy conservation. As will be seen in the particular court cases described later, it was these three events–the 1954 amendment and the *Abood* and *Central Hudson* cases–that led the way to today's marketing order challenges.

First Amendment Challenges to Generic Advertising

Abood and *Central Hudson* helped set the tone for a plethora of First Amendment cases destined to cause the Court to re-examine the speech (advertising) and association components of marketing orders for generic advertising. A brief listing of rulings in a few of the cases is instructive. In the same year *Abood* was decided, the Supreme Court ruled that the government cannot compel individuals to use their property to endorse an ideological message with which they disagree (*Wooley v. Maynard* 1977). The Court strengthened this message by adding that the government may not force individuals to respond to a hostile message when they prefer to remain silent (*Pacific Gas & Electric Co. v. Public Utilities Commission of California* 1986). Nor could the government require an individual to be identified or associated publicly with another's message (*Pruneyard Shopping Center v. Robins* 1980). The Supreme Court ruled in *Roberts v. United States Jaycees* (1984) that the government's interference in one's freedom of association could only be justified by "compelling state interests, unrelated to the suppression of ideas, that cannot be achieved through means significantly less restrictive of associational freedoms."[7]

Beginning with *Abood*, the Supreme Court in the 1980s signaled that it would observe governmental involvement in the First Amendment, especially governmental involvement in "commercial speech," with a high level of scrutiny. It did not take long for opponents of marketing orders to take notice. With the *Abood* and *Central Hudson* cases to guide them, opponents of generic advertising argued, often convincingly, that their First Amendment rights were violated by advertising campaigns that forced them to associate with their competitors and fund a promotional message with which they did not agree.

United States v. Frame (1989)

Throughout most of the twentieth century, U.S. beef consumption steadily increased. Annual beef demand reached its zenith in 1975 at around ninety pounds per person. Ten years later, demand had fallen to 1960s levels–around seventy pounds per person.[8] In 1985, Congress amended the Beef Promotion and Research Act of 1976 (7 USC) to strengthen and expand the foundering beef market through advertising and promotion.[9] The newly amended act required cattle producers and importers to finance a national beef promotion campaign by paying an assessment of one dollar per head of cattle sold. L. Robert Frame, Sr., who operated a cattle auction business and raised cattle in Pennsylvania, refused to pay his assessments. The U.S. Secretary of Agriculture brought action against him, winning in the district court. Frame appealed, and the case was heard in the Court of Appeals for the Third Circuit in July 1988. Frame argued that his First Amendment rights of free speech and association had been violated by a generic advertising program compelling him to associate with his competitors and pay for advertising when he preferred to remain silent.

In its 1989 ruling, the circuit court found that Frame's First Amendment rights had, in fact, been implicated by the forced association and advertising. Nevertheless, according to the court, this did not, in and of itself, justify nullifying the Beef Act. The rationale the court used was insightful. The government had argued that the promotion and advertising was "government speech," having been put forward by The Cattlemen's Board and Operating Committee, an instrument created by the U.S. Secretary of Agriculture to further the goals of the act. ("Government speech" is held to a much lesser degree of scrutiny than speech protected under the First Amendment.) Nevertheless, the court rejected this argument and ruled that the beef promotion program, though compelled by the government, was in fact a self-help, "commercial speech" program. Moreover, citing *Abood* and *Roberts*, the court observed that this particular type of commercial speech had to be held to a higher standard of scrutiny than other commercial speech because Frame was compelled to associate with the Cattlemen's Board. Writing for the court, Judge Scirica set forth the framework the court would use to scrutinize the act.

> Accordingly, we will sustain the constitutionality of the Beef Promotion Act only if the government can demonstrate that the Act was adopted to serve compelling state interests, that are ideologically neutral, and that cannot be achieved through means significantly less restrictive of free speech or associational freedoms. (*Frame* 1988, p. 1134)

The court was impressed with several features of the Beef Act. First, the act was put forward as a "self-help program" to insure the integrity of independent cattlemen; second, the promotion was ideologically neutral; and third, the act "expressly prohibits spending for political activity" (*Frame* 1988, p. 1134). Again citing *Abood*, Judge Scirica wrote, "Frame has failed to characterize his objection to the advertisements in a manner that would allow a reviewing court to reasonably infer a dispute over anything more than mere strategy" (*Frame* 1988, pp. 1135–1137). Interestingly, the court noted that the Beef Act would pass the three-prong test of *Central Hudson* and that those three prongs were in fact less restrictive than the *Abood* and *Roberts* tests used in *Frame*: the "*Central Hudson* test bolsters, rather than undermines, our conclusion that the Act is constitutional" (*Frame* 1988, p. 1137). Thus, the circuit court rejected Frame's free speech and association argument, finding instead that the Beef Promotion Act did implicate the First Amendment rights of those compelled to participate but that clearly the government "enacted this legislation in furtherance of an ideologically neutral compelling state interest, and has drafted the Act in a way that infringes on the contributors' rights no more than necessary to achieve the stated goal" (*Frame* 1988, p. 1137). Frame's appeal to the Supreme Court was denied.[10]

Cal-Almond, Inc. v. U.S. Department of Agriculture (1993)

The federal generic-advertising marketing order for almonds, established in 1950, differs slightly from many other such orders in its provisions for collecting the advertising assessments. The almond order is administered by the Almond Board of California, which is composed of ten members from the industry. Assessments are collected by the handler (processor) when a grower brings crops to the handler. Many of the almond handlers are also growers. The assessments are based on volume and administered by the board to cover its activities, which include research and development, quality control, volume regulation, and generic advertising and promotion. The biggest difference between this order and others is that during the 1980s handlers could receive a credit against their assessment when they undertook their own consumer advertising programs as long as their advertising efforts met requirements set by the board.

Some handlers felt that the board's requirements in the credit-back provision favored certain handlers and the largest almond cooperative, Blue Diamond Almonds (Blue Diamond), in particular over others. In 1987, when the first round of litigation in the Cal-Almond case began, Blue Diamond was the largest handler in the industry, having a 92 percent share of the almonds sold in grocery stores, and also sold almonds through its own retail stores. Most of the other almond handlers sold their almonds to food processors such as cereal makers. Also, the almond marketing order allowed for bloc voting, whereby an almond cooperative could vote on behalf of its grower members; the larger the cooperative, the more votes it could cast. Through such voting, Blue Diamond, representing more than half of the growers in the industry, maintained a majority position on the Almond Board.[11]

In 1984, Saulsbury Orchards and Almond Processing, Inc. (hereafter called Saulsbury) refused to pay its annual advertising assessments. Most of Saulsbury's sales were to cereal manufacturers, and Saulsbury helped fund advertisements for some of the almond-containing cereals. Saulsbury also provided almond-related advertising for a chain of mini-markets in the Boise, Idaho, area that carried Saulsbury almonds. However, when Saulsbury tried to recoup those advertising expenditures, it was denied credit by the board. The cereal expenditures were denied because of board regulations that the advertised consumer product must contain at least 50 percent almonds for the advertising to be creditable. The mini-market advertising expenditures were deemed unrecoverable because board regulations denied credit for advertisements at retail outlets not operated by a handler.[12] Another handler,

Cal-Almond, Inc., sold most of its almonds to an ice cream manufacturer and helped advertise the ice cream. Cal-Almond was also denied credit for its advertising expenditures because the ice cream was less than 50 percent almonds. Blue Diamond, on the other hand, was allowed to recoup its advertising expenditures. In 1987, Cal-Almond joined Saulsbury and another handler, Carlson Farms, in challenging the almond order. The matter finally appeared before the Ninth Circuit Court of Appeals in January 1993 as *Cal-Almond v. USDA* (*Cal-Almond*).

The *Cal-Almond* First Amendment attack on the almond order was similar to that proffered in *Frame*; the appellants argued that the generic advertising program infringed on their First Amendment rights of free speech and association. The district court from which this case was appealed had held that the almond marketing program neither implicated nor violated the appellants' First Amendment rights because no one was compelled to advertise and because handlers could get the assessments back if they participated in creditable advertising. The Ninth Circuit Court disagreed with the lower court's reasoning, however, precisely because the order does compel handlers to expend money on assessments or creditable advertising, which, as the Third Circuit Court had found in *Frame*, burdened the handlers' First Amendment rights.

The Third Circuit Court in *Frame* had cited *Abood* and *Roberts* in arriving at its decision. The Ninth Circuit Court, however, applied the three prongs of the *Central Hudson* test to the almond order's generic advertising program. For the first prong, the court asked whether the governmental interest was substantial. The court agreed with the language of the almond order that stimulating demand for almonds to increase returns to the almond industry was a substantial government interest. However, it balked at supporting the government under the second and third prongs.

Recall that the second prong of the *Central Hudson* test asks whether the restrictions directly advance the government's interest. The court first focused its attention on the creditable advertising restrictions and held that, to satisfy *Central Hudson*, USDA had to show that (1) the advertising that received credit was better at stimulating demand for almonds than the board's own advertising efforts, and (2) the advertising that was denied credit was worse than the board's efforts. In concluding that the generic promotion did not pass the second prong of the test, the court mainly relied on two observations. First, USDA could provide no evidence or studies showing that the board's advertising had any effect on almond demand. Thus, USDA had no basis for claiming that the generic advertising "directly advanced" the government's interest. Second, the court wondered whether handlers themselves could provide any evidence that the credit-back regulations helped them sell more almonds. Clearly Saulsbury's and Cal-Almond's experience suggested that the restrictions were detrimental. However, even Blue Diamond could provide no evidence that the regulations helped. Roger Baccigaluppi, Blue Diamond's president and chief executive officer, testified that Blue Diamond would "probably" continue to spend just as much money on advertising without the regulations, and Walter Payne, Blue Diamond's vice president for sales, marketing and distribution, testified that Blue Diamond would conduct the same amount of advertising in the same manner in the absence of the credit regulations. Judge Brunetti summarized the Ninth Circuit Court's ruling, stating that "the regulations do not 'directly advance' the government's asserted interest in increased almond sales and are therefore an unconstitutional restriction on appellants' First Amendment rights" (*Cal-Almond* 1999, p. 439).

The court then turned to the third and final prong of *Central Hudson*: Are the regulations more extensive than necessary to further the government's interest in selling almonds? The court agreed with the general notion that advertisements for which credit is permitted increase almond sales but chided USDA for offering no justifications for the restrictions that denied credit. USDA had argued that "the regulations reflect a reasonable judgment that the Board will make better use of those monies in its market promotion programs" than could be made by the handlers in their noncreditable advertising expenditures. Again, Judge Brunetti wrote,

> It is true that the fit between means and ends need not be perfect, but there seems to be no logical justification for these types of restrictions other than the restrictions are designed to benefit Blue Diamond, who overwhelmingly dominates the retail almond market, at the expense of smaller handlers such as appellants, who sell primarily to ingredient manufacturers....The creditable advertising regulations...are more extensive than necessary to serve the interest of increasing almond sales. (*Cal-Almond* 1993, p. 440)

Since the restrictions failed on two of the three prongs of the *Central Hudson* test, the Ninth Circuit Court ruled that the advertising portion of the almond order was an unconstitutional violation of the appellants' First Amendment rights.

The Fallout from *Frame* and *Cal-Almond*

In the aftermath of *Frame* and *Cal-Almond*, after years of stability, marketing orders now faced a real constitutional threat. In the years that followed, dozens of challenges arose. Some of the programs that subsequently came under attack (the vast majority of which were in California and hence under the jurisdiction of the Ninth Circuit Court) were the California Apple Commission, the California Cling Peach Growers Advisory Board, the California Cut Flower Commission, the California Grape Rootstock Improvement Commission, the California Kiwifruit Commission, the California Milk Advisory Board, the California Plum Marketing Board, the California Table Grape Commission, the Cattleman's Beef Board, Marketing Orders 916 and 917 (California peaches, plums, and nectarines), the Mushroom Council, and Promo-Flor (a national floral group). Some attacks succeeded; others failed. Court opinions from circuit to circuit showed seeming disagreement on the application of *Abood*, *Central Hudson*, and related Supreme Court decisions (see, for example, *Goetz v. Glickman* 1998). There is little need to present the many cases except to note that confusion among interested parties was growing. The Third Circuit Court in *Frame* and the Ninth Circuit Court in *Cal-Almond* had proffered different conclusions on very similar programs, and there was no clear indication which ruling would be applied by district judges. Sooner or later, the U.S. Supreme Court would have to get involved.

Glickman v. Wileman (1997)

In February of 1995, two years after its decision in *Cal-Almond*, the Ninth Circuit Court heard a similar argument from sixteen handlers of California nectarines, peaches, and plums. The tree-fruit case began as a dispute over size and quality regulations in Marketing Orders 916 and 917. The case had previously wound its way through the hands of an administrative law judge, who had ruled in favor of the handlers; a USDA judicial officer, who had ruled in favor of the secretary; and a district court that also had sided with the secretary. Adding a First Amendment argument, the handlers appealed the district court's ruling to the Ninth Circuit Court (*Wileman Bros. & Elliott, Inc. v. Espy*; later renamed *Wileman Bros. & Elliott, Inc. v. Glickman* (1995) and hereafter referred to as *Wileman*). Like the handlers in *Cal-Almond*, tree-fruit handlers in *Wileman* objected to the generic advertising assessments on First Amendment grounds. The case differed somewhat from the *Cal-Almond* case in that the tree-fruit marketing order did not allow for credits for the handlers' own promotion campaign expenditures. Still, many of the same arguments were made.

In particular, there was the charge that certain generic advertisements had favored products distributed by only some of the handlers. For example, the handlers argued that certain generic ads put forward the message that "Red is Better" and cited a promotional chart, financed by the assessments, that listed a proprietary variety of nectarine, the "Red Jim," that was owned by a member of the Nectarine Administrative Committee.

The Ninth Circuit Court used its own standard as laid out in *Cal-Almond* and applied the three prongs of *Central Hudson* to the *Wileman* case. On the first prong, the Ninth Circuit Court found that the government did have a substantial interest in enhancing tree-fruit growers' returns: "The handlers point to no reason why the government's interest in promoting peaches and nectarines is any less substantial than it is for almonds." On the second prong, the court agreed with the secretary's claim that generic advertising had increased sales but went further.

> However, according to *Cal-Almond*, the question is not whether the generic advertising program has increased peach and nectarine sales–it undoubtedly has. Rather, the question is whether the mandatory generic advertising program sells the product more effectively than the "specific, targeted marketing efforts of individual handlers."

The Ninth Circuit Court ruled that there was no evidence that the generic advertising program for tree fruit was better at increasing demand than individual handler advertising, so the generic program failed the second prong of *Central Hudson*. Third, the court concluded that the tree-fruit marketing orders for generic advertising were more restrictive than that of the almond order (because the tree-fruit orders lacked a credit-back provision for individual advertising). Thus, because the less restrictive almond order had failed the third prong of *Central Hudson* (i.e., it was not sufficiently narrowly tailored), the orders for tree fruits also failed this prong. Having failed two of the three prongs, the generic promotion aspect of the tree-fruit orders, like that of the almond order, was ruled an unconstitutional infringement on handlers' rights of free speech and association.[13]

The U.S. Secretary of Agriculture appealed the Ninth Circuit Court's ruling to the U.S. Supreme Court on the grounds that the Ninth Circuit's First Amendment decision in *Wileman* conflicted with the Third Circuit Court's decision in *Frame*. The Supreme Court granted the secretary's petition and heard testimony in December 1996 (*Glickman v. Wileman Brothers & Elliott, Inc. et al.* 1997). The Supreme Court's decision is well known: in a five-to-four ruling, it reversed the Ninth Circuit Court. In its decision, the Court put forward two main findings: (1) the handlers' disagreement with the content of some of the advertising had no bearing on the validity of the entire generic program, and (2) the Ninth Circuit Court had erred in using *Central Hudson* to test the constitutionality of the program.[14]

Writing for the Court, Justice Stevens repeatedly stressed the statutory context within which the generic promotion program had arisen and that generic campaigns had to be viewed in light of the regulatory scheme that Congress had put forward.

> The legal question that we address is whether being compelled to fund this advertising raises a First Amendment issue for us to resolve, or rather is simply a question of economic policy for Congress and the Executive to resolve.
>
> In answering that question we stress the importance of the statutory context in which it arises. California nectarines and peaches are marketed pursuant to detailed marketing orders that have displaced many aspects of independent business activity that characterize other portions of the economy in which competition is fully protected by the antitrust laws. The business entities that are compelled to fund the generic advertising at issue in this litigation do so as a part of a broader collective enterprise in which their freedom to act independently is already constrained by the regulatory scheme. (*Glickman* 1997, pp. 468–470)

The Court then pointed out that there were three characteristics of the generic advertising orders' regulatory scheme that distinguished the orders from other laws that had been found to violate the First Amendment. First, the marketing orders do not prevent producers from communicating any message to any audience, a fact that distinguished this case from the *Central Hudson* case. Second, the marketing orders do not compel handlers to engage in any actual or symbolic speech. Third, the marketing orders do not compel the handlers to endorse or to finance any political or ideological views, which distinguished this case from *Abood*. "Thus," Justice Stevens wrote, "none of our First Amendment jurisprudence provides any support for the suggestion that the promotional regulations should be scrutinized under a different standard from that applicable to the other anticompetitive features of the marketing orders."[15]

The Court stressed that the regulatory nature of the marketing orders required that the generic advertising be judged in a different light from that of other commercial speech cases. Congress had made a regulatory decision that, right or wrong, certain commodities should be marketed jointly.

> The basic policy decision that underlies the entire statute rests on an assumption that in the volatile markets for agricultural commodities the public will be best served by

compelling cooperation among producers in making economic decisions that would be made independently in a free market. It is illogical, therefore, to criticize any cooperative program authorized by this statute on the ground that competition would provide greater benefits than joint action. (*Glickman* 1997, pp. 474–475)

Thus, the Ninth Circuit Court's application of the three prongs of *Central Hudson*, especially that court's requirement that the secretary prove that demand had been increased more effectively by generic advertising than it would have been by the marketing efforts of independent producers, was "inconsistent with the very nature and purpose of the collective action program" (*Glickman* 1997, p. 475).

The Supreme Court did note the similarity with *Abood* and subsequent cases in that these cases showed that "assessments to fund a lawful collective program may sometimes be used for speech over the objection of some members of the group" (*Glickman* 1997, p. 472). Recall that the *Abood* test merely requires (1) that the assessments be spent on promotion relevant to statutory goals, and (2) that participants not be compelled to fund nonrelevant, ideological speech. The Court found that the tree-fruit orders satisfied both of these requirements. Under *Central Hudson*, the Ninth Circuit Court had ruled that the generic advertising had passed the first prong (the government had a substantial interest in increasing demand). With an *Abood* test, however, the Supreme Court indicated that the government need not show even this. The government's legislative action alone was enough to satisfy the necessity of the program. Simply put, as long as the regulatory means furthered the goals of the acts and did not compel ideological speech, that was all that was necessary to satisfy the acts' constitutionality. The question of whether individual handlers were hurt was not under consideration since they had chosen to operate in a regulated environment in the first place.

In sum, what we are reviewing is a species of economic regulation that should enjoy the same strong presumption of validity that we accord to other policy judgments made by Congress. The mere fact that one or more producers "do not wish to foster" generic advertising of their product is not a sufficient reason for overriding the judgment of the majority of the market participants, bureaucrats, and legislators who have concluded that such programs are beneficial. (*Glickman* 1997, p. 477)

Rulings in all of the pending First Amendment marketing order challenges had been postponed until the Supreme Court's ruling in *Wileman*. After the Supreme Court's decision, appellants in other cases tried to argue that the marketing orders in those cases were, in some sense, different from the tree-fruit orders. For example, the handlers in *Cal-Almond* contended that *Wileman* did not apply because the credit-back assessment mechanism of the almond order was different than that of the tree-fruit orders. In light of *Wileman*, the Ninth Circuit Court rejected this argument, reversing its previous ruling in *Cal-Almond*. In all of the cases that had been put on hold pending the Supreme Court's decision, the federal appeals courts applied *Wileman* to uphold the constitutionality of the promotion programs—all except one.

Here We Go Again

In November of 1999, the Sixth Circuit Court of Appeals ruled the Mushroom Promotion Act of 1990 unconstitutional because the act that created it was not in the same spirit as the broader, collective regulation embodied in the AMAA.[16] The Mushroom Promotion Act, like the Beef Act scrutinized in *Frame*, had been termed a "stand-alone" program because it was not formed under the original 1937 AMAA. However, given that the Supreme Court saw no distinction between the stand-alone promotion program in the Beef Act and the promotion program in the tree-fruit orders, the Sixth Circuit Court's ruling surprised many on both sides of the marketing order debate.

United Foods, Inc., a Tennessee food processor, challenged the 1990 Mushroom Act on the grounds that the assessments were compelled commercial speech and that the mushroom industry differed from that of the tree-fruit industry in the *Wileman* case. A U.S. district judge in Tennessee had ruled against United Foods, citing the *Wileman* decision, and United Foods appealed.

The attorneys for United Foods used a very interesting argument to distinguish the mushroom industry from the tree-fruit industry in *Wileman*. Focusing on the language of Justice Stevens' opinion concerning regulation and compelled association, they emphasized that the regulatory environment that justified the tree-fruit order was almost completely absent in the mushroom industry. Further, they argued that tree fruits need a generic program because the grading and standards regulations that are also part of the tree-fruit orders foster product homogeneity. Mushrooms, on the other hand, are not subject to such grading regulations and truly are heterogeneous, so a generic program provides no real benefit since "mushroom producers can freely differentiate their product and stimulate demand through individual competitive advertising."[17] The government argued that *Wileman* was not based on the degree of regulation in an industry, but the Sixth Circuit Court found the limited-regulation argument persuasive. Writing for the majority, Judge Merritt stated the court's opinion that "The Court's holding in *Wileman*, we believe, is that nonideological, compelled, commercial speech is justified in the context of the extensive regulation of an industry but not otherwise."[18] Therefore, without the extensive regulation present in the tree-fruit marketing orders, there was no justification for any further limits on compelled speech.

Arguing that the Sixth Circuit Court's ruling conflicted with the U.S. Supreme Court's ruling in *Wileman*, the government appealed. The U.S. Supreme Court agreed to hear the case and, in a surprise to many who were following the marketing order debates, ruled in a six-to-three decision to uphold the Sixth Circuit Court's ruling. Writing for the majority, Justice Kennedy stated,

> The program sustained in [*Wileman*] differs from the one under review in a most fundamental respect. In [*Wileman*] the mandated assessments for speech were ancillary to a more comprehensive program restricting marketing autonomy. Here, for all practical purposes, the advertising itself, far from being ancillary, is the principal object of the regulatory scheme.[19]

Because of this, the Court could not uphold "compelled subsidies for speech in the context of a program where the principal object is speech itself." In other words, as long as the generic advertising is part of a broader regulatory scheme (like the marketing orders for tree fruit), the assessments pass constitutional muster, but if generic advertising is the primary purpose for collecting the assessments, the assessments then violate the First Amendment. Recall how, in the *Wileman* case, the tree-fruit industry was found to be different from other industries (where compelled association would be illegal).

Although the Court left a great deal of room in determining what is or is not "a broader regulatory scheme," it is important to note exactly how Justice Kennedy and the majority viewed the mushroom market as different from the tree-fruit market.

> Beyond the collection and disbursement of advertising funds, there are no marketing orders that regulate how mushrooms may be produced and sold, no exemption from the antitrust laws, and nothing preventing individual producers from making their own marketing decisions.

Neither had the mushroom market been "subjected to a uniform price, or otherwise subsidized through price supports or restrictions on supply."[20] For both opponents and supporters of generic advertising programs, these distinctions must be considered carefully. It seems that the almond order, which includes supply controls, would pass the *United Foods* test, but what about beef, pork, or cotton? In fact, most state-authorized orders and commissions exist primarily to fund collective research and promotion and do not authorize additional industry collectivization or other regulations (Lee et al. (1996) identified thirty-six state-authorized promotion programs in California alone). Further, what would happen if Congress adjusted the regulations in an industry? Would the advertising programs be implicated if, for example, a grading regulation were abandoned? Would the Court's ruling create incentives for generic promotion supporters to seek more regulations for their industries to protect their advertising programs?

So far, it appears that these questions will be answered on a case-by-case basis and that courts will determine whether an individual check-off program is more like the one for tree fruits or the one for mushrooms. For example, in a challenge to a state check-off, the Ninth Circuit Court of Appeals used the *United Foods* decision to invalidate the California Table Grape Commission's promotion program (*Delano Farms v. California Table Grape Commission*, 318 F. 3d 895, 2003) because of the lack of extenuating regulations and collectivization in the program. But anticipating how courts will determine whether an industry is highly regulated is difficult. Milk, for example, seems to be one of the most highly regulated food industries given the grading standards and complicated regulations that determine prices. A federal district court judge in Pennsylvania made such a determination and ruled that *Wileman* was applicable in the case of the national Dairy Promotion Program because of the extent of regulation in dairy marketing. However, the Third Circuit Court of Appeals (*Cochran v. Veneman*, 252 F. Supp. 2d 126, 2003) ruled that milk did not in its view seem highly regulated and reversed the district court, citing *United Foods*.

A Revival of the Government Speech Argument?

Citing the *United Foods* case, U.S. District Court Judge Charles Kornmann in South Dakota ruled in June 2002 that the beef check-off program was unconstitutional. In October 2002, a U.S. district judge in Michigan, Richard Enslen, ruled that the pork check-off program was not only unconstitutional but "rotten" as well (*Michigan Pork Producers Association v. Campaign for Family Farms*, 229 F. Supp. 2d 772, 2002) and struck down the entire pork check-off, including portions for research and education. Both judges followed the high court's rulings in *United Foods* to make their decisions, but what set these cases apart from previous ones was the government's argument. Rather than basing its case only on the degree of regulation in the industry, the government revisited a rationale that had proven fruitless in the *Frame* case a decade earlier, arguing that the promotion programs did not infringe on an individual's free speech at all because the speech is not being made by an individual. Rather, it is *government* speech. Clearly the government-speech argument carried as little weight in the new beef and pork decisions as it had in *Frame* a decade earlier. In November 2002, a district court judge in Montana, Richard Cebull, likewise found that the beef program should be judged in light of *United Foods* and not *Wileman*. What differed in this case, however, was that Cebull ruled that the beef check-off "creates programs where the government utilizes private cattlemen to disseminate a single message, a message prescribed by Congress and the USDA" (*Charter v. USDA* 2002). In other words, the government is making the speech *through* the cattlemen rather than *for* the cattlemen. As such, the speech is not governed by the First Amendment and, therefore, not germane to constitutional scrutiny.

It is not surprising that each of the cases was appealed. At this writing, appeals courts have heard all three but ruled on only two of them. The lower court's ruling in the Michigan pork case was upheld on October 22, 2003, by the Sixth Circuit Court of Appeals (*Michigan Pork Producers Association v. USDA* 2003). The South Dakota beef case was appealed to the Eighth Circuit Court on July 8, 2003 (*Livestock Marketing Association v. USDA* 2003), whereupon a three-judge panel unanimously rejected the government-speech argument and upheld the district court's use of the *United Foods* case. This case was appealed to the U.S. Supreme Court, and the Supreme Court has agreed to hear the case in the fall of 2004. A decision will likely come early in 2005. At this writing, the Ninth Circuit Court has yet to rule on the Montana beef case but, given the U.S. Supreme Court's decision to hear the related South Dakota case, the Ninth Circuit Court is likely to wait to give its ruling.

Conclusion

Although the conventional wisdom is that farmer collectivities such as those seen in marketing orders and other check-off programs are "time warped," as Justice Scalia stated, out of the Great Depression, in fact, the ideals underlying these programs are much older than that. I have also argued that the challenges to marketing orders, especially those in the last ten years, could not have arisen at any other time. Today's challenges arose when they did because a handful of Supreme Court decisions in the early 1980s implicated government involvement in collective promotion. Once *Abood* and *Central Hudson* had been decided, it was inevitable that generic advertising would be looked at more closely. *Wileman* and *United Foods* are simply logical outgrowths of these earlier decisions.

Given the evolution of the court cases, the natural question to have asked after *Wileman* was "How much industry regulation is needed for a program to be legal?" The Court's response came with the ruling in *United Foods*. However, since the Court did not set out a specific formula by which to determine the degree of regulation, the number of challenges to commodity promotion that have since arisen is not surprising. In fact, unless it provides further clarification, the Court likely introduced a strange incentive into the battle. The *United Foods* case sends a signal to growers who support generic advertising that more regulation is better than less. And the seeds of this incentive may be starting to bear fruit. On February 4, 2004, for example, a bill passed the Washington House of Representatives that would make the Washington Apple Commission a part of the state's regulatory structure as a state board, rather than an independent commission, under the State Department of Agriculture (House Bill 2367). It will be interesting to see how many other industries halt plans to alter regulations or even seek to extend their programs to include more regulations like those found in the tree-fruit industries. A decision next year (2005) by the U.S. Supreme Court in the beef case should clear up the vagaries that were created in the wake of *Wileman* and *United Foods*. The Court is no doubt aware of the confusion that exists, thus I expect the Court to try and clarify its previous rulings. Of greatest interest will be how the Court comes down on the government-speech argument. If it accepts this rationale, generic advertising will likely be around for a very long time. If the Court rejects this premise, then expect more challenges and a most uncertain future for mandated promotion programs.

Notes

This chapter is based in part upon the author's 2003 article entitled "The Generic Advertising Controversy: How Did We Get Here and Where Are We Going?" published in the *Review of Agricultural Economics*, Volume 25, pp. 294–315. The author is grateful to Julian Alston, Tim Barron, David Barton, Giacomo Bonanno, Bradley MacLean, Barbara McDowell, Roger McEowen, Susan Offutt, and Richard Sexton for very helpful and instructive discussion and comments and to Natalie Karst for careful editing. The author takes full responsibility for all errors.

1 The terms "marketing order" and "check-off" are often used interchangeably. In fact, however, while all marketing orders are check-offs, not all check-offs are marketing orders. The distinction is a statutory one–the term marketing order is used for programs established under the specific marketing acts discussed in this chapter. However, courts have tended to view the two types of programs as equivalent, making the distinction a subtle one.

2 See the discussion in Woeste 1998, pp. 17–36.

3 Hovenkamp (1991, pp. 301–304) documents how support among economists for "trusts" also began growing at the turn of the century, though much of that support had eroded by the 1920s. For an example of such support, the interested reader is referred to Jenks (1892) and to Galbraith and Black (1938).

4 See especially the discussion of Depression-era legislation in Farrell and Runge (1983) and in Paarlberg (1983).

5 For brevity, much discussion on the formation of the Agricultural Adjustment Act was left out, but the interested reader is also referred to the discussion in Iron's (1982) Chapters 6–8, which shows that the views of both the farmers themselves and the lawyers working on the act were hardly unanimous regarding the type of programs to put in place. See also the discussion of New-Deal agricultural-policy reform by Gilbert (2000), Harl (1999, p. 388), and Paarlberg and Paarlberg (2000).

6 This voting provision is often neglected in press accounts, leading some to believe that marketing orders are simply imposed on growers by the government. For example, a fifteen-paragraph *Wall Street Journal* article that attempted to answer the question "Should Government Market Pork and Peaches?" (Felsenthal 1996) mentions neither the original grower referenda that started the programs nor the subsequent grower votes that keep them running.

7 Justice Brennan wrote the majority opinion in *Roberts v. United States Jaycees* (1984, p. 623).

8 Current beef demand is a little more than sixty pounds per person (USDA, ERS 2000, p. 11).

9 The Beef Promotion and Research Act is not part of the 1937 Agricultural Marketing Agreement Act, which covered the dairy, fruit, and vegetable industries. The Beef Act is termed a "stand alone" marketing program. Nevertheless, the administration, promotion, and assessment aspects of such stand-alone programs are nearly identical to those of the marketing orders.

10 *Frame v. United States* 1990.

11 Market-share figures came from endnote 9 of *Cal-Almond, Inc. v. U.S. Dept. of Agriculture*, 14 F. 3d 429 (9th Cir. 1993). The court documents, however, are not accurate. Blue Diamond's share refers to its branded consumer snack products only. In fact, Blue Diamond faced significant competition from Planters and Dole, especially in baking almonds, for which Dole's market share was larger than Blue Diamond's (Barron 2001).

12 *Saulsbury Orchards and Almond Processing, Inc. v. Yeutter* 1990.

13 *Wileman Bros. & Elliott, Inc. v. Espy* 1995, p. 439.

14 Petition granted in 517 U.S. 1232, *Glickman v. Wileman Brothers Inc. et al.*, 521 U.S. 457 (1996). It is also important to note that the Supreme Court felt inclined to rule on what was seen as a contradiction between the Ninth Circuit Court's *Wileman* ruling and the Third Circuit Court's *Frame* ruling. After the Supreme Court's *Wileman* decision, some attorneys argued that *Wileman* had no bearing on stand-alone programs (like the Beef Act and the Mushroom Promotion Act). However, the Supreme Court's decision to take the *Wileman* case (given that there was a conflict with *Frame*) signaled that the Court did not view the First Amendment issues involved in stand-alone programs as distinct from those in marketing orders.

15 See also Breyer 1982, pp. 156–161.

16 *United Foods, Inc. v. United States of America* 1999.

17 The author is grateful to Bradley A. MacLean of the Nashville firm of Farris, Warfield & Kanaday, PLC, for providing excerpts from the brief.

18 *United Foods, Inc. v. United States of America* 1999, p. 224.

19 *United States v. United Foods, Inc.* 2001, p. 6.

20 *United States v. United Foods, Inc.* 2001, p. 6.

References

Abbots Dairies Div. of Fairmont Foods, Inc. v. Butz. 1975. 389 F. Suppl. (E. Dist. Penn. 1975).

Abood v. Detroit Board of Education. 431 U.S. 209 (1977).

Agricultural Act of 1954, U.S. Code. Vol. 7, §401(c).

Agricultural Marketing Agreement Act of 1937. U.S. Code. Vol. 7, §601, et seq.

Barron, T. Managing Director, Cornucopia, LLC and former President of Dole Dried Fruit and Nuts. Personal communication, 12 November 2001.

Bonnen, J.T., and D.B. Schweikhardt. "The Future of U.S. Agricultural Policy: Reflections on the Disappearance of the 'Farm Problem'." *Review of Agricultural Economics* 20 (1998):2–36.

Breimeyer, H.F. "Conceptualization and Climate for New Deal Farm Laws of the 1930s." *American Journal of Agricultural Economics* 65 (1983):1153–1157.

Breyer, S. *Regulation and Its Reform.* Cambridge MA: Harvard University Press, 1982.

Cal-Almond, Inc. v. U.S. Department of Agriculture. 14 F. 3d 429 (9th Cir 1993), 521 U.S. 1113 (1997), 98-16921 9th Cir. (1999).

California Marketing Act of 1937, California Food and Agriculture Code, Division 21.

Carnation Co. v. Butz. 372 F. Supp. 883 (Dist. of D.C. 1974).

Cecelia Packing v. U.S. Dept. of Agriculture. 10 F. 3d 616 (9th Cir. 1993).

Central Hudson Gas & Electric v. Public Service Commission of New York. 447 U.S. 557 (1980).

Charter v. USDA. 230 F. Supp. 2d 1121 (2002).

Cline, H. "U.S. Supreme Court Agrees to Hear Market Order Case." *California Farm Press* 6 July 1996, p. 4.

Cochran v. Veneman. 252 F. Supp. 2d 126 (2003).

Consolidated-Tomoka Land Co. v. Butz. 498 F. 2d 1208 (5th Cir. 1974).

Delano Farms v. California Table Grape Commission. 318 F. 3d 895 (2003).

Ezekiel, M., and L.H. Bean. *Economic Bases for the Agricultural Adjustment Act.* Washington DC: United States Department of Agriculture, 1933.

Farrell, K.R., and C.F. Runge. "Institutional Innovation and Technical Change in American Agriculture: The Role of the New Deal." *American Journal of Agricultural Economics* 65 (1983): 1168–1173.

Federal Agricultural Improvement Act of 1996. Title V – Agricultural Promotion. Subtitle A – Commodity Promotion and Evaluation. Sec 501(c).

Felsenthal, E. "Should Government Market Pork and Peaches?" *The Wall Street Journal.* 26 November 1996, p. B1.

Forker, O.D., and R.W. Ward. *Commodity Advertising: The Economics and Measurement of Generic Programs.* New York NY: Lexington Books, 1993.

Frame, L. Robert, Sr., et al. v. United States. 110 U.S. 1168 (1990). See also *United States v. L. Robert Frame, Sr. and Vintage Sales Stables, Inc.* 885 F. 2d 1119 (1988).

Freeman v. Hygeria Dairy Co. 326 F. 2d 271 (5th Cir. 1964).

Galbraith, J.K., and J.D. Black. "The Maintenance of Agricultural Production during the Depression: The Explanations Reviewed." *Journal of Political Economics* 46 (1938):305–323.

Gardner, B.L. "Changing Economic Perspectives on the Farm Problem." *Journal of Economic Literature* 30 (1992):62–101.

Gilbert, J. "Eastern Urban Liberals and Midwestern Agrarian Intellectuals." *Agricultural History* 74 (2000):162–180

Glickman v. Wileman Brothers & Elliott, Inc. 117 Sup. Ct. 2130, 138 L. Ed. 2d 585, 95–1184 (1997). See also *Glickman* 521 U.S. 457 (1997) and 517 U.S. 1232 (1996).

Goetz v. Glickman. C.A. 94-1299-FGT, Dist. of Kansas (1996), C.A. No 96-3120 (10th Cir. 1998).

Harl, N.E. *Agricultural Law*. Vol. 9, Ch. 70. New York NY: Matthew Bender & Company, Inc., 1999.

Hovenkamp, H. *Enterprise and American Law*. Cambridge MA: Harvard University Press, 1991.

Iron, P.H. *The New Deal Lawyers*. Princeton NJ: Princeton University Press, 1982.

Jenks, J.W. "Trusts in the United States." *The Economic Journal* 2 (1892):70–99.

Keller, M. *Regulating a New Economy: Public Policy and Economic Change in America, 1900–1933*. Cambridge MA: Harvard University Press, 1990.

Knutson, R.D., J.B. Penn, and B.L. Flinchbaugh. *Agricultural and Food Policy*. 4th ed. Upper Saddle River NJ: Prentice Hall, 1998.

Lee, H., J.M. Alston, H.F. Carman, and W. Sutton. *Mandated Marketing Programs for California Commodities*. Berkeley CA: University of California Division of Agriculture and Natural Resources, Giannini Foundation Information Series 96-1, August 1996.

Livestock Marketing Association v. USDA. 335 F. 3d 711 (2003).

Michigan Pork Producers Association v. Campaign for Family Farms. 229 F. Supp. 2d 772 (2002).

Michigan Pork Producers Association v. USDA. Nos. 02-2337/2338 (6th Cir. 2002).

Paarlberg, D. "Effects of New Deal Farm Programs on the Agricultural Agenda a Half Century Later and Prospect for the Future." *American Journal of Agricultural Economics* 65 (1983): 1163–1167.

Paarlberg, R., and D. Paarlberg. "Agricultural Policy in the Twentieth Century." *Agricultural History* 74 (2000):136–161.

Pacific Gas & Electric Co. v. Public Utilities Commission of California. 475 U.S. 1 (1986).

Panama Refining Co. v. Ryan. 293 U.S. 388 (1935).

Pruneyard Shopping Center v. Robins. 447 U.S. 74 (1980).

Rasmussen, W.D. "The New Deal Farm Programs: What They Were and Why They Survived." *American Journal of Agricultural Economics* 65 (1983):1158–1162.

Richards, T.J. "Marketing Orders and Game Theory." *National Food and Agricultural Policy Project Newsletter*. Mesa AZ: National Food and Agricultural Policy Project, Arizona State University East, 1995. Available online at *http://nfapp.east.asu.edu/Newsletters/1995/aug95.htm*.

Roberts v. United States Jaycees. 486 U.S. 623 (1984).

Saulsbury Orchards and Almond Processing, Inc. v. Yeutter. 917 F. 2d 1190 (9th Cir. 1990).

Schecter v. United States. 295 U.S. 495 (1935).

Sequoia Orange Co. v. Yeutter. 973 F. 2d 752 (9th Cir. 1992).

United Foods, Inc. v. United States of America. 197 F. 3d 221 (6th Cir. 1999).

United States v. Butler. 297 U.S. 1 (1936).

United States v. David Buthick Co. 15 F. Supp. 655 (Dist. of Mass. 1936).

United States v. Mills. 315 F. 2d 828 (4th Cir. 1963).

United States v. Rock Royal CO-OP, Inc. 307 U.S. 533 (1939).

United States v. United Foods, Inc. 533 U.S. 00-276 (2001).

United States Department of Agriculture, Economic Research Service. *Major Trends in US Food Supply, 1909–99*. FoodReview 23, Washington DC, April 2000.

United States Department of Agriculture, Economic Research Service. *A History of American Agriculture, 1607–2000*. Washington DC, 2000.
Wileman Bros. & Elliott, Inc. v. Espy. 58 F. 3d 1367 (9th Cir. 1995).
Woeste, V.S. *The Farmer's Benevolent Trust: Law and Agricultural Cooperation in Industrial America, 1865–1945*. Chapel Hill NC: The University of North Carolina Press, 1998.
Wooley v. Maynard. 430 U.S. 705 (1977).

Part II

California Promotion Case Studies

4

The Benefits and Costs of Promotion Programs for California Table Grapes

Julian M. Alston, James A. Chalfant, Jason E. Christian, Erika Meng & Nicholas E. Piggott

The California Table Grape Commission was established in 1968. Almost thirty years later, Alston et al. (1997) published a report on the economic impacts of the commission's market development and promotion activities. The central question addressed in that study was whether the program resulted in an increase in demand and sales and whether, as a result, industry net revenue increased enough to cover the costs of the program. To answer this question, econometric models were developed using time-series data on the key economic variables affecting consumption of table grapes. The central element was a model of annual per capita demand for California table grapes in the United States and Canada. The results from this model were reinforced with models of monthly demand in selected U.S. cities and models of demand in selected Asian export markets. This chapter summarizes the main elements of the aggregate analysis in that report. The models have not been updated. Consequently, the information presented refers to the period analyzed in the study, which was prior to 1997.

Markets for California Table Grapes

California has consistently been responsible for approximately 90 percent of total grape production in the United States and around 97 percent of fresh grape production. Grape production in California is classified by end use into three categories: table grapes, raisin grapes, and wine grapes. Classification is based on the most significant use of the grape variety, although many of the varieties are suitable for multiple uses.[1] In particular, although their primary use continues to be in the dried market, Thompson Seedless grapes have been utilized widely for blending by the wine industry and have been popular for fresh market consumption. Specific varieties of grapes grown for table use and directed mostly to the fresh market continue to comprise the smallest segment of the California grape industry in terms of both acreage and production.

Consumption, Imports, and Exports

Total domestic consumption of table grapes consists of two components: U.S.-produced grapes utilized in the fresh market and grapes imported for fresh-market consumption.[2] The overall trend in consumption of table grapes from 1950 to 1968 was clearly one of steady decline, with sharper drop-offs occurring until 1971, a year in which total consumption reached a low of 352 million pounds (1.7 pounds per capita). The declining trend then reversed dramatically, and total consumption climbed to more than one billion pounds during the first half of the 1980s. Domestic production accounted for a large portion of the increased consumption, particularly through 1988; however, rapidly increasing quantities of imports after 1980 also played an important role. Grape imports in the 1950s and 1960s were primarily Concord grapes from Canada. Import quantities remained fairly constant and low during the 1950s, 1960s, and much of the 1970s, during which time the majority of imported grapes came from Chile. Imports from Chile increased from 77 million pounds in 1980 to more than 619 million in 1994. Imports from Mexico have also increased dramatically since 1980.

Trends in per capita consumption for the United States were similar to those in total consumption. After declining to a low of 1.7 pounds per person in 1971, per capita consumption of domestically produced grapes peaked at almost five pounds by 1988. Total per capita consumption of both domestic and imported grapes also reached its maximum of more than seven pounds in that year. Annual per capita consumption of imported grapes likewise increased markedly during the 1980s and attained its high of more than three pounds in 1990. Total per capita consumption declined somewhat in the first half of the 1990s but generally remained at more than 6.5 pounds per person per year.

On the export side, the overall trend has been less dramatic. The very gradual trend of increasing exports did not change noticeably until after 1985, when export quantities began to increase at a much faster rate. Canada remains the primary importer of U.S.-produced grapes, although its share decreased from around 80 percent of total U.S. exports in the 1960s and 1970s to less than 50 percent of total exports in 1994. Much of the growth in exports to countries other than Canada resulted from newly developed markets in Asia, and Mexico also has proven significant since implementation of the North American Free Trade Agreement (NAFTA).

The California Table Grape Commission

The California Table Grape Commission (CTGC) was established in 1968 to carry out promotional activities on behalf of the state's table grape industry.[3] It was the first agricultural commission organized in the state. Its activities include developing and expanding domestic and foreign consumer demand for fresh market grapes and overseeing operation of the necessary distribution channels, particularly during peak periods of supply. CTGC is funded by an assessment on growers on each pound of grapes shipped.[4] The rate of assessment is determined annually by the commission's board. In 1997, the rate was set at $0.006087 per pound, almost 13 cents per twenty-one-pound box.

CTGC has allocated the majority of its budget to promotion activities. Budget shares for these activities have ranged from approximately 80 percent of the total budget for the first ten to fifteen years of its existence to an average of more than 90 percent since the 1980s. Promotion activities include advertising, merchandising, and public relations for both domestic and foreign markets. Advertising alone has accounted for at least 50 percent of the total budget with at least another 25 percent set aside for merchandising; the balance of the budget has been distributed among administration, food service, public relations, research, and other special projects. CTGC has engaged in export promotion since the late 1970s and was undertaking promotional activity in more than twenty countries by the mid-1990s. In addition to the goal of expanding exports in existing markets, attention also has focused on obtaining export access to countries with prohibitive trade barriers, particularly in Asia and Latin America.

During the past half a century, much has changed in the table grape industry, especially in the period following founding of CTGC. Since many causal factors tended to change together, it is difficult to identify specific causes for particular changes in the industry and to isolate the elements attributable specifically to particular efforts by CTGC. Increased consumption resulted partly from population growth, but per capita consumption also increased. Some of the growth in per capita consumption may be attributable to changes in other demographic variables, such as the age distribution and ethnic mix of the population, or per capita incomes. Also, the growth may reflect a general shift toward healthier diets that include more fruit. In addition, improved technology, both in production and postharvest handling, has meant that table grapes have become less expensive in real terms while consumers enjoy improved quality and a wider range of varieties that are available virtually all year long. These changes have almost certainly contributed to rising per capita consumption. Moreover, significant quantities of table grapes are exported. Various factors may have contributed to the growth in export markets. CTGC is engaged in marketing activities, including market promotion programs, related to these markets. Some support for these activities has been provided by the U.S. government under various export promotion programs administered by the United States Department of Agriculture (USDA).

Promotional efforts of CTGC in both domestic and international markets have worked in parallel with other changes and in many ways have been intimately intertwined with them (for example, one role of promotion is to educate consumers about the availability of new varieties and the healthfulness of the product). The challenge is to measure what proportions of total past changes in consumption have been attributable to these various causes.

A Model of North American Demand for Table Grapes

In this section, we report the results of estimating time-series models of total U.S. and Canadian demand for table grapes, including promotion variables, to obtain estimates of the sales response to promotion. Domestic U.S. and Canadian data were combined, effectively treating the United States and Canada as a single market. We used aggregate annual data for the period 1968–1993.

The Conceptual Model

We used Q_t to represent the per capita quantity of table grapes (of uniform quality) demanded by a representative consumer during a particular year, *t*. The standard theory of consumer demand suggests a model in which the quantity demanded, Q_t, depends on the corresponding *real* (or deflated) values of the price of table grapes, RPG_t; the prices of all other goods (other fruits, in particular) that are substitutes or complements for table grapes, RPS_t;

and total money income or expenditure on all goods, $REXP_t$.[5] Algebraically, this model can be expressed as:

$$(1) \quad Q_t = f(RPG_t, RPS_t, REXP_t).$$

In this model, we would expect the own-price effect to be negative (a negative coefficient on RPG_t), such that an increase in the price of grapes would cause a decrease in consumption. The cross-price effects (the coefficients on other prices, RPS_t) can be positive or negative but are expected to be predominantly positive, especially for close substitutes, and the income effect (the coefficient on $REXP_t$) is probably positive and in the range for a normal good, corresponding to an income elasticity of demand for table grapes between 0 and 1. In other words, an increase in the price of grapes would lead to a decrease in grape consumption while an increase in the price of a substitute or in total money income would lead to an increase in consumption.

This demand equation implicitly assumes constant tastes and preferences for table grapes. To accommodate changes in preferences arising from promotion or anything else that may change and affect demand (such as demographic characteristics of the consumer), the model can be augmented with other demand-shift variables.[6] In a model of consumer demand for California table grapes, appropriate shift variables can be included to represent the effects of such things as (1) increased consumer health consciousness and rising consumer interest in natural foods; (2) other demographic changes, such as changes in the age structure of the population, a higher rate of labor-force participation by women, changes in the ethnic composition of the population, and the fact that more meals are eaten away from home; (3) promotion by CTGC and others and other changes in merchandising expenditures; (4) changes in the varietal mix and general quality of California table grapes; (5) the effects of year-round availability arising from a longer season for domestic varieties and increased imports; and (6) the increased availability of imported substitutes.

We included four demand-shift variables: (1) a variable to represent both rising year-round availability and increased import competition (the per capita quantity of imports of Chilean table grapes), $CHILE\text{-}IMP_t$; (2) a variable to represent the changing varietal mix (the fraction of Thompson Seedless in the total quantity of table grapes), $TS\text{-}SHARE_t$; (3) the quantity of generic promotion (represented by the total—not per capita—promotion expenditures by CTGC expressed in real terms by dividing by the U.S. Bureau of Labor's

Consumer Price Index (CPI)), $RPROMO_t$; and (4) a linear time-trend variable, $YEAR_t$, that was included to represent the effects of other trends, as previously described, that were not being modeled explicitly. Incorporating the shift variables led to an augmented model of demand as follows:

(2) $Q_t = f(RPG_t, RPS_t, REXP_t, RPROMO_t, CHILE\text{-}IMP_t, TS\text{-}SHARE_t, YEAR_t).$

Generic promotion by CTGC, $RPROMO_t$, was expected to have a positive effect on demand if CTGC has been successful in increasing demand for table grapes; otherwise, this variable would have no effect on demand. The other variables could have positive or negative effects, depending on which influences they represented and which of the influences (positive or negative) are most important. For instance, the effect of U.S. imports of Chilean grapes could be negative, reflecting a competitive effect, or positive, representing synergistic benefits from year-round availability. The effect of a decrease in the share of total domestic consumption that is of the Thompson Seedless variety could be positive if a declining share of Thompson Seedless grapes results from a rising share of more valuable or preferred varieties. It could be negative if the Thompson Seedless variety is relatively valuable or if its share is positively correlated with some other important quality factor that is not included in the equation. Such questions can be answered only empirically.

Data Used in the Demand Models

Data on the quantities of California table grapes consumed in North America were derived from tabulations by CTGC. The annual aggregate was calculated by subtracting crop-year exports (excluding Canada) from crop-year shipment totals for the entire California industry. CTGC's tabulations were on a lug basis; these were converted to pounds based on an average lug weight of twenty-two pounds. The models used per capita table grape quantities. To calculate these quantities, and to convert other variables to per capita terms, population data were collected for the United States and Canada from the tabulations by the International Monetary Fund (IMF) in *International Financial Statistics*. Alston et al. (1997) provide a detailed description of the data used in the analysis and the sources of those data.

Wholesale price reports from the various terminal markets covered by the Federal-State Market News Service through USDA were the primary source of data on prices of fresh grapes. For the annual aggregate model, an average of monthly prices from the Los Angeles wholesale market was calculated. The average of CPIs for apples, oranges, and bananas from the U.S. Bureau of Labor Statistics was used as a measure of the price of substitutes. The income variable was derived from annual data on total private domestic consumption expenditures for the United States and Canada that was taken from IMF's tables. Converting figures to real per capita terms involved several steps. First, aggregate Canadian consumption expenditures were converted to U.S. dollars using the annual average exchange rate, also from IMF. Total U.S. dollar expenditures were then converted to per capita terms by dividing by the North American population (United States plus Canada). Finally, the resulting per capita series was deflated using the U.S. CPI. Data were extracted from CTGC financial statements of expenditures on advertising, merchandising, and promotions/communications/public relations. The total promotion expenditure was defined as the sum of expenditures on those three activities. Data on imports of table grapes from Chile were obtained from the U.S. Department of Commerce.

Estimation Results and Selection of the Preferred Model

We focus on the results from a demand equation that is linear in all variables except promotion. We included the square root of promotion instead of the level of promotion and accordingly called it the *square root* model. The square root model was at least as good statistically as any other model we tried, and it was the preferred model from the standpoint of allowing diminishing marginal returns to promotion.[7] The equation for the square root model is

$$(3)\quad Q_t = \beta_0 + \beta_{PG}RPG_t + \beta_{PS}RPS_t + \beta_{EXP}REXP_t + \beta_{PRO}\sqrt{RPROMO_t} + \beta_{IMP}CHILE\text{-}IMP_t + \beta_{TS}TS\text{-}SHARE_t + \beta_T YEAR_t + e_t.$$

In this model, the β coefficients are multipliers that, holding other independent variables constant, translate changes in prices and other right-hand-side variables into changes in per capita quantities consumed. The final term, e_t, represents residual changes in per capita quantities consumed that are not accounted for by changes in the right-hand-side variables; e_t is sometimes referred to as the "error" term since it can be thought of as the error in predicting Q_t using only the right-hand-side variables. Under standard assumptions, these errors are normally distributed random variables with an expected value of 0 and a constant variance.

The strategy for estimation was to estimate the square root model first and then to (1) examine the estimated coefficients to see whether they satisfied our expectations based on the theory noted previously, and (2) at the same time, examine the residuals to see whether their behavior was consistent with conventional econometric assumptions. In addition, diagnostic tests were applied to see whether the validity of the model and its parameters could be rejected by the results from alternative models, using alternative functional forms, and making different assumptions about whether prices and promotion are statistically exogenous. Only if a model passes all of these tests–that is, it is consistent with economic theory and our expectations about the signs and sizes of the coefficients, has well-behaved residuals, and is not rejected by an alternative specification–can we confidently take the next step and use the estimated model to simulate alternative market scenarios as a basis for benefit-cost analysis.

Regression results are presented in Table 4.1. The first column shows the results for the first model, corresponding directly to Equation (3). In this table, the figures in parentheses are t-statistics for the coefficients (a value of less than −2.1 or greater than 2.1 indicates that the coefficient is statistically significantly different from 0 using the conventional 95 percent level of significance) and the figures in brackets are the elasticities of demand with respect to the corresponding variables computed at the mean of the sample data. The base equation fits the data well, accounting for about 90 percent of the variation in consumption (indicated by the R^2).

Most of the parameter estimates had signs consistent with the theory, but one exception was total expenditure, which had a negative coefficient, although it was not statistically significantly different from 0. A disappointing aspect of this model was that the coefficients measuring the effects of prices of both grapes and competing fruit, while of the expected signs and implying plausible magnitudes for elasticities, also were not significantly different from 0. Promotion had a statistically significant, positive effect on consumption of California table grapes, as did the quantity of Chilean imports and the share of Thompson Seedless grapes. The coefficient on the time trend was not statistically significant (and, in fact, was never significant in any variant of this model we tried).

The second column of Table 4.1 shows the results from estimating the same model without including the prices of substitutes or the time trend. We chose to drop the price of substitutes and retain the expenditure variable in the model, even though the latter was slightly less statistically significant and had

Table 4.1. North American Per Capita Demand for California Table Grapes

Independent Variables	(1)	(2)	(3)	(4)	(5)	(6)
RPG_t	−0.516	−0.455	−0.801	−0.968	−0.968	−1.281
	[−1.31]	[−1.22]	[−2.75]	[−3.37]	[−4.03]	[−5.41]
	(−0.15)	(−0.14)	(−0.24)	(−0.29)	(−0.29)	(−0.51)
RPS_t	0.591	–	–	–	–	–
	[1.02]	–	–	–	–	–
	(0.27)	–	–	–	–	–
$REXP_t$	−0.167	−0.188	−0.082	0.014	0.014	0.100
	[−1.16]	[−1.63]	[−0.91]	[0.14]	[0.17]	[1.61]
	(−0.72)	(−0.81)	(−0.36)	(0.06)	(0.06)	(0.51)
$\sqrt{RPROMO_t}$	0.649	0.564	0.767	0.567	0.567	0.519
	[2.15]	[2.62]	[4.55]	[2.98]	[4.65]	[5.45]
	(0.20)	(0.18)	(0.24)	(0.18)	(0.18)	(0.16)
$CHILE\text{-}IMP_t$	0.838	0.887	0.587	0.297	0.297	0.040
	[3.59]	[4.30]	[3.44]	[1.34]	[1.95]	[0.23]
$TS\text{-}SHARE_t$	4.363	4.142	2.955	2.344	2.344	1.541
	[2.39]	[2.34]	[2.18]	[1.78]	[2.75]	[2.29]
$YEAR_t$	−0.015	–	–	–	–	–
	[−0.35]	–	–	–	–	–
$D68_t$	–	–	1.117	–	–	–
	–	–	[4.10]	–	–	–
$D81\text{–}93_t$	–	–	–	0.423	0.423	0.599
	–	–	–	[1.89]	[2.72]	[3.08]
CONSTANT	31.199	2.939	2.055	1.533	1.533	1.158
	[0.37]	[1.75]	[1.61]	[1.25]	[1.80]	[1.73]
R^2	0.92	0.92	0.96	0.96	0.96	0.97
$\bar{R}^2$	0.89	0.90	0.94	0.95	0.95	0.96
Durbin Watson	1.64	1.66	2.24	2.62	2.62	2.59
Sample	1968–93	1968–93	1968–93	1969–93	1969–93	1969–93

Source: Alston et al. (1997, p. 25), Table 3.2.
Note: The t-statistics for values are shown in brackets; elasticities (at means) are shown in parentheses.

the "wrong" sign, for three reasons. First, the expenditure variable belonged in the model according to the theory while the price of a particular substitute did not need to be included since we had effectively included the price of all other goods by using the CPI as a deflator. Second, it was not clear that the particular prices we used appropriately represented the price of close substitutes for table grapes. Finally, the cross-price elasticity simply might not be very large, in which case it would be difficult even with an unbiased estimate to distinguish the true effect given our small sample size and the relatively larger effects of other variables. In any event, the model shown in the second column is only slightly different from the first one in terms of its overall fit and the values of the coefficients on the included explanatory variables.

An initial round of diagnostic tests indicated some potential problems with the model, suggesting a structural change in the relationship prior to 1975. To test for structural changes outside the range where a full "Chow" test could be performed (allowing all of the parameters of the model to shift), we estimated equations that included a dichotomous "dummy" variable for 1968, $D\text{-}68_t$ (set equal to 1 in 1968 and 0 otherwise) and then dummies for both 1968 and 1969 and so on until dummies were included for all years 1968 through 1974. A test for a significant coefficient on an individual dummy variable is an effective test of whether the consumption for that year fits with the general model. Based on a sequence of tests, we concluded that 1968, the first year of operation of CTGC, could be regarded as an outlier, and the model that included the dummy variable, $D\text{-}68_t$, was preferred to one without it. This was essentially the same as dropping the observation for 1968.

These results indicated that the model was consistent with the data only from 1969 forward; there was a significant structural change in demand between 1968 and the following years. The regression for 1968–1993, including a dummy variable for 1968, is reported in the third column of Table 4.1. This equation fits the data very well, accounting for about 94 percent of the variation in annual per capita consumption. However, an examination of the residuals from the model, shown in the third column, and some further formal diagnostic testing indicated that heteroskedasticity could be a problem (i.e., the assumption of a constant variance of the residuals may have been violated), possibly a symptom of specification error. For the diagnostic testing, models were estimated excluding the data for 1968. Since our tests involved splitting the sample, $D\text{-}68_t$ could not be included because this variable was always 0 in the later subsample. When we used the DIAGNOSTIC procedure in the econometric package, SHAZAM, the Goldfeldt-Quandt test indicated evidence of a statistically significant change in the error variance. Options to deal with this problem included applying generalized least squares to correct the estimates from the model or changes in the specification of the model. Both approaches were used. Based on the Goldfeldt-Quandt test results, we split the sample and estimated separate sets of parameters for the full model, first for the period 1969–1980 and then using data for 1981–1993. The parameter estimates for the two samples differed but generally were not statistically significantly different between the two periods or between the models for either subperiod or for the full sample. Subsequent tests led us to allow for a different intercept parameter between the two periods. Within the subsamples, the residuals were homoskedastic. These results led us to include a dichotomous intercept dummy variable (equal to 0 before 1981 and 1 after 1980) and re-estimate the model for the entire sample.

There remained evidence of heteroskedasticity, so we obtained generalized least squares estimates, corrected for different error variances for the periods before 1981 and after 1980. The fourth column in Table 4.1 reports the model with the intercept dummy added as estimated by ordinary least squares (OLS) without the heteroskedasticity correction. The fifth column reports the same OLS estimates of the βs but with revised standard errors that were computed according to White's (1980) procedure. A comparison of these two columns shows that allowing for possibly different error variances increased the precision of the estimates (reflected in larger calculated t-statistics). The sixth column in the same table refers to estimates computed by weighted least squares, allowing for a different error variance after 1980. The weighted least-squares parameter estimates were all slightly different from the OLS estimates and without important qualitative changes with two exceptions: (1) the coefficient on Chilean imports, which was positive and almost statistically significant in the models shown in the first three columns of Table 4.1, became statistically insignificant; and (2) the coefficient on expenditure changed sign but remained statistically insignificant once we added $D81\text{–}93_t$.

The Preferred Model

The model shown in the sixth column served as the starting point for the simulations and cost-benefit analysis.[8] The demand equation was written as

$$(4)\quad Q_t = 1.158 + 0.599\, D81\text{–}93_t - 1.281\, RPG_t + 0.100\, REXP_t + 0.519\sqrt{RPROMO_t} + 0.040\, CHILE\text{-}IMP_t + 1.541\, TS\text{-}SHARE_t + e_t.$$

The price coefficient was negative, significant, and of plausible magnitude and had an elasticity at the sample mean of −0.51: a 1 percent increase in price would lead to a 0.51 percent decrease in annual per capita quantity consumed. The estimated coefficient on total per capita expenditure was positive but not statistically significant. The corresponding elasticity of demand with respect to income at the sample mean was 0.51, which was plausible, indicating that a 1 percent increase in income would lead to a 0.51 percent increase in consumption of table grapes. The coefficient on promotion was large and statistically significant. Notably, this effect was relatively insensitive to the various changes in specification previously described: the elasticity of demand with respect to promotion ranged from 0.16 to 0.24 across the models shown in Table 4.1 and in the preferred model it was 0.16. The estimated effects of Chilean imports and the Thompson Seedless share were positive but not statistically significant. Overall, this model was plausible. It fits the data generally well and explains a high proportion of the variation in consumption of table grapes, as indicated by the $\bar{R}^2$ value of 0.96.

Evaluation of the Benefits from Promotion for 1969 through 1993

In this section, we use the estimated demand parameters from the previous section in a market simulation model to estimate gross and net benefits to the industry from CTGC promotions. Measuring these impacts required (1) a conceptual structural model of the industry market equilibrium, (2) estimates of supply and demand parameters that could be used to parameterize the structural model, (3) estimates of the demand response to promotion expenditures, and (4) information that would allow a transformation of the effects of promotion (through retail demand shifts) and assessments or check-offs (through commodity supply shifts) into measures of benefits and costs.[9]

Conceptual Model of Supply and Demand

The econometric work discussed in the previous section allowed us to estimate how much quantities of grapes sold would increase in response to a given increase in promotion expenditures while holding prices (and other variables) constant. This did not, however, tell us how much sales would actually increase when promotion changed, since prices could not be assumed to remain constant. Indeed, the increase in prices following a promotion-induced shift in demand is an important source of the benefits realized by growers and packers of table grapes. To evaluate the industry's demand-shifting activities

Figure 4.1. A Conceptual Supply and Demand Model

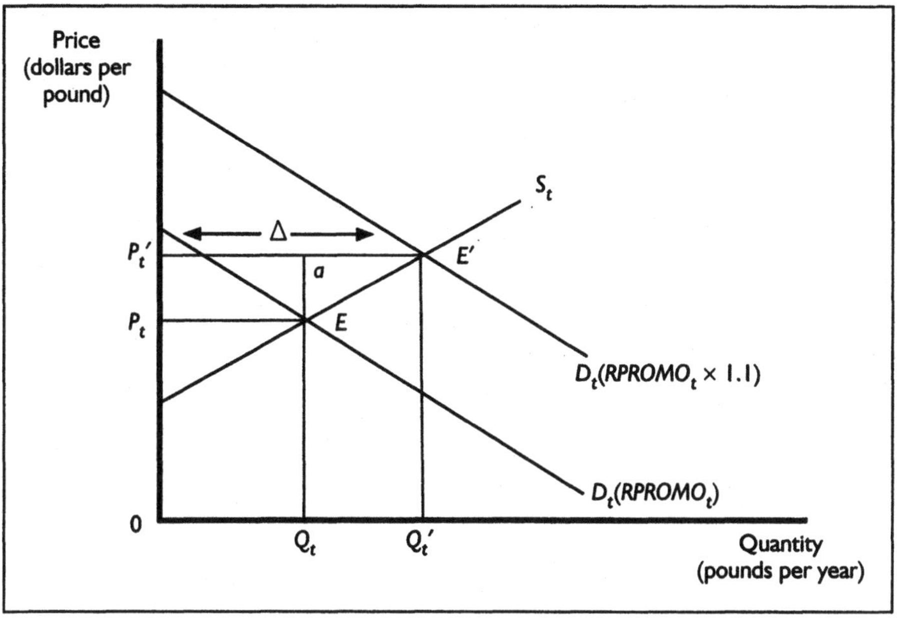

properly, therefore, we had to account for both demand effects and the response of supply to increased price.

Demand Shifts from Promotion. Figure 4.1 illustrates the conceptual supply and demand relationships for a typical year *t*. In the figure, the line labeled S_t represents the supply curve for table grapes. It traces the quantities available to domestic consumers at various prices. At higher prices, more fresh grapes are available domestically; at lower prices, larger quantities of grapes are diverted to other uses such as juice, wine, and raisins or to the export market or they are left unharvested. The line labeled $D_t(RPROMO_t)$ represents the demand curve: at higher prices, consumers purchase a smaller quantity of grapes than they would at lower prices if other factors were held constant. In particular, CTGC's promotion expenditure is held constant at its actual value, $RPROMO_t$, along this demand curve. The equilibrium price, the price at which quantities supplied and demanded are the same, is the price observed at point *E*: price P_t is consistent with the observed quantity Q_t.

The effects of a 10 percent increase in $RPROMO_t$ are illustrated by the outward shift in the demand curve. The new curve is labeled $D_t(RPROMO_t \times 1.1)$.

The econometric model allows us to estimate the horizontal distance of the demand shift, which is identified by Δ in the diagram. In our preferred model, the coefficient on $(RPROMO_t)^{0.5}$ was 0.519. Suppose the actual promotion expenditure was $5 million (in 1995 dollars). This means that a $0.5 million (i.e., 10 percent) increase in CTGC promotion expenditures would be expected to lead to an increase in per capita table grape consumption of $0.519(5.5^{0.5} - 5.0^{0.5}) = 0.057$ pounds if there were no change in price.

Multiplying by the population of the U.S. and Canada (POP_t = 286.5 million in 1993) yielded the total horizontal demand shift from a 10 percent increase in promotion expenditures–about a 2 percent increase in consumption at constant prices. However, this is greater than the actual increase in consumption that would result. An increase in price is needed to bring forth the additional quantities necessary to satisfy the increased demand. This is reflected in the fact that the supply curve slopes up. The new equilibrium point is where the new demand curve crosses the supply curve, E'. Price and quantity both increase to the new equilibrium values P_t' and Q_t'.

The evaluation of promotion expenditures requires both an estimate of increments in prices and quantities resulting from the expenditures and a measure of the costs of supplying the additional quantities to the market. In the diagram, increased prices call forth additional supplies. These supplies come at a cost that, in the case of a perennial crop, may largely be the earnings foregone from other uses of existing production (e.g., for wine or juice) rather than new production. The sellers who were already in the market at price P_t profit by the increase in price to P_t'; their gain is given by $(P_t' - P_t)Q_t$–the area $P_t'EaP_t$ in the diagram. The gains to the sellers of additional supplies are much smaller since they must be reduced by the cost (including revenues from foregone sales in other markets) of the new fresh supplies. This amount is given in the diagram by the area of the triangle aEE'. The total gain to table grape producers is given by the area of the trapezoid $P_t'E'EP_t$; this represents the gain in *producer surplus* resulting from the 10 percent increase in promotion expenditures. Changes in producer surplus coincide with changes in producers' economic profits from the production of grapes in such a situation. The only information requirement, in addition to the econometrically estimated demand parameters for responses to prices and promotion, is information on the supply response to price.

A similar exercise can be carried out to evaluate returns to the entire promotion budget. The position of the demand curve with zero promotion can be calculated–it is shifted to the left by $(0.519)(RPROMO_t)^{0.5}(POP_t)$–and then

Figure 4.2. Incidence of Assessments

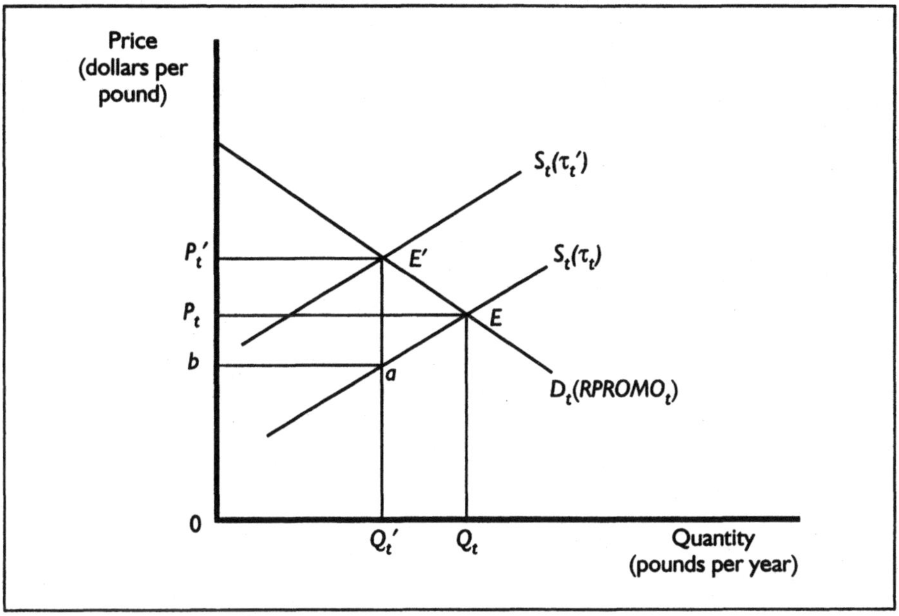

the zero-promotion prices and quantities can be identified and the size of the corresponding trapezoid can be calculated.

Costs of Assessments to Finance Promotion. The gain in producer surplus is not adjusted for the cost of the increase in promotion expenditures. To evaluate the profitability of these expenditures, the gain must be set against the cost. We used two measures of cost. One measure was the total cost of the program (or of the marginal increase in expenditure). However, when the promotion cost is financed by a per-pound assessment, some of the incidence of the assessment falls on consumers through the operation of supply and demand. Thus, the total cost may differ from the cost to producers. A second measure compared the benefits to producers with the producers' share of the cost, allowing that some of the costs of the assessment are borne by consumers.

Figure 4.2 shows the same initial supply and demand curves that appear in Figure 4.1. They are labeled $S_t(\tau_t)$ and D_t, where τ_t represents the actual assessment per pound in year *t*. The equilibrium is at point *E* with price P_t and quantity Q_t. An increase in the assessment per pound is reflected as a shift up of the supply curve to $S_t(\tau_t')$ by the amount of the increase (given by simply adding the additional assessment to the previous price at which producers would be willing to supply any given quantity along the supply curve). This leads to a new equilibrium at point E', a higher consumer price P_t', a smaller net producer price, and a smaller quantity produced and consumed, Q_t'.

The extent of the consumer price increase depends on the slopes of the supply and demand curves. If supply were fixed and unresponsive to price (so that the supply curve is a vertical line), there would be no increase in the consumer price and all of the additional assessment would be borne by producers. The proportion of the assessment borne by producers is smaller when supply is more price-responsive (flatter, more price-elastic).

The additional amount of assessment revenue, given by the increase in the assessment, equals $\tau_t'Q_t' - \tau_tQ_t$. For small changes in the assessment, this is approximately equal to the change in the assessment times the final quantity: $(\tau_t' - \tau_t)Q_t'$. In Figure 4.2, this is equal to area $P_t'E'ab$. This corresponds closely to the full social cost of the change in assessment to finance a change in promotion expenditure of that amount (it leaves out the area of the triangle $E'Ea$, which is negligible for the small changes in assessments considered here). The loss of producer surplus (or profit) associated with the same increase in the assessment equals area P_tEab, only a fraction of the total amount being spent. In the work that follows, we compare producer benefits and the two alternative measures of producer costs.

Aggregation over Time

The approach just described compares hypothetical alternative scenarios against the actual historical record. It could be used to evaluate the *marginal* gross or net benefits by simulating the effects of a 10 percent decrease or increase in promotion expenditure in a particular year, or in every year; or it could be used to evaluate the *average* gross or net benefits from the total investment in promotion by simulating the past under a scenario of no promotion at all. In each case, it is possible to simulate changes in the assessments jointly with corresponding changes in promotion.

If we simulate changes in more than one year, we must be able to aggregate benefits and costs over time. A natural impulse is simply to add them up. But this is appropriate only if past benefits or costs could not have been invested to earn some interest. If the relevant interest rate is not zero, past benefits and costs should be compounded forward to the present. We computed present values of benefits and costs using two alternative interest (compounding) rates: 0 percent (simply adding up over time) and 3 percent (a reasonable value for the long-term, risk-free, *real* rate of interest).

The Supply Model

To conduct the benefit-cost analysis, we combined the preferred demand model with an assumed supply function. First, from the demand side, the *predicted quantities*, $\hat{Q}_t$, were calculated for each year by substituting the actual values for each of the explanatory variables in the demand equation with the point estimates of the parameters:

$$(5)\quad \hat{Q}_t = 1.158 + 0.599\, D81\text{–}93_t - 1.281\, RPG_t + 0.100\, REXP_t + 0.519\sqrt{RPROMO_t} + 0.040\, CHILE\text{-}IMP_t + 1.541\, TS\text{-}SHARE_t.$$

Next, the supply function was defined to be of the constant elasticity form and to pass through the points defined by the predicted quantities from the demand model. That is, the supply function for each year was of the form

$$(6)\quad Q_t = A_t R_t^{\varepsilon} \text{ where } A_t = \hat{Q}_t / R_t^{\varepsilon} \text{ and}$$

where R_t is the producer return per pound in year t, defined as $R_t = (1 - \tau_t)P_t$ where τ_t is the actual promotion expenditure per pound consumed in year t expressed as a fraction of the market price in year t (i.e., the rate of assessment required to finance the actual promotion expenditure). A_t is a parameter that varies from year to year to ensure that, given actual values of prices and the other exogenous variables, each year the supply equation passes through the points defined by the predicted quantities from the demand model. This meant that we could combine the calibrated supply model and the estimated demand model to simulate past actual prices and predicted quantities. Supply functions were calibrated using alternative supply elasticities (ε) of 0, 0.5, 1, 2, and 5. Changes in producer surplus were calculated by integrating the function over the range of a price change. In practice, this translated to the following formula for the change in producer surplus:

$$(7)\quad \Delta PS_t = \frac{R_t' Q_t' - R_t Q_t}{1 + \varepsilon}.$$

Simulations

Using these definitions of supply and demand equations, we could replicate the past: by equating the equations for supply and demand and solving for the market equilibrium, we obtain values of actual prices and predicted quantities (from the demand model) given the actual values for the exogenous variables. In addition, we could simulate counterfactual scenarios using hypothetical values for the exogenous variables. We conducted six types of counterfactual simulations:

- using hypothetical values for the promotion expenditure in every year (either 1.1 times actual values or zero promotion expenditure) with actual assessment rates.
- using hypothetical values for the assessment rate in every year (either 1.1 times actual values or zero assessments) with actual promotion expenditure.
- changing both the promotion expenditures and the assessments (setting both at either 1.1 times actual values or zero).

For each simulation we calculated two measures of producer costs of promotion: (1) the cost of the promotion expenditure, and (2) the producer surplus loss associated with an assessment sufficient to generate the same amount of expenditure. Combining these two measures with the six simulations and the two alternative discount rates (0 or 3 percent) yielded a total of twenty-four benefit-cost comparisons for every value of the supply elasticity. Using five supply elasticity values means we have a total of 120 estimates of benefits, costs, or both.

Benefit-Cost Ratios

The results from the benefit-cost analysis conducted using the point estimates of the parameters are summarized in Table 4.2. The upper half of Table 4.2 refers to estimates using 0 percent to compound the benefits forward over time; the lower half uses a 3 percent compounding rate. For now, we focus on the measures using 0 percent compounding (the upper half), which are smaller but not much different from their 3 percent counterparts. We will review the center column, which was derived using a supply elasticity of 1.

The first entry in the center column indicates that, over the twenty-six-year period, total producer benefits from promotion were $4,322.6 million. The next entry down shows that the total producer incidence of assessments to finance that promotion was $29.4 million. The next entry down is simply the amount spent on promotion during that same twenty-six year period, $115.2 million. The mean value of the ratio of the total producer benefit to the producer incidence of the assessments is 147.1. This benefit-cost ratio indicates that, for every dollar borne by producers in assessments to fund promotion, producers gained $147 in additional profit. This is a very high benefit-cost ratio. Any ratio greater than 1 indicates profitability. The fifth and final entry in this set shows that the mean value of the ratio of the total producer benefit to the total cost of promotion is 37.5, which is still a very high benefit-cost ratio.

Table 4.2. Benefits and Costs of Promotion of California Table Grapes in North America

	Supply Elasticity				
	0	0.5	1	2	5
0 Percent Compounding					
Average Benefits and Costs					
Present Value of Producer Benefits	19,876.5	6,589.8	4,322.6	2,606.5	1,205.8
Present Value of Producer Cost Incidence	115.2	46.1	29.4	17.1	7.6
Present Value of Total Program Expenses	115.2	115.2	115.2	115.2	115.2
Producer Benefits/Producer Costs	172.5	143.0	147.1	152.1	157.9
Producer Benefits/Total Expenses	172.5	57.2	37.5	22.6	10.5
Marginal Benefits and Costs					
Present Value of Producer Benefits	970.1	380.1	240.2	139.0	61.6
Present Value of Producer Cost Incidence	11.5	4.6	2.9	1.7	0.8
Present Value of Total Program Expenses	11.5	11.5	11.5	11.5	11.5
Producer Benefits/Producer Costs	84.2	82.2	81.5	80.8	80.3
Producer Benefits/Total Expenses	84.2	33.0	20.8	12.1	5.3
3 Percent Compounding					
Average Benefits and Costs					
Present Value of Producer Benefits	25,964.1	8,992.2	5,939.9	3,603.2	1,676.0
Present Value of Producer Cost Incidence	153.3	63.7	41.0	24.1	10.8
Present Value of Total Program Expenses	153.3	153.3	153.3	153.3	153.3
Producer Benefits/Producer Costs	169.4	141.2	144.9	149.6	155.2
Producer Benefits/Total Expenses	169.4	58.7	38.8	23.5	10.9
Marginal Benefits and Costs					
Present Value of Producer Benefits	1,267.3	516.0	329.5	192.2	85.7
Present Value of Producer Cost Incidence	15.3	6.4	4.1	2.4	1.1
Present Value of Total Program Expenses	15.3	15.3	15.3	15.3	15.3
Producer Benefits/Producer Costs	82.7	80.8	80.1	79.5	79.0
Producer Benefits/Total Expenses	82.7	33.7	21.5	12.5	5.6

Source: Alston et al. (1997, p. 40), Table 3.4.
Note: Present values are in millions of constant (1995) dollars.

The next set of five entries in the center column refers to the same measures of benefits and costs when considering a marginal increase of 10 percent rather than looking at total benefits relative to total promotion expenditure. The first entry in this group indicates that, in total over the twenty-six-year period, producers would have benefited by $240.2 million if the promotion expenditure had been increased 10 percent over the actual value in each year. The next entry down shows that the total producer incidence of assessments to finance the additional promotion would have been $2.9 million over the same period. The next entry down shows that the cost of spending an additional 10 percent on promotion each year would have been $11.5 million. The mean value of the ratio of the marginal producer benefit to the producer incidence of the assessments is 81.5, a little more than half the corresponding average benefit-cost ratio. The fifth and final entry in this set shows the mean value of the ratio of the marginal producer benefit to the total cost of an additional 10 percent expenditure on promotion in every year. The ratio is 20.8, which is also a little more than half the value of its counterpart considering average rather than marginal benefits and costs, for which the ratio was 37.5.

What is the effect of different assumed values for the supply elasticity? This can be seen by comparing entries across columns in a given row. As the supply elasticity increases from 0 to 5, the total producer benefit falls from $19,876.5 million to $1,205.8 million. The more elastic the supply function, the more the expansion in demand is satisfied by an increase in quantity supplied and the less it is satisfied by an increase in price. If supply were perfectly inelastic ($\varepsilon = 0$), there would be no increase in production and the increase in demand would have to be completely met by an increase in price. In this case, since demand is inelastic, any increase in demand generates a proportionately greater increase in price, which accounts in part for the very large measures of producer benefits when supply is fixed ($\varepsilon = 0$).

Likewise, a similar pattern can be seen in the mean value of the producer incidence of assessments to finance the promotion, which ranged from $115.2 million when supply was perfectly inelastic to much less than one-tenth of that value, $7.6 million, when the supply elasticity was 5. Indeed, for small shifts in supply due to assessments and in demand due to promotion, the patterns of the producer benefits from a demand shift and producer incidence of the costs of an assessment should match perfectly. Hence the benefit-cost ratio should not be affected much by the supply elasticity, especially for smaller shifts. This can be seen in the table. The estimate of the average benefit-cost ratio varied between 172.5 and 143. For the marginal benefit-cost ratio, considering only a 10 percent change in assessments and promotion, the ratio was more nearly constant, ranging from 84.2 to 80.3.

When the producer benefit, which decreases as the supply elasticity increases, is divided by the total promotion expenditure, which remains constant regardless of the supply elasticity, the ratio declines substantially with increases in the supply elasticity. The benefit-cost ratio (using average benefits) when the supply elasticity was 0 was the same as the value for the ratio using the producer incidence of the assessments (since, when the supply elasticity is 0, all of the burden of the assessments falls on producers). However, it fell from this value of 172.5 to 10.5 when the supply elasticity increased to 5. A similar pattern held for the marginal benefit-cost ratio, which fell from a maximum of 84.2 when supply was fixed to a value of 5.3 when the supply elasticity was increased to 5.

Interpretation

The benefit-cost ratios calculated in this study are high both in absolute terms and when compared to estimates in the literature for other commodity promotion programs, which could lead some to doubt the results. We addressed this possibility by providing results for models of domestic per capita demand using monthly data and models of export demand and by examining the implications of the estimates in more detail.

Additional Models of Demand Response to Promotion

Two sets of demand models were estimated using monthly per capita data: (1) using aggregate per capita data for the period 1972–1993, and (2) using per capita data for individual U.S. cities for two years, 1992 and 1993. The primary purpose of the aggregate monthly model was to serve as a check on the annual model. We used the same basic specification, augmented with monthly intercept dummy variables to capture seasonal shifts in demand, and tried the promotion variable in square root and in linear form.

The results generally conformed to those from the aggregate model. The elasticities of demand with respect to price and promotion from the preferred square root model, estimated using weighted least squares, were −0.34 and 0.22, respectively, which is in a range that implies a benefit-cost ratio greater than the ratio of 150 implied by the elasticities of −0.51 and 0.16 generated by the annual model. The city-month models included a variable to represent the effects of spot radio and television advertising expenditures by CTGC by city and month, included linearly and in square root form with an allowance for lagged response. In the square root model, the elasticities of demand with respect to price and advertising were −0.24 and 0.40, which are comparable to those from the other models but imply an even more favorable benefit-cost ratio.

Export demand models were estimated using data on exports to four Asian markets—Hong Kong, Malaysia, Singapore, and Taiwan—both in aggregate and individually for each country. In the preferred model of aggregate export demand, the elasticities of demand with respect to price and promotion from the square root model were −0.83 and 0.21, respectively. The corresponding average benefit-cost ratio was in the range of 9, and the marginal benefit-cost ratio was about 4.2.[10]

An Alternative Interpretation

We conducted some further simulations based on a hypothetical increase of 10 percent in promotion expenditure in 1993 with producer levies increased just enough to pay for that increase in promotion. This was an effort to see how much we would have to change our best estimates of key parameters to reduce the benefit-cost ratio to 1. We considered two of the parameters in the aggregate domestic demand model to be critical: the own-price elasticity (measured by the slope coefficient β_{RPG}) and the elasticity of demand with respect to promotion (measured by the slope coefficient β_{RPROMO}); we also looked at the supply elasticity, ε. First, we fixed each of these key parameters at its point estimate (with a value of 1 assumed for the supply elasticity) and found that a 10 percent increase in promotion expenditures implied a nearly $12 million increase in producer surplus. Then we varied each parameter in turn, holding the others constant, to find the value for the parameter in question that would imply no increase in producer surplus. This is the value that would have to be true for the change in promotion expenditures to represent a break-even proposition—a value such that an increase in promotion would increase producer revenues by no more than the increase in promotion cost and therefore equivalent to a marginal benefit-cost ratio of 1 when comparing the producer benefit to the producers' share of the total cost of the promotion expenditure. According to our results, even an increase in the size of the demand elasticity to more than −20 (compared with the 1993 value of −0.25) or a supply elasticity greater than 10 would still generate a small profit from increased promotion expenditures. These values are quite extreme, supporting the conclusion that increases in promotion expenditures would indeed be profitable using our best point estimates. Alternatively, using the best estimate of the price elasticity of demand and the assumed supply elasticity of 1, the elasticity of demand with respect to promotion would have to be reduced by a factor of one one-hundredth, to 0.0016, before the 10 percent increase in promotion expenditures would not be profitable.

Conclusion

Our econometric results provide strong evidence supporting the view that promotion by CTGC has significantly expanded demand for California table grapes. Those results were used in a market simulation model, along with a range of assumed values for the elasticity of supply, to compute estimates of benefits from promotion and compare them to the costs. The estimated benefits were many times greater than the total costs or the producer incidence of costs of a check-off, even when we used very large assumed values for supply elasticities that resulted in smaller estimates of producer benefits. To obtain a benefit-cost ratio of 1, just sufficient for the program to break even, we would have to believe that the true measure of demand response to promotion was one one-hundredth of our best estimate. Statistically, our estimates indicate that such a small value is highly improbable.

Notes

1 In 1993, 19 percent of grapes that were classified as raisin-type and 22 percent of grapes classified as table-type were crushed for use in the wine industry. In the period 1984–1993, an average of 11.3 percent of raisin-type grape production was consumed in the fresh market (U.S. Department of Agriculture and California Department of Food and Agriculture 1993).

2 We calculated the first component, consumption of U.S.-produced grapes as table grapes, from fresh shipments of table- and raisin-type grapes less total fresh-market exports.

3 At the time of CTGC's establishment, growers of Tokay grapes were already covered under an existing marketing order and were not included under the CTGC mandate. However, in 1972 Tokay-grape-grower representatives voted for inclusion of Tokay grapes in CTGC's promotion programs. The Tokay federal marketing order was terminated in 1995.

4 In addition to funds raised by the assessments, in some years CTGC has used some Market Access Promotion Program (MAPP) funds provided by the U.S. government to support export promotion activities.

5 To remove the effect of inflation, we divided all prices and incomes by a general price index, the Consumer Price Index, thereby expressing all of the monetary variables in the demand equation in *real* terms.

6 Blaylock and Smallwood (1986) document some of the general trends in consumer demand for food that may be reflected in shift variables of these types. See also Frazao (1992).

7 The statistical results were essentially identical between the square root model and a model in which the only difference is that the promotion variable entered linearly rather than in square root form.

8 We subjected this model to extensive further diagnostic tests as described in detail in Alston et al. (1997). The tests included (1) a battery of tests for heteroskedasticity and omitted variables, (2) tests for serial correlation of the residuals, (3) tests against alternative specifications of the functional form for the demand equation, and (4) tests for simultaneity and endogeneity. The model stood up well against these tests; again, see Alston et al. (1997) for details.

9 As well as reporting the point estimate of a benefit-cost ratio, Alston et al. (1997) report the corresponding 99 percent confidence interval: a lower bound and upper bound between which the true benefit-cost ratio would be found with 99 percent confidence.

10 Note, however, that these benefit-cost ratios were calculated without allowing for the fact that much of the funding for export promotion–about 75 percent–was provided by the state and federal governments rather than from assessments.

References

Alston, J.M., and J.A. Chalfant. "Can We Take the Con Out of Meat Demand Studies?" *Western Journal of Agricultural Economics* 16 (1991):36–48.

Alston, J.M., J.A. Chalfant, J.E. Christian, E. Meng, and N.E. Piggott. *The California Table Grape Commission's Promotion Program: An Evaluation.* Berkeley CA: University of California Division of Agriculture and Natural Resources, Giannini Foundation Monograph 43, November 1997.

Blaylock, J.R., and D.M. Smallwood. *U.S. Demand for Food: Household Expenditures, Demographics, and Projections.* Washington DC: U.S. Department of Agriculture, Economic Research Service Technical Bulletin 1713, 1986.

California Table Grape Commission. *Marketing the Natural Snack: The First Twenty-Five Years.* Sacramento CA, 1995.

Frazao, E. *Food Spending by Female-Headed Households.* Washington DC: U.S. Department of Agriculture, Economic Research Service Technical Bulletin 1806, 1992.

International Monetary Fund. *International Financial Statistics.* Washington, DC, various issues.

Jones Putnam, J., and J.E. Allshouse. *Food Consumption, Prices, and Expenditures, 1996: Annual Data, 1970–1994.* Washington DC: U.S. Department of Agriculture, Economic Research Service Statistical Bulletin 926, 1996.

Prescott, R. *Charts on U.S. Food Consumption.* Washington DC: U.S. Department of Agriculture, Economic Research Service, February 1983.

U.S. Department of Agriculture/California Department of Food and Agriculture. "Marketing California Grapes for Fresh Use, 1992 and 1993 Season." Federal-State Market News Service, 1993.

U.S. Department of Labor, Bureau of Labor Statistics. "Consumer Price Index: Average Price" data. Washington DC.

White, H. "A Heteroskedasticity-Consistent Covariance Matrix Estimator and a Direct Test for Heteroskedasticity." *Econometrica* 48 (1980):817–838.

White, K.J., S.A. Haun, N.G. Horsman, and S.D. Wong. *SHAZAM Econometrics Computer Program: User's Reference Manual*, Version 6.2. New York NY: McGraw-Hill Book Company, 1990.

5

An Economic Analysis of Generic Egg Advertising in California

Todd M. Schmit, J. Carlos Reberte & Harry M. Kaiser

Since 1984, the California Egg Commission (CEC) has financed egg advertising, promotion, education, and research aimed at increasing egg consumption and enhancing returns for California egg producers. California produces more eggs than any other state–more than 25 million laying hens producing more than six billion eggs per year–approximately 10 percent of the national total. CEC generates revenue for operation of its programs by assessing egg handlers 1 cent per dozen shell eggs sold within the state.[1] CEC's assessment and marketing program is completely independent of the national program operated by the American Egg Board (AEB). As such, the assessments levied on California producers and handlers by CEC are in addition to assessments at the national level.[2] Eggs that are sold outside of California, exported, consumed by the military, or transported to breaking plants within the state for further processing are exempt from CEC's assessments.

CEC has an annual budget of about $4 million. Approximately 90 percent of this budget is directed to advertising, public relations, promotion, and education and research purposes; media advertising comprises the largest share. Annual expenditures for advertising have consistently been more than $3 million, approximately 80 percent in the form of media advertising and the remaining 20 percent corresponding to the production costs of advertising (CEC unpublished data). Since 1985, CEC advertising expenditures have totaled more than $37 million.

Figure 5.1. Quarterly California Egg Consumption Per Capita

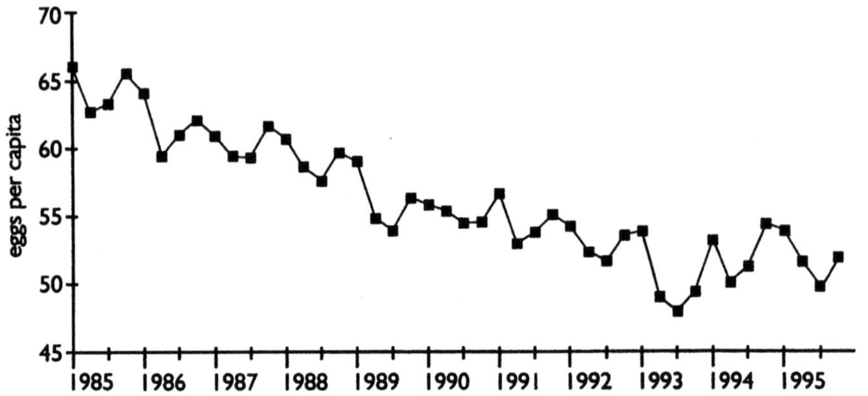

Source: California Egg Commission unpublished data, 1996.

Figure 5.1 displays per capita egg consumption in California from 1985 through 1995. Between 1985 and mid-1995, quarterly per capita consumption of eggs and egg products dropped from sixty-six eggs to forty-eight eggs; since then, it has increased to its current level of nearly fifty-four eggs. This represents an 18 percent decrease overall in consumption and compares to a decrease at the national level of approximately 10 percent for the same period. The downward trend in demand for eggs has been attributed to the abundance of information about heart disease and cholesterol intake. This occurred most predominantly from 1985 to late 1990; consumption has tended to stabilize since then, which may be attributable to a saturation of cholesterol-related information in the market and to distinctions made between so-called "good" and "bad" cholesterol in certain food products. In fact, recent consumer tracking studies have found that consumers' negative attitudes toward eggs are no longer increasing (Smith 1993). Much of past generic egg advertising was developed under a defensive strategy to counter negative publicity about cholesterol; however, efforts currently are focused on convenience and the nutritional value of eggs in a well-balanced diet.

Compared to consumption, producer prices have been much more variable. As Figure 5.2 indicates, real producer prices for eggs have generally trended downward during the period, and in nominal terms, prices are virtually equivalent to where they were in 1985. Prices trended downward into late

1988, increased into early 1991, dropped substantially over the next two years, and then leveled out over the remainder of the sample period.

Given the slow but steady decline in per capita egg consumption and highly variable producer prices, an evaluation of the effectiveness of generic egg promotion in California is both timely and important. Accordingly, the purpose of this study was to measure the impact of generic egg advertising in California on producer prices and net revenue from 1985 through 1995. To measure these impacts, an econometric model describing the supply and demand for California eggs was estimated. The estimated model was then used to simulate the impact of two advertising scenarios on producer prices, production, and net revenues for the 1985 through 1995 period. In the first scenario, generic egg advertising expenditures were set equal to historical levels; in the second scenario, expenditures were increased by 1 percent. Based on the results of the simulation, the marginal rate of return to advertising expenditures (i.e., benefits to costs of additional generic advertising) was computed.

Previous Research

Various egg price-forecasting models and egg-industry models were common in the 1970s and 1980s (e.g., Miller and Masters 1973; Roy and Johnson 1973; Salathe, Price, and Gadson 1983; Blaylock and Burbee 1985; Chavas and Johnson 1981; Schrader et al. 1978), but little research has focused on the impact of generic advertising on farm-level prices and per capita consumption.

Figure 5.2. California Producer Egg Prices, 1985–1995

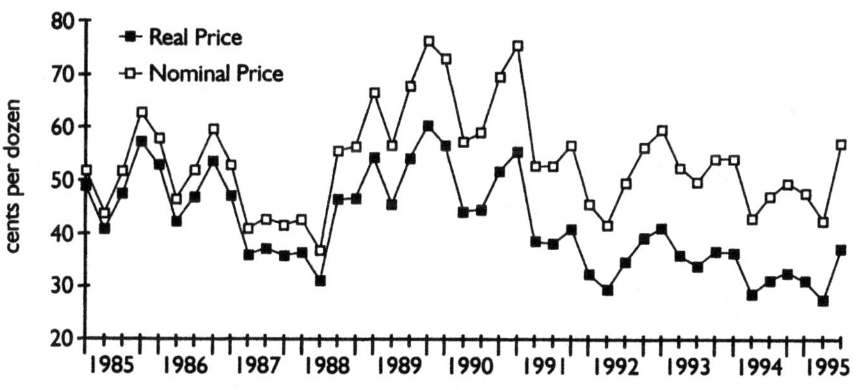

Source: Bell 1996.

Chavas and Johnson provided the most detailed model of the egg industry in the literature, encompassing estimation of production, consumption, and storage components. In 1990, Brown and Schrader estimated an econometric model for the egg industry from 1955 through 1987 and found that information on links between cholesterol and heart disease had a significant negative impact on consumer demand for eggs. However, the effectiveness of generic advertising to maintain consumption levels was not addressed in either of those studies.

Generic egg advertising was studied more recently by McCutcheon and Goddard (1992) and Chyc and Goddard (1994); however, their studies dealt with the Canadian egg sector, which is supply-managed; imports and egg production are restricted to maintain producer prices. Although evaluation of advertising in light of government intervention is more problematic, both studies determined that generic egg advertising had a positive impact on demand.

The only recent study to examine generic egg advertising impacts in the United States is Reberte, Schmit, and Kaiser (1996), who estimated a comprehensive model of the U.S. egg sector, including production and consumption equations for both whole and processed egg products. Although similar in form to Chavas and Johnson (1981), the model incorporated national AEB advertising expenditures since 1990. Advertising was shown to significantly influence farm prices and, ultimately, producer net revenues.

The Conceptual Model

Production of eggs is a fairly straightforward process. Approximately five months after chicks have been placed into the hatchery supply flock, hens are released into the laying flock. The laying process lasts twelve to fourteen months, after which the hens are either slaughtered or force-molted. In the latter case, the hens receive a two-month rest before resuming the laying cycle in the production flock. The supply equation in this study was specified as follows:

$$(1)\quad PROD = f(PEGG{-}1, FEED{-}1, LAY, PTVTY, QTR1, QTR2, QTR3, TIME, PROD{-}1),$$

where *PROD* is production of eggs in California; *PEGG*−1 is the real producer price for large eggs in Southern California lagged one quarter;[3] *FEED*−1 is the real feed price in Southern California lagged one quarter; *LAY* is a four-quarter moving average of the size of the California laying flock; *PTVTY* is the average layer's productivity for the quarter; *QTR*1, *QTR*2, and *QTR*3 are quarterly dummy variables to account for seasonal variation in supply; *TIME* is a time trend as a proxy for structural change; and *PROD*−1 is egg production lagged one quarter. Lagged producer prices reflect a delay in farmers' production responses to a current change in price. Feed costs are for "typical" feed at delivered net prices in Southern California.[4] Lagged production is incorporated as in previous egg-industry studies (Salathe, Price, and Gadson 1983; Chyc and Goddard 1994). Egg prices and production costs for the time period evaluated were obtained from a study by Bell (1996).

Demand for California eggs was modeled in price-dependent form and included the following variables: farm egg demand, income, price of substitutes, generic advertising, a time trend, three quarterly dummy variables, and the percent of women in the labor force. The price of egg substitutes was represented in the model by a composite retail price index for red meat for the western region. There are conflicting views about what is a proper substitute product for eggs. Like previous studies, our model included red meat prices as possible substitutes. Although the composite price index for red meat imparts some rigidity in the model, it eliminates possible collinearity problems associated with separate red-meat price series. A price index for cereals and bakery products was included in early model specifications but was later removed due to poor performance and a high standard error relative to the coefficient estimate. The time trend was included to represent changes in consumer tastes and preferences for eggs over the preceding ten years. Consumer demand was included as an identity in the model as farm production less net exports (including military consumption).

Similar to Reberte, Schmit, and Kaiser (1996) and to Brown and Schrader (1990), we included the percent of women in the labor force in the demand equation. Conflicting views exist regarding the relationship of this variable to egg consumption. On one hand, Stillman (1987) suggests a negative relationship; as more women work outside the home, less of their time is available to prepare eggs for breakfast. On the other hand, Brown and Schrader assert that increasing numbers of working women increases the number of breakfasts eaten away from home, which may increase overall egg consumption. Other consumer demographic variables were included in earlier specifications (e.g., the proportion of expenditures that were for food eaten away from home for breakfast and lunch), but they performed poorly and were removed prior to the final specification given here.

Following numerous previous empirical studies on generic advertising evaluations, this study measured advertising effort in terms of advertising expenditures. As mentioned earlier, AEB's advertising and promotion program operates independently of CEC's advertising efforts. AEB and CEC advertising expenditures were converted to a per capita basis and then summed to account for all advertising in California. Although actual per capita advertising expenditures in California may be higher or lower than those at the national level, the approach followed here provides a reasonable approximation. The average quarterly advertising expenditure in California per one hundred people from 1985 through 1995 was $2.46. Of that amount, $2.31—nearly 94 percent—came from CEC expenditures. Both total and per capita CEC advertising expenditures are detailed in Figure 5.3.

Two cholesterol indexes were separately included in preliminary estimations; however, neither was incorporated in the final specification due to poor performance and unexpected signs. The first index was constructed by Ward (1992) based on national survey data and measured the percent of consumers expressing strong or moderate concern about cholesterol in their diets. The second index we constructed was similar to the one in McGuirk et al. (1995) and measured the number of articles dealing with dietary cholesterol issues in national publications.[5] Both Brown and Schrader (1990, eggs) and McGuirk et al. (1995, meat) found a negative and significant relationship between cholesterol "concern" and consumption of eggs and meat. However, both studies used earlier and longer time periods than the one covered by this study. To the extent that dietary cholesterol concern can be viewed as a structural change in the demand for eggs, its impact should have been captured by the time trend.

Based on the preceding discussion, the farm-level demand for eggs in California was modeled as the following equation:

$$(2)\quad PEGG = f(PROD, PMEAT, INC, ADV, QTR1, QTR2, QTR3, TIME, WOMEN),$$

where *PEGG* is the real producer price for large eggs in Southern California; *PROD* is farm-level production in California available for state consumption and net exports; *PMEAT* is the real retail price index for red meat in the western region of the U.S.; *INC* represents real disposable income per capita in California; *ADV* represents per capita generic egg advertising expenditures in California; *QTR1*, *QTR2*, and *QTR3* are quarterly dummy variables to account

Figure 5.3. Real CEC Advertising Expenditures by Quarter, 1985–1995

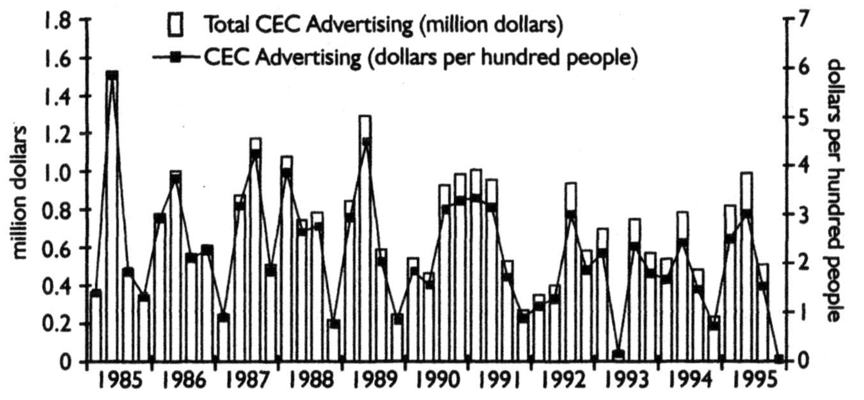

Source: California Egg Commission.

for seasonal variation; *TIME* is a time trend; and *WOMEN* is the proportion of women employed in the labor force.

Econometric Results

Both the supply and demand equations were specified on a per capita basis using a double-logarithmic functional form (Table 5.1). The model was estimated using quarterly data from January 1985 through June 1995. The procedure suggested by Godfrey (1993, pp. 116–117) was used to test for autocorrelation. This procedure requires first obtaining each equation's vector of residuals and then re-estimating the model, including in each equation the lagged residual as an additional regressor. The null hypothesis of no autocorrelation for a given equation is rejected if the coefficient on the lagged residual is significant based on a t-test. This procedure indicated that the residuals of the demand equation exhibited first-order autocorrelation. Autocorrelation correction procedures were subsequently applied.

The supply equation explained more than 98 percent of the variation in production, and all variables had expected signs. As expected, there exists a positive relationship between producer prices and production; however, the own-price elasticity of supply is relatively low. The short-run supply elasticity of 0.03 is comparable with previous research (e.g., Schrader et al. 1978; Chavas and Johnson 1981) and the long-run elasticity increased only to 0.05.

Table 5.1. California Supply and Demand Specification Results

Supply Equation

$$\ln(PROD/POP) = -8.055 + 0.033 \ln(PEGG)_{-1} - 0.063 \ln(FEED)_{-1} + 0.465 \ln(LAY)$$

(−2.02) (1.19) (−1.19) (2.15)

$$+ 0.686 \ln(PTVTY) - 0.029\, QTR1 - 0.041\, QTR2 - 0.014\, QTR3 - 0.005\, TIME$$

(5.52) (−2.90) (−4.25) (−1.36) (−2.53)

$$+ 0.360 \ln(PROD/POP)_{-1}$$

(2.89)

$R^2 = 0.98$

Demand Equation

$$\ln(PEGG) = -170.678 - 0.579 \ln(PROD/POP) + 5.019 \ln(PMEAT) + 14.149 \ln(INC)$$

(−4.78) (−1.23) (4.67) (4.97)

$$-0.007\, QTR1 - 0.201\, QTR2 - 0.118\, QTR3 - 1.246 \ln(TIME) + 1.387 \ln(WOMEN)$$

(−0.23) (−4.37) (−3.56) (−5.17) (2.27)

$$+ 0.026 \ln(ADV/POP) + 0.038 \ln(ADV/POP)_{-1} + 0.038 \ln(ADV/POP)_{-2}$$

(1.76) (1.76) (1.76)

$$+ 0.026 \ln(ADV/POP)_{-3}$$

(1.76)

$R^2 = 0.86$

$\rho_1 = 0.416$ (1.95)

Note: *POP* is the average quarterly state population, t-ratios are in parentheses, and ρ represents the autoregressive parameter. Production, egg price, production cost, disposable income, and advertising expenditure data were provided by CEC. Data for *PMEAT* and *WOMEN* were obtained from an online computer file from the U.S. Department of Labor, Bureau of Labor Statistics, Washington, DC, 1985–1995.

Due to the advancement of vertical integration in the poultry sector, some would argue that the supply of eggs is quite sensitive to changes in price; however, a considerable share of egg production is under contract to market firms or is carried out as only one phase in vertically integrated operations (Salathe, Price, and Gadson 1983). In addition, empirical estimates of supply elasticities are generally lower for products with few alternative uses (Tomek and Robinson 1990); as such, large capital investment suitable only for egg production may prohibit production adjustments in response to price changes beyond altering the number of laying hens or their productivity through feed ration adjustments or other measures. Furthermore, the availability of poultry from breeding flocks and other biological constraints can restrict production response (Salathe, Price, and Gadson 1983). To address the implications of these assumptions specifically, sensitivity analysis with regards to supply elasticities is reported in the following section to evaluate the resulting impacts on producer returns to advertising.

The number and productivity of laying hens were also found to have a positive effect on production. Lagged feed prices were negatively related to production and exhibited a fairly inelastic response. The negative signs on the three seasonal dummy variables suggested that production is highest in the fourth quarter. Finally, given the steady decline in per capita production over the time period evaluated, the negative trend variable was not unexpected.

The demand equation explained 86 percent of the variation in producer prices, and all variables had expected signs. The results indicated an elastic response to price of approximately −1.7 (reciprocal of the price flexibility coefficient). This result differed from the results of two earlier studies (Brown and Schrader 1990; Chavas and Johnson 1981) that calculated price elasticities of −0.2 and −0.3. The more elastic results, while somewhat surprising, may have been due to several factors. California consumer responses may not be typical of responses by consumers nationwide, especially for the relatively short time horizon evaluated. In addition, retail prices for eggs in California were more than 40 percent higher on average than national retail prices, so one would expect a higher elasticity in California.

The positive and significant coefficient on the meat price index indicated that red meats are substitute products for eggs. The positive coefficient on income suggested that eggs are a normal good. As noted in other studies, both cross-price and income parameters showed inelastic responses to price (0.20 and 0.07, respectively). Farm-level demand was higher in the fourth quarter, which may have been due in large part to the holiday season. Not surprisingly, given the steady decline in consumption since 1985, the trend variable displayed a negative and statistically significant relationship.

Lagged as well as current generic advertising expenditures were included to account for delays in the demand response to advertising (see, for example, Forker and Ward 1993, p. 169). To mitigate the impact of multicollinearity among the lagged advertising variables, the lag weights were approximated using a second-degree polynomial lag structure with both endpoints restricted to zero. In this way, only one advertising parameter had to be estimated. Several lag lengths were considered, for as long as a full year (four quarters). Based on the significance of the lagged coefficients, three lags were included in the final specification. The estimated coefficients indicated that generic advertising had a positive and statistically significant (at the 10 percent significance level) impact on the producer egg price. The long-run elasticity of price with respect to generic advertising, obtained by summing the advertising coefficients over all lags, was 0.13; i.e., a 1 percent increase in advertising expenditures results in an average increase in the producer egg price of 0.13 percent.

Producer Returns to Advertising

A dynamic, in-sample simulation was conducted to measure how well the model replicated historical values of the endogenous variables. Root mean square percentage errors (%RMSEs) were computed to measure how predicted values deviated from actual values. The %RMSEs for price, production, and consumption were 1.98 percent, 0.41 percent, and 1.74 percent, respectively. In addition, acceptable mean-square-error decomposition proportions existed for all endogenous variables with small proportions corresponding to the bias (U^m) and regression (U^r) components and proportions close to 1 corresponding to the disturbance component (U^d) (Table 5.2). Thus, the model was deemed acceptable for simulation purposes.

While the econometric results indicated that generic egg advertising had a positive impact on price, the bottom line to California egg producers is the impact on net revenue. To ascertain the impact of advertising on producer revenue, the model was simulated under two alternative scenarios: (1) with actual, inflation-adjusted advertising expenditures, and (2) with a 1 percent increase in expenditures. The change in net economic benefits to producers was then computed for each quarter in the sample as the difference in producer surplus:

(3) $\Delta PS_t = \Delta GR_t - \Delta C_t,$

Table 5.2. Simulation of Root Mean Square Percentage Errors (%RMSE) and Mean Square Error (MSE) Decomposition Proportion Inequality Coefficients

Variable	%RMSE	Bias (U^m)	Reg (U^r)	Dist (U^d)
LFARMP	1.979	0.000	0.048	0.952
LPRODPC	0.414	0.004	0.000	0.996
CONSPC	1.741	0.005	0.015	0.980

Note: *LFARMP* is producer price and *LPRODPC* is per capita production, both converted to their natural logarithms; CONSPC is per capita consumption.

where ΔPS is the change in producer surplus for each time period t, ΔGR is the change in gross revenues, and ΔC is the change in production costs. The simulation procedure accounts not only for the production response due to changes in producer prices from advertising but also for the impact of check-off assessments on producer costs.[6] This was accomplished by subtracting the per-unit check-off from producer prices generated through the simulation procedure. Therefore, the simulated gross revenues are net of check-off payments.[7]

The results of this simulation indicated that generic egg advertising had a substantial impact on egg producers' net revenue. A 1 percent change in real advertising expenditures resulted in a 0.13 percent increase in producer prices.[8] Furthermore, a 1 percent change in real advertising expenditures, which is equivalent to $304,000, resulted in a change in producer net revenue of approximately $2.1 million over the sample period. This translates into a marginal rate of return of 6.9 (i.e., an additional dollar added to existing advertising expenditure levels generated $6.90 in producer profits). By comparison, Liu et al. (1990) computed a marginal rate of return of 4.8 for the national dairy advertising campaign and Ward (1992) estimated a marginal rate of return of 6.7 for the beef check-off program.

This result has two important implications. First, California egg producers are benefiting from generic egg advertising because the benefits exceed the cost of the program. Second, since the marginal benefits of advertising exceed marginal costs, more money should be spent on advertising California eggs.

Sensitivity analysis was conducted on producer net returns by changing the price elasticity of supply and indicated that net returns were fairly insensitive to alternative producer responses. Specifically, supply elasticity would need to increase nearly thirtyfold before the marginal rate of return would drop below 1. Producer returns were also evaluated around a 90 percent confidence interval of the advertising-coefficient parameter estimates. Based on this interval, resulting marginal rates of return varied from 1.25 to 8.88. Even at the lower range of advertising-parameter estimates, advertising increased resulting producer net returns over the sample period (i.e., 1.25 > 1.00), increasing the confidence in our conclusion that generic egg advertising in California has benefited egg producers' net returns.

Finally, the internal rate of return (IRR) to advertising expenditures was calculated. The IRR method is common in *ex post* evaluations of research projects and it allows ranking of alternative programs in terms of their relative profitability (Alston, Norton, and Pardey 1995). The procedure is similar to the rate-of-return method previously discussed; however, here the returns to advertising are calculated by discounting the stream of benefits and costs over time. The IRR to generic advertising is the solution to:

$$(4) \quad 0 = \sum_{t=0}^{n} \frac{\Delta PS_t - \Delta AE_t}{(1 + IRR)^t},$$

where AE denotes advertising expenditures. This equation generated a monthly IRR of 58.9 percent. A program is considered profitable if its IRR exceeds the opportunity cost of the funds invested.

Summary

A supply-and-demand econometric model of the California egg industry was estimated to evaluate the impact of generic egg advertising in the state on producer prices and returns over a ten-year period. Econometric estimation indicated that advertising efforts have had a positive impact on producer prices and net profits. The model was simulated with existing advertising expenditure levels and with expenditures 1 percent higher than actual levels. A 1 percent change in advertising expenditures resulted in a 0.13 percent average increase in producer prices and a marginal rate of return to advertising of 6.9. In other words, each additional dollar spent on advertising generated $6.90 in producer profits. The resulting impact on net returns was evaluated over a 90 percent confidence interval of the advertising-parameter estimates, indicating that returns exceeded costs to producers over the entire interval. The estimated returns to generic egg advertising in California indicate that advertising efforts in the state have been quite profitable. Since consumer reaction to future advertising is unpredictable, future efforts to model the California egg industry will be important in evaluating the program's continuing level of effectiveness.

Notes

Funding for this study was provided by the California Egg Commission and the American Egg Board.

This chapter originally appeared as an article entitled "An Economic Analysis of Generic Egg Advertising in California: 1985–1995" by Schmit, Reberte, and Kaiser in *Agribusiness: An International Journal* (© 1997 John Wiley and Sons). This material is used by permission of John Wiley and Sons, Inc.

1 Producers handling less than 26,000 cases of eggs (or liquid equivalent) per year are exempt from the assessment. There are currently nineteen exempt handlers, but their combined marketing constitutes less than 2 percent of the total volume handled. Starting in 1993, separate assessments were also levied on distributors of eggs and egg products

produced outside the state and imported into California. Assessments on egg products are based on a thirty-eight-pound liquid equivalent weight of thirty dozen eggs (one case) and are currently 7 cents per case with a mandatory limit of 30 cents. The CEC producer shell-egg assessment of 1 cent per dozen currently averages approximately 2 percent of farm price.

2 Since February 1995, producers with more than 75,000 laying hens have paid a mandatory national assessment of 10 cents per thirty-dozen case. Prior to this, the assessment was 5 cents and applied to producers with more than 30,000 laying hens.

3 All monetary measures (except generic advertising expenditures) in the supply and demand equations were deflated by the Consumer Price Index for the Western Region of the United States (1982–84 = 100). Generic egg advertising expenditures were deflated by the Media Cost Index provided by Leo Burnett Media.

4 "Typical" refers to a feed mixture of corn and soybean meal blended in a ratio of approximately 85:15. Feed price was measured in real dollars per hundredweight.

5 The publication index was constructed by counting the number of articles each month regarding nutrition impacts of dietary cholesterol reported in the *Reader's Guide to Periodical Literature* from 1985 through 1995. The index was weighted by the circulation levels of the various magazines and converted to a quarterly basis for inclusion in the model.

6 Since generic advertising programs must also cover overhead costs, overhead costs should be included in the return calculation; however, data on those costs were not available. In addition, some assessment revenues are directed toward education, public relations, and research programs that cannot be directly modeled here due to data limitations.

7 See Reberte, Schmit, and Kaiser (1996) for a further explanation of this procedure. The model was simulated in SAS using the simulation procedure in PROC MODEL.

8 Similar simulations were run at advertising changes of 10 percent and 50 percent, resulting in percentage increases in average producer prices of 1.22 percent and 5.30 percent, respectively.

References

Alston, J.M., G.W. Norton, and P.G. Pardey. *Science Under Scarcity: Principles and Practice for Agricultural Research Evaluation and Priority Setting*. Ithaca NY: Cornell University Press, 1995.

Bell, D. *Southern California Egg Production Costs and Income: 1979–1995*. Upland CA: California Egg Commission, 1996.

Blaylock, J., and C. Burbee. *Consumer Demand for Eggs and Market Implications*. United States Department of Agriculture, Economic Research Services Staff Report AGES850924, November 1985.

Brown, D.J., and L.F. Schrader. "Cholesterol Information and Shell Egg Consumption." *American Journal of Agricultural Economics* 72 (1990):548–555.

Chavas, J.P., and S.R. Johnson. "An Economic Model of the U.S. Egg Industry." *Applied Economics* 13 (1981):321–335.

Chyc, K.M., and E.W. Goddard. "Optimal Investment in Generic Advertising and Research: The Case of the Canadian Supply-Managed Egg Market." *Agribusiness: An International Journal* 10 (1994):145–166.

Forker, O.D., and R.W. Ward. *Commodity Advertising: The Economics and Measurement of Generic Programs*. New York NY: Lexington Books, 1993.

Godfrey, R.G. *Misspecification Tests in Econometrics: The Lagrange Multiplier Principle and Other Approaches.* New York NY: Lexington Books, 1993.

Liu, D.J., H.M. Kaiser, O.D. Forker, and T.D. Mount. "An Economic Analysis of the U.S. Generic Dairy Advertising Program Using an Industry Model." *Northeastern Journal of Agricultural and Resource Economics* 19 (1990):37–48.

McCutcheon, M.L., and E.W. Goddard. "Optimal Producer and Social Payoff from Generic Advertising: The Case of the Canadian Supply Managed Egg Sector." *Canadian Journal of Agricultural Economics* 40 (1992):1–24.

McGuirk, A., P. Driscoll, J. Alwang, and H. Huang. "System Misspecification Testing and Structural Change in the Demand for Meats." *Journal of Agricultural and Resource Economics* 20 (1995):1–21.

Miller, B.R., and G.C. Masters. "A Short-Run Price Prediction Model for Eggs." *American Journal of Agricultural Economics* 55 (1973):484–489.

Reader's Guide to Periodical Literature. Bronx NY: H.W. Wilson Company, 1985–1995.

Reberte, J.C., T.M. Schmit, and H.M. Kaiser. *An Ex Post Evaluation of Generic Egg Advertising in the U.S.* Ithaca NY: Cornell University National Institute for Commodity Promotion Research and Evaluation, 1996.

Roy, S.K., and P.N. Johnson. "Econometric Models for Quarterly Shell Egg Prices." *American Journal of Agricultural Economics* 55 (1973):209–213.

Salathe, L.E., J.M. Price, and K.E. Gadson. "The Food and Agricultural Policy Simulator: The Poultry- and Egg-Sector Submodel." *Agricultural Economics Research* 35 (1983):23–34.

Schrader, L.F., H.E. Lanzelere, G.B. Rogers, and O.D. Forker. *The Egg Subsector of U.S. Agriculture: A Review of Organization and Performance.* West Lafayette IN: Purdue University Agricultural Experiment Station, North Central Region Research Publication 258, June 1978.

Smith, R. "AEB, ENC Getting OK for Consumers to Eat More Eggs." *Feedstuffs* 65(13) (1993): 1,4.

Stillman, R.P. "A Quarterly Forecasting Model of the U.S. Egg Sector." United States Department of Agriculture, Economic Research Service Technical Bulletin 1729, September 1987.

Tomek, W.G., and K.L. Robinson. *Agricultural Product Prices.* 3d ed. Ithaca NY: Cornell University Press, 1990.

U.S. Department of Labor, Bureau of Labor Statistics. "Consumer Price Index: Average Price" data. Washington DC, 1985–1995.

Ward, R.W. *The Beef Checkoff: Its Economic Impact and Producer Benefits.* Gainesville FL: University of Florida, January 1992.

6

Evaluation of Prune Promotion by the California Dried Plum Board

Julian M. Alston, Hoy F. Carman, James A. Chalfant, John M. Crespi & Richard J. Sexton

The California prune industry, through the California Dried Plum Board (CDPB, previously called the California Prune Board)[1] and its various packers, especially Sunsweet Growers, the largest marketer of California prunes, has invested substantially in promotion of prunes to consumers. Under a state marketing order program first established in 1952, all California prune producers pay mandatory assessments that are administered by CDPB to promote and increase demand for California prunes. These efforts are in addition to promotion activities undertaken privately by Sunsweet. In 1998, Alston et al. published a study of the economic impacts of industry-financed market development and promotion activities for California prunes. The central questions addressed by that study were: (1) how have CDPB marketing expenditures affected the demand for and sales of California prunes, and (2) have net revenues to producers from CDPB marketing programs increased enough to offset program costs? This chapter summarizes the main elements of that report.

Marketing California Prunes

California leads the world in prune production, accounting for 99 percent of U.S. production and 70 percent of the world's supply. Since 1949, domestic shipments of California prunes have generally declined, while exports of California prunes have tended to increase, in both absolute and percentage terms. In the 1990s, about 60 percent of California's annual prune shipments were consumed in the United States and about 40 percent were exported; this compares with average shares of 83 percent to domestic use and 17 percent to exports between 1949 and 1953. Domestic prune shipments dipped to 85,000 tons in 1978 and remained below 100,000 tons annually until 1987, when they reached 103,000 tons.

Acreage of prunes began to increase in the mid-1980s as California producers switched land that generally had been planted in annual crops to perennial crops in search of higher returns. Bearing acreage rose from a low of 65,000 acres in 1985 to projections of 94,000 acres for 2002. Although highly variable due to weather conditions and alternate bearing tendencies, yields have risen consistently over time. For example, the average yield in 2000 was about 2.55 tons per acre, in contrast to about 2 tons per acre in 1980. (See Table 6.1 for the chronology of acreage, yields, and grower prices for California prunes.)

The rate of prune consumption has been stagnant, so rising acreages and yields have meant generally declining returns to the industry during the 1990s. The 2000 crop consisted of 200,128 saleable tons with 65,153 tons carried forward from the 1999 crop. The industry projected shipments for the 2000/01 crop year of 158,256 tons, implying a carry-out from 2000/01 of 107,000 tons, or roughly one-half a year's production. The projected price per ton for 2001 was $759, compared to a high of $1,120 per ton in 1993.

Promotion of Prunes by CDPB and Sunsweet Growers

A state marketing order for prunes with provisions for promotion and research was approved by California prune producers in 1952. The objective of CDPB, the administrative committee for the marketing order, is to increase worldwide demand for California prunes. CDPB administers domestic and international generic (nonbranded) programs that encompass advertising, sales promotion, public relations, and educational activities. CDPB also funds production and processing research. About 50 percent of its annual domestic budget is spent on television advertising (CDPB annual reports). Advertising can be made more effective by targeting certain groups, and CDPB has, for example, targeted women aged forty-five and older who either rarely or have not recently consumed prunes. Sales promotion activities include supermarket displays that tie into television-advertising themes for prunes or nutrition attributes. CDPB also makes links with national health campaigns, such as the National Cancer Institute's *5-a-Day* program, which recommends eating five servings of fruits and vegetables every day (Peterson 1994).

Table 6.1. California Prunes: Bearing Acreage, Yields, and Grower Prices, 1981–2000

Year	Bearing Acres	Yield in Short Tons	Grower Price in Dollars per Ton
1981	65,400	2.44	654
1982	66,600	1.89	679
1983	67,700	2.14	667
1984	68,700	2.15	693
1985	71,700	1.97	680
1986	72,600	1.36	819
1987	75,600	3.03	734
1988	77,200	1.96	782
1989	77,400	2.92	779
1990	80,100	1.84	873
1991	80,200	2.33	940
1992	80,400	2.29	1,030
1993	83,000	1.46	1,120
1994	84,000	2.30	1,090
1995	83,500	2.17	1,040
1996	85,000	2.62	839
1997	82,000	2.61	883
1998	83,000	1.30	764
1999	83,000	2.14	906
2000	86,000	2.55	845

Source: U.S. Department of Agriculture, Economic Research Service, *Fruit and Tree Nuts Situation and Outlook Yearbook*, 2000; California Agricultural Statistics Service 1996.

CDPB also conducts international promotion activities, such as in-store demonstrations, publicity, trade education, and advertising. Support for international promotion increased sharply in 1987 and has increased in real terms since then. Approximately 50 percent of funding for international promotion activities has been provided by the United States Department of Agriculture's (USDA's) Market Access Promotion Program (MAPP). A rise in export sales coincided with CDPB's participation in the MAPP and its predecessor programs. In 1992, CDPB promoted consumption of California prunes in thirteen nations, but funding reductions in the MAPP led CDPB to eliminate promotion programs in several of those countries. In 1994, CDPB conducted promotion campaigns in Germany, Italy, the United Kingdom, Japan, and Mexico (Peterson 1994).

CDPB's funding for production research in real dollars has remained roughly constant. In the 1990s, about 4 percent of CDPB's budget was invested in production research. All CDPB generic advertising and promotion support for California prunes was discontinued from 1975 through 1978, when the Sunsweet members of CDPB voted to eliminate CDPB's assessment for generic promotion. CDPB reinstated a public relations program in 1979 and resumed an assessment for generic advertising in 1980, subsequently spending an average of about $3 million per year.

Processing and marketing of California's annual prune crop is divided almost evenly between Sunsweet Growers, the industry's only marketing cooperative, and a group of independent growers and packers. Sunsweet members' returns are based on the selling price of the processed fruit. Independent growers sell to independent packers or handlers on a contractual basis. All California prune producers and handlers pay mandatory assessments to support CDPB's generic promotion efforts. Members of Sunsweet also support substantial brand advertising of Sunsweet prune products. Unlike the case of almonds (discussed in this volume), under the prune marketing order there is no assessment credit-back. That is, none of the funds a packer such as Sunsweet expends independently to promote its own brand is deducted from the assessment the packer must pay to CDPB.

Accounting for Changes in U.S. Prune Consumption

Economic theory was used as a guide in specification of the model of demand, identification of variables that were used to explain changes in consumption patterns, and interpretation of the results from estimation. This section summarizes the relevant theory as it applies to models of U.S. per capita demand for prunes.

A Consumer Demand Model

Suppose we use Q_t to represent the per capita quantity of prunes (of uniform quality) demanded by a representative consumer during a particular year, t. The theory of consumer demand suggests a model in which the quantity demanded, Q_t, depends on the corresponding price of prunes, PP_t; the prices of all other goods that are substitutes or complements for prunes (such as other dried fruits and fresh fruits in particular), PS_t; and total money income or expenditure on all goods, EXP_t. This model can be expressed as:

(1) $Q_t = f(PP_t, PS_t, EXP_t)$.

To make this model operational, we must specify a particular functional form for $f(\bullet)$–for instance, a linear functional form, which we use later. In this model, we expect the own-price-effect to be negative (a negative coefficient on PP_t)–higher prune prices will reduce per capita consumption. The cross-price effects (the coefficients on other prices, PS_t) can be positive or negative but are expected to be predominantly positive, especially for close substitutes, and the income effect (the coefficient on EXP_t) is probably positive and in the range for a normal good, corresponding to an income elasticity of demand for prunes between 0 and 1. In sum, we expect an increase in the price of prunes to lead to a decrease in the quantity of prunes demanded, while an increase in the price of a substitute or in total money income leads to an increase in prune demand.

In addition, the theory of consumer demand implies that the demand equation should be homogeneous of degree zero in money income and prices–doubling money income and all prices should leave consumption unaffected, since nothing real has changed. This homogeneity condition is commonly imposed by dividing all of the prices and income by a general price index, such as the Consumer Price Index (CPI), thereby expressing all of the monetary variables in the demand equation in *real* terms (denoted RPP_t, RPS_t, and $REXP_t$). The resulting model is:

(2) $Q_t = f(RPP_t, RPS_t, REXP_t)$.

Both of these demand equations (1 and 2) implicitly assume constant tastes and preferences for prunes. To accommodate changes in preferences arising from promotion or anything else that may affect demand (such as demographic characteristics of consumers), the model can be augmented with other demand-shift variables.[2] Clearly, promotion is one such variable. To obtain reliable estimates of the influence of the factors that were most important for this study–in particular, the responsiveness of demand to price and promotion–it was necessary to take into account the influence of other demand-shift variables as well. Otherwise, there was a risk that the effects of omitted shifters would be attributed falsely to the variables included in the model.

In a model of consumer demand for California prunes, appropriate shift variables can be included to represent the effects of such things as

(1) increased consumer health consciousness and rising interest among consumers in natural foods; (2) other demographic changes, such as changes in the age structure of the population (likely to be especially important for prunes since they are relatively more frequently consumed by older people), a higher rate of labor-force participation by women, changes in the ethnic composition of the population, and more meals being eaten away from home; (3) generic promotion by CDPB, brand promotion by Sunsweet and others, and other changes in merchandising expenditures; and (4) changes in the quality of California prunes.

To deal with all of these individual variables explicitly in a model was impossible given our limited data set and the difficulty of identifying their individual effects. Many variables change smoothly and together over time. Thus, we included three types of shift variables that we believed would have the most important effects: (1) the quantity of promotion, represented by total–not per capita–promotion expenditures by both CDPB (*RPROCDPB*) and Sunsweet (*RPROSUN*) expressed in real terms by dividing expenditures by the CPI;[3] (2) a linear time-trend variable, $TIME_t$, included to represent the effects of other trends previously described that were not being modeled explicitly; and (3) quarterly intercept indicator variables (SPR_t, WIN_t, and SUM_t) to reflect seasonal shifts in demand. Note that fall is the base season–the estimated coefficients for SPR_t, WIN_t, and SUM_t show the change in demand from the fall base season.

Incorporating the shift variables led to the following augmented demand model:

$$(3)\quad Q_t = f(RPG_t, RPS_t, REXP_t, RPROCDPB_t, RPROSUN_t, SUM_t, WIN_t, SPR_t, TIME_t).$$

The direction of the effects of the demand-shift variables was not as easy to predict as that of the more conventional variables. Successful promotion was expected to have an unambiguously positive effect on demand, whereas unsuccessful promotion was expected to have little or no effect on demand. The effect of the time trend was likely to be negative, reflecting a general shift of domestic consumer demand away from prunes over time, as previously discussed. It was expected that per capita consumption of prunes would increase with increases in the fraction of the population in older age categories. However, it was likely to be difficult to separate age effects from cohort effects, especially in a monthly model that covered only a few years (indeed, as noted previously, an important question was whether declining per capita consumption is an age effect, reversible as the population becomes older again, or a cohort effect, which is not reversible).

Econometric Estimation of Aggregate Monthly Domestic Demand Models

This section summarizes the regression models used to estimate the effects of promotion on monthly[4] per capita prune consumption in the United States. The estimates indicated that promotion expenditures had a statistically significant, positive effect on monthly prune consumption at the retail level. Before presenting the demand models and the empirical results, we provide an overview of the data and data sources for the models.

Data

The time period for this part of the study was from September 1992 through July 1996, comprising fifty-one observations of four weeks each. The monthly regressions presented here were based on data collected from four sources. The data describing quantities of prunes consumed in the U.S. came from Information Resources, Inc. (IRI) Infoscan retail market profiles representing U.S. consumption based on sixty-four cities/regions. We used "total" prune consumption over this period, which was total sales of all brands in retail stores. During this period, retail sales accounted for approximately 30 percent of all prune sales, and Sunsweet Growers represented about 70 percent of the retail sales market (CDPB 1996a). The quantities were converted into pounds per capita using an estimate of the U.S. midyear population that was compiled by the International Monetary Fund (IMF) in *International Financial Statistics*. We converted this population statistic into population per observation period using a growth formula provided by the U.S. Census Bureau. The prices used in the demand models were the average retail price (dollars per pound) of prunes purchased and the average price of dried fruit other than prunes (dollars per pound) purchased from retail food stores.[5] These prices also came from Infoscan IRI data and were converted into 1996 dollars using the CPI. Deflation by the CPI has the effect of treating an aggregate of all other goods as a general substitute for prunes. Although it is difficult to identify a particular index to represent those goods that are close substitutes for prunes, the models presented in this section used the real price of an aggregate of other dried fruit (RPS_t). It must be stressed that there was very little price movement for prunes and other dried fruit during the observation period for the monthly demand models; the average real price of prunes ranged from $2.05 to $2.26

per pound per month. Thus, we were attempting to correlate a consumption variable with price variables that simply did not move very much. A data set covering a longer time period and with more variation in prices might provide improved estimates of price effects on demand.

Promotion included generic promotion expenditures by CDPB and brand promotion by Sunsweet Growers. There is little or no branded promotion (besides display units and other trade promotions like "buy one, get one free") by other packers, and Sunsweet is the only packer that advertises its prunes nationally. The aggregate monthly promotion variables (*PROCDPB* and *PROSUN*) each included monthly sales promotions and monthly television advertising.

PROMOCDPB represented monthly sales promotion expenditures by CDPB for prunes. Specifically, *PROMOCDPB* was the sum of the following budgetary items: coupon programs, public relations, sampling, and merchandising/sales promotion. Throughout this section, "sales promotion" refers to the sum of these budgetary items. All figures were converted into monthly data using daily averages for each month to match the Infoscan IRI quantity and price series. *PROMOSUN* represented sales promotion expenditures by Sunsweet Growers for prunes and consisted of the cooperative's monthly total expenditures for dried fruit promotion.

ADCDPB represented television advertising expenditures for the specific observation point, as billed to CDPB. Unlike the data on sales promotion expenditures, which came from crop-year budgets, these data accurately reflected the timing of television advertisements. The figures were converted to monthly expenditures using daily averages. *ADSUN* represented television advertising expenditures billed to Sunsweet. Those data, reported on a weekly basis, were summed to provide monthly figures for the observation period.

Finally, aggregate promotion variables were constructed for CDPB and Sunsweet that included both sales promotion and television advertising expressed in millions of dollars: *PROCDPB* = *ADCDPB* + *PROMOCDPB* and *PROSUN* = *ADSUN* + *PROMOSUN*.

The timing of sales promotion expenditures posed a potential problem, as only the two television advertising variables, *ADCDPB* and *ADSUN*, accurately reflected when advertising was seen by viewers. The two sales promotion variables were off by an indeterminate amount since they came directly from CDPB's and Sunsweet's accounting records and aggregated various important sales promotion expenditures (coupon payments, public relations, sampling, and merchandising/sales promotion) that occurred at various times. Thus, while the *AD* variables reflected dollar figures that corresponded to the actual occurrence of the promotion (e.g., television commercials), the *PROMO* variables did not. Since some promotion expenses were paid when promotion occurred, others were paid ahead of time, and still others were paid after promotion, it was not clear whether these variables should be lagged in the regressions.

Ultimately, we decided to use the variables as presented in the budgets. Promotion expenditures were expressed in real terms to reflect the view that there would be no real change in the amount of promotion undertaken if CDPB's and Sunsweet's promotion budgets doubled and the cost of promotion also doubled. Lacking specific information on unit costs of promotion, we assumed that promotion costs rose with prices generally and used the CPI to deflate promotion expenditures.

A total expenditure variable representing consumer income was also included. As monthly data were not available, the expenditure variable was based on quarterly data on total private domestic consumption expenditures (in billions of dollars) for the U.S. taken from *International Financial Statistics*, published by IMF. These figures were then deflated by a quarterly CPI (August 1996 = 1.00) to put them in real terms and divided by a quarterly estimate of U.S. population to put them in per capita terms. A time-trend variable was also used in the demand models. Over the period of study, there was a slight decrease in per capita prune consumption, so we expected the coefficient of the trend variable to be negative if this decline in consumption was not attributable to other variables in the model. Conversations with CDPB staff suggested that prune consumption varies seasonally. To investigate this possibility, three quarterly indicator variables (*WIN* = winter, *SPR* = spring, and *SUM* = summer) were specified and fall quarter (October, November, and December) was used as the base season. The seasonal variables were equal to 1 for each month during a given season and 0 otherwise. When the four-week period for a particular observation spanned two seasons, it was assigned to the season that applied for the majority of days in the four-week period.

Table 6.2 summarizes the definitions of the variables used in the monthly demand models. Summary statistics for the monthly demand variables are shown in Table 6.3.

Estimation Results and Selection of the Preferred Monthly Models

In this section, we present regression equations that represent the monthly demand for prunes in the United States for the period spanning the months

Table 6.2. Description of Variables in the Monthly Demand Model

Variable	Definition	Units	Data Source
$TIME_t$	"Month."	One observation is four weeks long.	NA
Q_t	"Monthly" per capita consumption of dried prunes in the U.S. from the week ending September 6, 1992, to July 7, 1996. Recall, one "month" here is exactly four weeks long.	Pounds of dried prunes per person per month.	Prune Consumption from Infoscan IRI retail market profiles. U.S. population from IMF, *International Financial Statistics.*
RPP_t	Real average retail price of prunes.	Real dollars (August 1996 = 1) per pound of processed prunes.	Prices came from Infoscan IRI and were deflated by CPI data from IMF, *International Financial Statistics.*
RPS_t	Real average retail price of all other dried fruit.	Real dollars per pound.	Prices came from Infoscan IRI and were deflated by CPI data from IMF, *International Financial Statistics.*
$REXP_t$	Real, average, quarterly per capita private domestic consumption expenditures.	Real dollars per person per quarter.	IMF.
$RPROCDPB_t$	Aggregate real promotion expenditures by CDPB for prunes in the U.S. market.	Real dollars per month.	CDPB.
$RPROSUN_t$	Aggregate real promotion expenditures by Sunsweet for prunes in the U.S. market.	Real dollars per month.	Sunsweet Growers.
WIN, *SPR*, *SUM*	Winter, spring, and summer dummy variables, respectively.	NA	NA

Note: *RPS* consists of two components: raisins, *RPR*, and other dried fruit, *RPF*.

Table 6.3. Summary Statistics of the Variables in the Monthly Demand Model

Variable	N	Mean	Standard Deviation	Minimum	Maximum
Q	51	0.015	0.002	0.012	0.020
RPP	51	2.179	0.045	2.054	2.264
RPS	51	2.394	0.042	2.305	2.527
RPR	51	2.010	0.041	1.916	2.067
RPF	51	4.599	0.363	3.984	5.258
REXP	51	18,691	499.52	17,785	19,483
RPROMO	51	1.512	0.483	0.516	2.397
TIME	51	26	14.866	1	51

Note: In this table, *RPROMO* = *RPROCDPB* + *RPROSUN* represents total monthly promotion in millions of dollars (aggregated to maintain Sunsweet's confidentiality). *RPR* and *RPF* are the real prices of raisins and other dried fruit, respectively.

from September 1992 through July 1996. The choice of the functional form for the demand equation, which was guided by diagnostic tests, was important because it could influence the results of the econometric estimation (e.g., see Chalfant and Alston 1988; Alston and Chalfant 1991). A variety of models was examined in determining an adequate specification. To investigate timing issues associated with the sales-promotion variable, several models that included lagged promotion variables with a variety of polynomial restrictions on the lag structure were tested. These models were all unsatisfactory, and they are not presented here. Further, in the course of testing various specifications, we decided to use the aggregate *PROCDPB* and *PROSUN* variables instead of disaggregating each of them into *AD* and *PRO*.[6]

In what follows, we focus on the results from a demand equation that was linear in all variables except that we included the square root of promotion–the *square root* model–instead of the level of promotion. This model allowed for diminishing marginal returns to promotion.[7] We wished to include separate variables representing generic promotion by CDPB and brand promotion by Sunsweet. Accordingly, we considered two types of models that include promotion variables in square root form. The first type consisted of two models that included the square root of all promotion. Model 1 (Equation (4a)) allowed the marginal effects of CDPB and Sunsweet promotion expenditures to differ, while Model 2 (Equation (4b)) set equivalent promotion parameters in Equation (4a) so that the marginal effects of CDPB and Sunsweet promotion expenditures were forced to be the same.

Model 1

$$(4a)\quad Q_t = b_0 + b_{PP}RPP_t + b_{PS}RPS_t + b_{EXP}REXP_t + (b_{CDPB}RPROCDPB_t + b_{SUN}RPROSUN_t)^{1/2} + b_{SPR}SPR_t + b_{WIN}WIN_t + b_{SUM}SUM_t + b_T TIME_t + e_t.$$

Model 2

$$(4b)\quad Q_t = b_0 + b_{PP}RPP_t + b_{PS}RPS_t + b_{EXP}REXP_t + b_{PRO}(RPROCDPB_t + RPROSUN_t)^{1/2} + b_{SPR}SPR_t + b_{WIN}WIN_t + b_{SUM}SUM_t + b_T TIME_t + e_t.$$

We also considered two other models based on separate square roots of promotion expenditures. In Equation (4c), the marginal effects of CDPB and Sunsweet promotion expenditures were allowed to differ.

Model 3

$$(4c)\quad Q_t = b_0 + b_{PP}RPP_t + b_{PS}RPS_t + b_{EXP}REXP_t + b_{CDPB}(RPROCDPB_t)^{1/2} + b_{SUN}(RPROSUN_t)^{1/2} + b_{SPR}SPR_t + b_{WIN}WIN_t + b_{SUM}SUM_t + b_T TIME_t + e_t.$$

A special case of Equation (4c) was given by assuming that Sunsweet's promotion activities had no effect on the total *demand* for prunes. This is shown by Equation (4d). Note that Equation (4d) can also be seen as a special case of Equation (4a) by setting $b_{SUN} = 0$.

Model 4

$$(4d)\quad Q_t = b_0 + b_{PP}RPP_t + b_{PS}RPS_t + b_{EXP}REXP_t + b_{CDPB}(RPROCDPB_t)^{1/2} + b_{SPR}SPR_t + b_{WIN}WIN_t + b_{SUM}SUM_t + b_T TIME_t + e_t.$$

We can test Models 2 and 4 as special cases of Model 1, and Model 4 can also be tested against Model 3. Models 1 and 3, however, are not nested as special cases and cannot be tested against one another using conventional nested tests.

In all of these models, the *b* coefficients are multipliers that, holding the other independent variables constant, translate changes in the right-hand-side variables into changes in quantities consumed. For the promotion variables, *RPROCDPB* and *RPROSUN*, however, the multipliers translate changes in the square root transformations of the underlying variables into changes in quantities consumed. The term e_t represents residual changes in per capita quantities consumed that were not accounted for by changes in the right-hand-side variables. This term is sometimes referred to as the "error" term since it can be thought of as the error in predicting Q_t using only the right-hand-side variables. These residuals are typically assumed to be normally distributed random variables, with an expected value of 0 and a constant variance.

The results from estimating Models 1 through 4 (Equations (4a) through (4d)) are shown in Table 6.4, in which the elasticities are displayed in brackets beneath the coefficient estimates and have been calculated at the means of the economic variables. The figures in parentheses are t-statistics. The adjusted R^2 statistics indicate the proportion of the variation in consumption that has been accounted for by the explanatory variables included in the equation. They ranged from 76.1 percent for Model 4 to 81.1 percent for Model 2.

A battery of diagnostic tests for model specification and variable endogeneity was performed on these models, and the reader is referred to the original report by Alston et al. (1998) for details. For the models shown in Table 6.4, Model 4 proved to be a misspecification, indicating that omission of Sunsweet's promotion was a serious problem. This is an important finding because it lends credence to the belief that any significant private-brand promotion in an industry must be accounted for in a model of demand. Thus, Models 1, 2, and 3 were all acceptable models, but Model 4 was not.

Models 1, 2, and 3 generally fit the data well and explained about 81 percent of the variation in consumption of prunes. For demonstration, the close correspondence between actual per capita consumption and the estimated (fitted) values for Model 2 over the sample period are shown in the upper panel of Figure 6.1. The lower panel of the figure shows the fraction of the fitted values accounted for by all of the variables other than promotion in Model 2. In other words, it is the fitted values net of the estimated effects of promotion (calculated by subtracting $0.535 \times 10^{-2} \times (RPROCDPB_t + RPROSUN_t)^{1/2}$ from the fitted value in each month). The importance of promotion in overall demand for prunes is clearly illustrated in Figure 6.1.

Table 6.4. Ordinary Least Square Estimates Comparing Different Specifications of Promotion Expenditures in the Preferred Monthly Model

Independent Variables	(1)	(2)	(3)	(4)
Constant	0.019	0.015	0.019	0.083
	(0.330)	(0.283)	(0.340)	(1.367)
RPP	−0.291E-2	−0.305E-2	−0.229E-2	−0.153E-2
	(−0.740)	(−0.794)	(−0.579)	(−0.346)
	[−0.418]	[−0.437]	[−0.329]	[−0.220]
RPS	−0.904E-2	−0.910E-2	−0.953E-2	−0.817E-2
	(−2.180)	(−2.228)	(−2.324)	(−1.787)
	[−1.424]	[−1.433]	[−1.501]	[−1.287]
RPROCDPB	0.305E-4	0.535E-2	0.280E-2	0.499E-2
	(2.25)	(6.339)	(2.498)	(4.824)
	[0.052]	[0.048]	[0.050]	[0.089]
RPROSUN	0.270E-4	0.535E-2	0.449E-2	
	(2.13)	(6.339)	(3.426)	
	[0.161]	[0.168]	[0.159]	
REXP	0.102E-5	0.129E-5	0.999E-6	−0.264E-5
	(0.34)	(0.505)	(0.343)	(−0.866)
	[1.248]	[1.585]	[1.230]	[−3.244]
WIN	0.182E-2	0.181E-2	0.183E-2	0.219E-2
	(3.76)	(3.797)	(3.823)	(4.176)
SPR	0.179E-2	0.173E-2	0.178E-2	0.301E-2
	(3.11)	(3.862)	(3.152)	(6.169)
SUM	−0.405E-3	−0.452E-3	−0.380E-3	0.220E-3
	(−0.83)	(−1.105)	(−0.791)	(0.438)
TIME	−0.826E-4	−0.911E-4	−0.820E-4	0.355E-4
	(−0.83)	(−1.047)	(−0.840)	(0.347)
Adjusted R^2	0.806	0.811	0.810	0.761
Log-likelihood	287.092	287.071	287.617	281.198
Durbin Watson	2.118	2.114	2.124	2.078

Notes: The t-statistics are in parentheses; elasticities (at means) are in brackets. See the definitions of Models 1–4 (i.e., Equations (4a)–(4d)) to interpret the coefficients on the promotion variables.

Models 1, 2, and 3 all said essentially the same things about demand for prunes. The own-price elasticity of demand was inelastic, falling in a range of −0.3 to −0.5. The period of study was one of very little price movement. Not surprisingly, it was difficult to find significant effects on quantity demanded from changes in prices. Low price elasticities are not uncommon for

Figure 6.1. U.S. Per Capita Monthly Prune Consumption (Retail) – Actual versus Fitted Values, September 1992 through July 1996

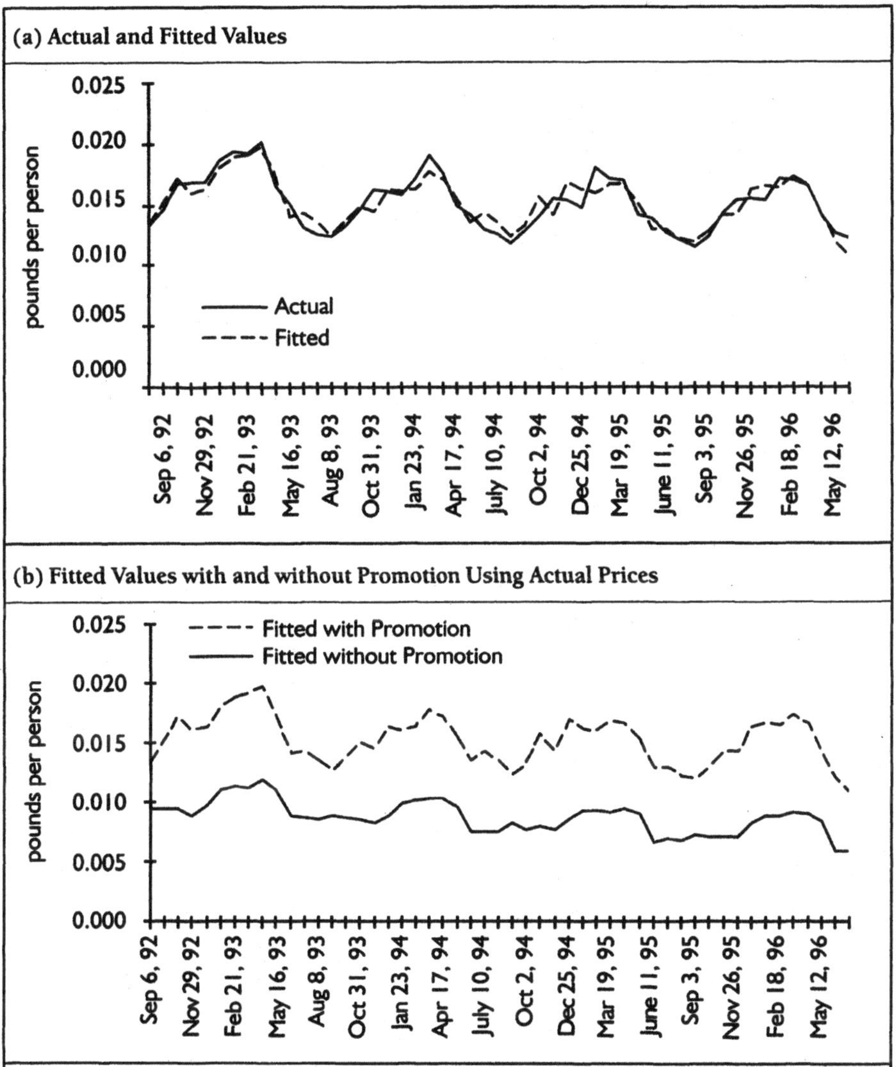

Source: Actual values from IRI. Fitted values from authors' calculations.

agricultural goods. Comparatively, a study of the prune market in the United Kingdom also suggested a rather low price elasticity of −0.7 (Thorogood & Associates 1994). Seasonality was also present in all of the estimated monthly demand models.

The income elasticity of demand was about 1.2 to 1.6, and the cross-price elasticity of demand with respect to the price of substitutes was about −1.4. Note that a negative cross-price elasticity could imply that the "substitute" good was, in fact, a complement. While it may be the case that dried fruit consumption moves in a complementary fashion with prune consumption, we suspect that the negative cross-price elasticity was capturing some of the downward trend in prune consumption that was not otherwise reflected in the time-trend variable.

The coefficients on the promotion variables were statistically significant at the 99 percent confidence level in each of the models shown in Table 6.4. In addition, the estimated coefficients on promotion and corresponding elasticities were very stable across model specifications. The elasticity of demand with respect to CDPB promotion was estimated at between 0.048 and 0.052, while the elasticity of demand with respect to Sunsweet promotion ranged from 0.159 to 0.168. It is important to note that these elasticities refer to the effect of CDPB and Sunsweet promotion on total sales of prunes. In particular, the Sunsweet elasticity cannot be used to evaluate the effectiveness or profitability of Sunsweet's promotion. To do so would require an elasticity of demand for Sunsweet prunes with respect to Sunsweet promotion.

Simulation Model and Benefit-Cost Analysis

In this section, we use the estimated monthly demand parameters from the previous section to estimate the gross and net benefits to the California prune industry from its expenditures on promotion. The preferred models of monthly retail demand are used to show a range of estimated values for gross and net benefits.

Measuring the welfare impacts of promotion funded by check-offs required (1) a conceptual structural model of the industry market equilibrium, (2) estimates of supply and demand parameters that could be used to define the values for the parameters in the structural model, (3) estimates of the demand response to promotion expenditures, and (4) information to allow a transformation of the effects of promotion (through retail demand shifts) and assessments or check-offs (through commodity supply shifts) into measures of benefits and costs.

Conceptual Model of Supply and Demand

The econometric work discussed in the previous section allowed us to estimate how much the quantities of prunes sold would increase in response to a given

Figure 6.2. Conceptual Supply and Demand Model

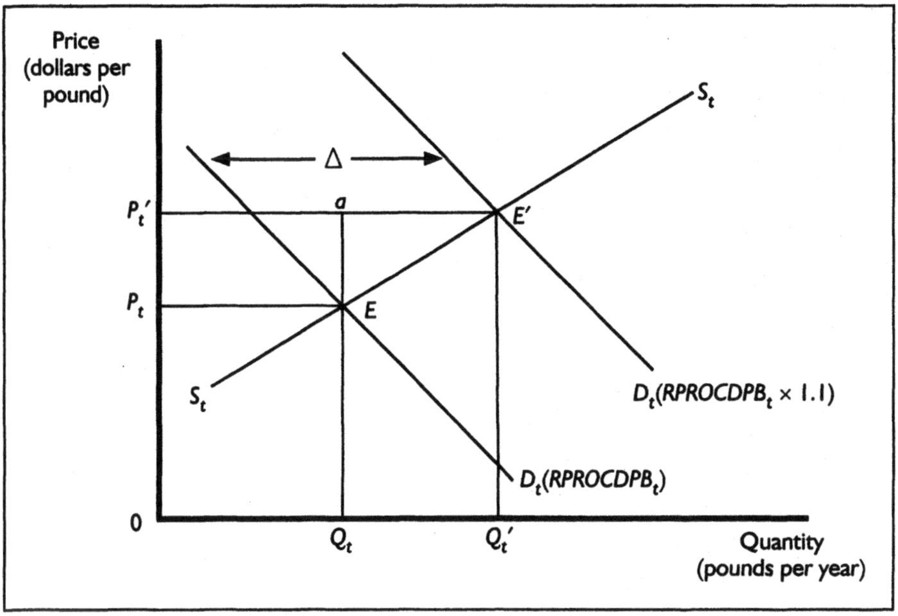

increase in promotion expenditures, holding prices (and other variables) constant. This did not, however, tell us how much sales would actually increase when promotion was changed since prices could not be assumed to remain constant. Indeed, the increase in prices following a promotion-induced shift in demand is an important part of the benefits of promotion realized by growers and packers. To evaluate the industry's demand-shifting activities properly, therefore, we had to account for both demand effects and the response of supply to increased price.

The diagram in Figure 6.2 illustrates the conceptual supply and demand relationships for a typical year *t*. In the figure, the line labeled S_t represents the supply curve for prunes. It shows the quantities available to domestic consumers at various prices; at higher prices, more fresh prunes are available domestically, while at lower prices, larger quantities of prunes are diverted to other uses, such as to the export market, or are left unharvested. The line labeled $D_t(RPROCDPB_t)$ represents the demand curve: at higher prices, consumers purchase a smaller quantity of prunes than they do at lower prices, holding other factors constant.[8] In particular, the promotion expenditure by CDPB was held constant at its actual value, $RPROCDPB_t$, along this demand curve. The equilibrium price at which quantities supplied and demanded are the same is the price observed at point *E*: price P_t is consistent with the observed quantity Q_t.

In this example, the effects of a 10 percent increase in CDPB promotion, $RPROCDPB_t$, are illustrated by the outward shift in the demand curve. The new curve is labeled $D_t(RPROCDPB_t \times 1.1)$. The econometric model allowed us to estimate the horizontal distance of the demand shift, identified by Δ in the diagram. In Model 2, as given in Equation (4b) (the second column of Table 6.4), for example, the coefficient on $(RPROCDPB_t)^{0.5}$ was 0.00535. Suppose the actual monthly promotion expenditure was $1 million (in 1996 dollars). This would mean that a $0.1 million (i.e., 10 percent) increase in total promotion expenditures would be expected to lead to an increase in per capita prune consumption of $0.00535 \times (1.1^{0.5} - 1.0^{0.5}) = 0.00026$ pounds per month if there was no change in price.

Multiplying by the population (POP_t = 265.5 million in 1996) yielded the total horizontal demand shift from a 10 percent increase in promotion expenditures, which was about a 2.1 percent increase in consumption at constant prices. However, this was greater than the actual increase in consumption that would result. An increase in price was needed to bring forth the additional quantities necessary to satisfy increased demand. This is reflected in the fact that the supply curve to the domestic market slopes up. The new equilibrium was given by the point where the new demand curve crossed the supply curve, *E′*. Price and quantity both increased to the new equilibrium values P_t' and Q_t'.

Evaluation of promotion expenditures required both an estimate of the increments in prices and quantities due to the expenditures and a measure of the costs of supplying the additional quantities to the market. In the diagram, increased prices call forth additional supplies; these supplies come at a cost, which, in the case of a perennial crop, may largely be the earnings foregone from other uses of existing production (e.g., exports) rather than from new production. The sellers who were already in the market at price P_t profit from the increase in price to P_t'; their gain is $(P_t' - P_t)Q_t$, or the area $P_t'aEP_t$ in the diagram. The gains to sellers of additional supplies are much smaller as they must be reduced by the cost (including revenue from foregone sales in other markets) of the additional quantities. This benefit is given by the area of the triangle *aEE′* in the diagram. The total gain to prune producers is given by the area of the trapezoid $P_t'E'EP_t$; this represents the gain in *producer surplus* resulting from the 10 percent increase in promotion expenditures. Changes in producer surplus coincide with changes in producers' economic profits from the production of prunes in such a situation. The only information required, in addition to the econometrically estimated demand parameters for responses to prices and promotion, was for the supply response to price.

To evaluate the profitability of these expenditures, the gain in producer surplus had to be set against the cost. We used two measures of cost. One measure was the total cost of the marginal increase in expenditures. However, when the promotion cost was financed by a per-pound assessment, some of the incidence of the assessment fell on consumers as a result of increased retail prices. Thus, the total cost could differ from the cost to producers. A second measure compared the benefits to producers with the producers' share of the cost, allowing that some of the costs of the assessment were borne by consumers.

Figure 6.3 shows the same initial supply and demand curves as in Figure 6.2, labeled $S_t(\tau_t)$ and $D_t(RPROCDPB_t)$, where τ_t represents the actual assessment per pound in year t and the equilibrium is at point E with price P_t and quantity Q_t. An increase in the assessment per pound is reflected as a shift up of the supply curve to $S_t(\tau_t')$ by the amount of the increase (given by simply adding the additional assessment to the previous price at which producers would be willing to supply any given quantity along the supply curve). This

Figure 6.3. Incidence of Assessments

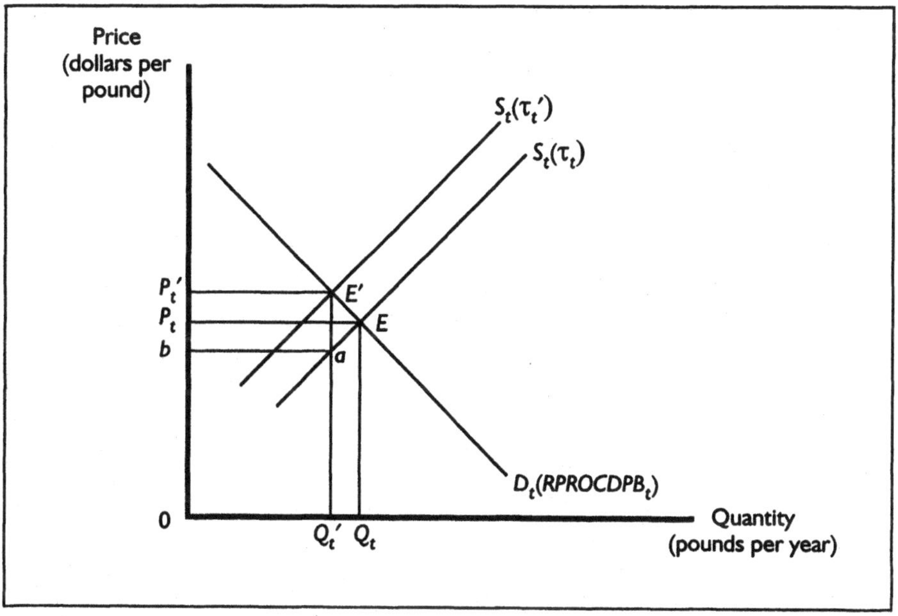

leads to a new equilibrium at point E' with a higher consumer price P_t', a smaller net producer price ($b = P_t' - \tau_t'$), and a smaller quantity produced and consumed, Q_t'. The extent of the consumer price increase depends on the slopes of the supply and demand curves. The more price-responsive (elastic) supply is, the smaller is the proportion of the assessment borne by producers.

The additional amount of assessment revenue is equal to $\tau_t'Q_t' - \tau_t Q_t$. For small changes in the assessment, this was approximately equal to the change in the assessment times the final quantity: $(\tau_t' - \tau_t)Q_t'$. In Figure 6.3, this is equal to area $P_t'E'ab$. This area corresponds closely to the full social cost of the change in assessment to finance a change in promotion expenditure of that amount (it leaves out the area of the triangle $E'Ea$, which is negligible for the small changes in assessments considered here). The loss of producer surplus (or profit) associated with the same increase in the assessment is equal to area P_tEab, only a fraction of the total amount spent. In the work that follows, we compare producer benefits and the two alternative measures of producer costs.

An Approximation Using Elasticities

Before turning to the specification of the supply side and the simulation models for computing benefit-cost ratios, we report results from an approximation procedure using just the estimated elasticities of demand with respect to price and promotion and the promotion intensity (promotion expenditure per dollar of sales). We derived and explained that approximation procedure in Alston et al. (1998), showing that μ, the benefit from a marginal increase in promotion expenditure (for an increase in promotion financed by a check-off), can be approximated using

$$\mu = Q\frac{\partial P}{\partial a} = Q\frac{\partial Q}{\partial a}\left|\frac{\partial P}{\partial Q}\right| = \frac{\alpha}{|\eta|}\frac{v}{a} = \frac{\alpha}{|\eta|\iota},$$

where a is the expenditure on promotion, v is the value of sales (the product of price, P, and the quantity sold, Q), α is the elasticity of demand with respect to promotion, η is the absolute value of the elasticity of demand with respect to the price, and $\iota = a / v$ is the intensity of promotion. This result is intuitively clear in the case where supply is fixed and the producer benefits from advertising are exactly equal to the price increase induced by the demand shift, multiplied by the quantity supplied. It also applies more generally when advertising is financed by a check-off. Using the results from the preferred monthly regression models, we computed values for μ at every monthly data point. In all of the instances examined, the estimated benefit from a marginal increase in the promotion expenditure exceeded 1.0.

Simulation Model

The simulation model combined results from the monthly demand analysis with assumed supply parameters. We compared the pattern of consumption and prices predicted by the estimated model, given actual historical promotion activities, with corresponding values predicted by the model following a counterfactual 1 percent increase in promotion expenditure in every month in the sample period (September 1992 to July 1996). We also simulated changes in the assessments jointly with corresponding changes in promotion. We used the differences between these actual and counterfactual scenarios to calculate measures of the *marginal* gross and net benefits. We computed present values of benefits and costs using an annual interest rate (compounding) of 3 percent (a reasonable value for the long-term, risk-free, *real* rate of interest).

To conduct the benefit-cost analysis, we combined the preferred demand model with an assumed supply function. First, from the demand side, *predicted quantities* were calculated by substituting the actual values for each of the explanatory variables in the estimated equation. For instance, for Model 2 (the second column in Table 6.4),

$$(5)\quad \hat{Q}_t = \hat{b}_0 + \hat{b}_{PP}RPP_t + \hat{b}_{PR}RPS_t + \hat{b}_{EXP}REXP_t + \hat{b}_{PRO}(RPROCDPB_t + RPROSUN_t)^{1/2} + \hat{b}_{SPR}SPR_t + \hat{b}_{WIN}WIN_t + \hat{b}_{SUM}SUM_t + \hat{b}_T TIME_t,$$

where the hats (^) on parameters denote their estimated values and the hat on Q_t denotes the predicted value given the estimated coefficients and the values of the exogenous variables. We replicated this step for each of the three reliable models. Next, the supply function was defined to be of the constant elasticity form and to pass through the points defined by the predicted quantities from the demand model. That is, the supply function was of the form

$$(6)\quad Q_t = A_t R_t^{\varepsilon} \text{ where } A_t \equiv \hat{Q}_t / R_t^{\varepsilon}$$

and R_t was the producer return per pound in year t, defined as $R_t = (1 - \tau_t)P_t$, where τ_t was the actual promotion expenditure per pound consumed in year t

expressed as a fraction of the market price in year *t* (i.e., the rate of assessment required to finance the actual promotion expenditure). A_t was a parameter that varied from month to month to ensure that, given the actual values of prices and the other exogenous variables, for each month the supply equation passed through the point defined by the predicted quantity from the demand model and the actual price. This meant that we could combine the calibrated supply model and the estimated demand model to simulate past actual prices and predicted quantities.

Supply functions were calibrated using alternative supply elasticities (ε) of 0, 0.5, 1.0, 2.0, and 5.0. This range of elasticities reflected a range of lengths of run; it also reflected our uncertainty about the exact implications of international trade in prunes for the elasticity of the *residual* supply to the domestic market. Changes in producer surplus were calculated by integrating the function over the range of a price change. In practice, this translated to using the following formula for the change in producer surplus:

$$(7) \quad \Delta PS_t = \frac{R_t' Q_t' - R_t Q_t}{1 + \varepsilon}.$$

Using these definitions of supply and demand equations, we first replicated the past: by equating the equations for supply and demand and solving for market equilibrium, we obtained values of actual prices and predicted quantities (from the demand model) given the actual values for the exogenous variables. In addition, we simulated counterfactual scenarios by using hypothetical values for the exogenous variables. Counterfactual simulations were conducted by:

- using hypothetical values for CDPB's promotion expenditure in every year (1.01 times the actual values) with actual assessment rates (in practice, we defined "actual" assessment rates by expressing the total promotion expenditure as a fraction of the total value of production);
- using hypothetical values for the assessment rate in every year (1.01 times the actual values) with actual promotion expenditures; and
- changing both the promotion expenditures and the assessments (setting both at 1.01 times the actual values).

For each simulation, we calculated two measures of producer marginal costs of promotion: (1) the cost of the marginal promotion expenditure, and (2) the producer surplus loss associated with an assessment sufficient to generate the same amount of additional expenditure. Only selected summary results are reported here.

An important issue was determining how much confidence could be placed in the particular values of the benefit-cost measures. How confident could we be, for example, that the net benefits were greater than, say, $1 million given a "best" estimate of $2 million? The precision of our estimates of the benefits depended on the precision of our estimates of the underlying parameters but in ways that were not easy to see clearly. To evaluate the precision of our measures of benefits and costs, we conducted Monte Carlo simulations, following the approach of Alston et al. (1997). The simulations yielded confidence intervals on our welfare measures, permitting us to make statements such as a 95 percent confidence interval for the benefit-cost ratio was formed by the interval from *a*:1 to *b*:1, where *a* was a lower confidence limit and *b* was an upper confidence limit. In practice, this required an iterative process where first a particular set of values of the parameters was drawn at random from the estimated joint statistical distribution. This set of parameters was substituted into the demand equation and used to generate predicted quantities, which were in turn used to set parameters for the supply equation for each supply elasticity. These supply equations were then used with the demand equation to conduct the counterfactual scenarios and then to evaluate the scenarios. The detailed results from the Monte Carlo study are reported in Alston et al. (1998).

Table 6.5 reports marginal benefits and costs of prune promotion implied by each of the three preferred demand models using five alternative values for the supply elasticity. In each case, we report both producer benefits relative to producer costs associated with a change in assessment to finance a change in promotion, along with producer benefits relative to the total costs of the change in promotion expenditure (not just the producer share). For example, the first row in Table 6.5 refers to Model 1 and the first column represents a supply elasticity of 0. Thus, using a supply elasticity of 0 and a demand equation based on Model 1, the benefit-cost ratio for a 1 percent increase in promotion expenditure was 3.00, regardless of whether we distinguished between producer cost and total cost (they were the same when the supply elasticity was 0). Looking across the columns in this row, we can see the effects of the increase in the supply elasticity. We can see that the supply elasticity did not affect the benefit-cost ratio when we considered only the producer share of the cost of the check-off—the benefit-cost ratio was essentially 3.00 regardless of the value of the supply elasticity. However, as the supply elasticity rose, producers received progressively smaller benefits from a given demand increase. Hence, the benefit-cost ratio computed using total promotion expenditures declined with increases in the supply elasticity, going from 3.00 when supply was fixed to 0.22 when the supply elasticity was 5. This pattern was repeated across all the models (and, looking across the rows and comparing within a column, we can see the effects of the different demand models). Thus, a primary issue is whether incremental promotion is financed by incremental changes in assessments or in a lump-sum fashion (see Goddard, Griffith, and Quilkey 1994; Alston, Carman, and Chalfant 1994; and Alston et al. 1997). In the following, we emphasize the results obtained by comparing producer benefits to producer costs rather than to total expenses, where total expenses were partly borne by producers and partly by consumers to the extent that the costs of the assessment were passed on through higher consumer prices.

Table 6.6 reports the point estimates for the ratio of producer benefits to producer costs, calculated using a supply elasticity of 1.0 (but representative of all supply elasticities), with the corresponding results from the Monte Carlo simulation. We report the mean and the upper and lower limits from a 95 percent confidence interval. Comparing the point estimates for the three models, the estimated benefit-cost ratios fell within a relatively narrow range, from 2.65 to 3.85. (As an alternative test of our model specification, we calculated

Table 6.5. Marginal Benefit-Cost Ratios for Prune Promotion: A Comparison of Estimates from the Preferred Models

	Benefit-Cost Ratios for a Supply Elasticity of				
Series	0.0	0.5	1.0	2.0	5.0
Model 1					
Producer Benefits/ Producer Costs	3.00	2.99	2.99	2.99	2.99
Producer Benefits/Total Expenses	3.00	1.32	0.85	0.50	0.22
Model 2					
Producer Benefits/ Producer Costs	2.65	2.65	2.65	2.65	2.65
Producer Benefits/Total Expenses	2.65	1.20	0.78	0.46	0.20
Model 3					
Producer Benefits/ Producer Costs	3.82	3.85	3.86	3.86	3.87
Producer Benefits/Total Expenses	3.82	1.48	0.92	0.52	0.23

Note: Present values are in millions of constant (August 1996) dollars using 3 percent (annual) compounding.

the benefit-cost for the misspecified Model 4 as well. The calculation revealed that an analyst who had inadvertently omitted Sunsweet promotion from the analysis (Model 4) would have estimated a benefit-cost ratio of 10.3–a clear example of omitted-variable bias leading to a biased benefit-cost ratio.)

Conclusion

The two main sources of expenditures on promotion of California prunes are CDPB and Sunsweet. Results from analysis of monthly retail data strongly support the proposition that prune advertising and promotion has been an effective mechanism for increasing demand for prunes. Across alternative model specifications, prune promotion had a consistently statistically significant, positive effect on per capita domestic prune consumption. For the various models estimated, the elasticity of prune demand with respect to CDPB promotion generally ranged from 0.048 to 0.052, meaning that a 10 percent increase in expenditures on generic promotion would have induced about a 0.5 percent increase in consumption, holding price and other explanatory variables constant. The models performed well against diagnostic tests, causing us to have reasonable confidence in the specification and, in turn, in the statistical results pertaining to the effects of promotion.

We used a simulation approach to translate the effects of promotion on demand into estimates of marginal benefits to prune growers. The simulation analysis required a complete model of the industry, including supply response. Since a supply analysis was not a component of the present study, simulations were conducted for a range of alternative synthetic supply functions. These simulations, buttressed by some complementary algebraic derivations, enabled us to estimate the marginal returns to the industry from expenditures on advertising and promotion.

Table 6.6. Marginal Benefit-Cost Ratio for a Supply Elasticity of 1.0

Model	Point Estimate	Simulation Results		
		Mean	95 Percent Lower Bound	95 Percent Upper Bound
1	3.0	10.2	0.3	41.6
2	2.7	9.0	0.6	33.2
3	3.9	18.4	0.4	49.5

The marginal benefit-cost ratio for advertising and promotion can hinge on the value of the price elasticity of supply, depending on how expenditures are funded. Marginal returns referred to the revenues generated from an incremental expenditure on advertising and promotion. The ratio of producer benefits to producer incidence of the check-off, however, did not depend on the supply elasticity. We emphasized this measure.

Optimal allocation of expenditures to advertising and promotion calls for expanding expenditures until the marginal dollar spent just yields a dollar back in benefits. The simulation analyses suggested that the industry stopped short of this optimizing condition during the 1992–1996 period covered by the monthly data. Based on our preferred models, the marginal benefit of an additional dollar of expenditure, given the amounts actually expended, ranged upward from $2.65, suggesting that additional expenditures on advertising and promotion would have generated positive net revenue to producers.

We conclude that promotion of California prunes conducted by CDPB has increased demand for prunes and returns to prune producers. Over the four-year period analyzed, the results suggest that investments by prune growers in promotion through CDPB yielded them marginal returns of at least $2.65 for every dollar spent. Moreover, marginal benefit-cost ratios in the range of 2.7:1 and higher indicate that the industry could have profitably invested even more in promotion during this period.

Notes

1 In 2001 the U.S. Food and Drug Administration approved a request by the industry to change the name of its product from "prune" to "dried plum." We utilized the widely recognized term "prune" in this chapter as it was the official name of the fruit at the time the study was conducted.

2 Blaylock and Smallwood (1986) documented some of the general trends in consumer demand for food that can be reflected in shift variables of these types.

3 The choice of whether to include promotion in per capita terms or in total was discussed by Alston et al. (1997).

4 Throughout this section, we use the term "monthly" to refer to the length of the observation period. In reality, the observations were four weeks long.

5 Prices of dried fruit excluding raisins and prices of raisins alone were also considered, but neither was included in any of the models.

6 When *AD* and *PRO* were included in a linear model, their coefficients were essentially equal. Thus, there was no information gained by separating *PROCDPB* or *PROSUN* into their components and we lost degrees of freedom in doing so.

7 Including the square root of promotion rather than the quantity of promotion imposed diminishing marginal returns on the demand response for promotion, whereas the linear model was characterized by constant marginal returns. Diminishing marginal returns

meant that each incremental dollar spent on promotion brought forth a smaller benefit than the last. It was preferable to have a structure that imposed (or at least permitted) diminishing returns for two related reasons. First, it would not be optimal to choose quantities of promotion in a range of constant or increasing marginal returns. Second, to solve for optimal promotion, we required a model that allowed for diminishing returns. Similar conditions did not apply to the other variables in the model since they were not controlled by the prune industry.

8 The econometric work discussed in the previous section provided estimates of the shape and position of this line. In particular, the inverse of the price coefficient was the line's slope (the price coefficient told the change in quantity demanded in response to a unit increase in price) while the horizontal position of the curve was given by the sum of the products of the values of each of the other variables, in year or month *t*, times their corresponding coefficients

References

Alston, J.M., H.F. Carman, and J.A. Chalfant. "Evaluating Primary Product Promotion: The Returns to Generic Advertising by a Producer Cooperative in a Small, Open Economy." *Promotion in the Marketing Mix: What Works, Where and Why*. Proceedings from the NEC-63 Spring 1994 Conference. E.W. Goddard and D.S. Taylor, eds., pp. 145–167. Guelph Ontario: University of Guelph, 1994.

Alston, J.M., H.F. Carman, J.A. Chalfant, J.M. Crespi, R.J. Sexton, and R.J. Venner. *The California Prune Board's Promotion Program: An Evaluation*. Berkeley CA: University of California Division of Agriculture and Natural Resources, Giannini Foundation Research Report 344, March, 1998.

Alston, J.M., and J.A. Chalfant. "Can We Take the Con Out of Meat Demand Studies?" *Western Journal of Agricultural Economics* 16 (1991):36–48.

Alston, J.M., J.A. Chalfant, J.E. Christian, E. Meng, and N.E. Piggott. *The California Table Grape Commission's Promotion Program: An Evaluation*. Berkeley CA: University of California Division of Agriculture and Natural Resources, Giannini Foundation Monograph No. 43, November 1997.

Blaylock, J.R., and D.M. Smallwood. *U.S. Demand for Food: Household Expenditures, Demographics, and Projections*. Washington DC: U.S. Department of Agriculture, Economic Research Service Technical Bulletin 1713, 1986.

California Agricultural Statistics Service. *1995 California Prune Acreage Survey*. Sacramento CA, 1996.

California Dried Plum Board (formerly California Prune Board) and the Prune Marketing Committee. *California Prune News: Annual Report*. Sacramento CA, reports from 1986 to 1992.

California Dried Plum Board (formerly California Prune Board). *California Prune News: Annual Report*. Sacramento CA, 1996a.

——. *PruneTec*. Sacramento CA, 1996b.

——. *The California Prune Buyer's Guide*. Sacramento CA, 1996c.

Chalfant, J.A., and J.M. Alston. "Accounting for Changes in Tastes." *Journal of Political Economy* 96 (1988):391–410.

Goddard, E.W., G.R. Griffith, and J.J. Quilkey. "Optimal Advertising Investment Decisions." Guelph Ontario: Department of Agricultural Economics and Business, University of Guelph, AEB 92/05, November 1992 (revised April 1994).

Information Resources, Inc. (IRI). Infoscan grocery market data on fruit juices and dried fruits, July 1992 to August 1996. Chicago IL, 1996.

International Monetary Fund. *International Financial Statistics*. Washington DC, monthly reports for July 1992 through January 1997.

Peterson, R. *California Marketing & Advertising Report*. Sacramento CA: California Prune Board, 1994.

Thorogood & Associates. *The Impact of Price on UK Market Size*. Sacramento CA: California Prune Board, 1994.

U.S. Bureau of the Census. *Statistical Abstract of the United States*. Washington DC, various editions.

U.S. Department of Agriculture, Agricultural Marketing Service. *How Prunes are Marketed*. Agricultural Marketing Service Bulletin 68, Washington DC, 1979.

U.S. Department of Agriculture, Economic Research Service. *Food Consumption, Prices, and Expenditures, 1970–94*. Economic Research Service Statistical Bulletin 928, Washington DC, 1996.

——. *Fruit and Tree Nuts Situation and Outlook Yearbook*. Washington DC, 2000.

U.S. Department of Agriculture, National Agricultural Statistics Service. *Agricultural Statistics*. Washington DC, various years.

U.S. Department of Commerce. *Statistical Abstract of the United States 1995*. Washington DC, 1995.

7

Evaluation of Avocado Promotion by the California Avocado Commission

Hoy F. Carman & R. Kim Craft

The California avocado industry conducted generic advertising programs under a state marketing order from 1961 through 1977 and has continued such programs under the California Avocado Commission (CAC) since 1978. A review of annual reports of the marketing order and commission programs indicates that California avocado producers spent more than $116 million on advertising, promotion, and related services ("marketing expenditures") through October 1995. This chapter summarizes a study of the effectiveness of California avocado industry advertising and promotion programs from 1961 through 1995. For a more detailed report, see Carman and Craft (1998).

The study focused on two interrelated questions: (1) what impact have marketing expenditures had on demand and supply for California avocados, and (2) have the marketing programs increased net revenues to producers enough to offset the program's costs? Answering these questions required that we specify and develop a detailed econometric model of the California avocado industry that included components for market demand and supply response over time. Most studies of the effectiveness of commodity advertising and promotion programs have estimated benefit-cost ratios by combining estimated demand relationships with assumptions concerning the supply elasticity or a range of supply elasticities. An important contribution of this study is its specification and estimation of a recursive perennial-crop supply-response model for California avocados that traces the estimated effects of advertising and promotion over time. The resulting estimated supply and demand relationships were used to simulate the economic benefits and costs of the avocado industry's advertising and promotion program.

The first step was to use estimates of annual avocado prices both with and without advertising to derive short-run net returns to advertising and promotion. We then used the acreage-response relationship to derive an estimate of long-run returns that accounted for production adjustments over time in response to the impacts on price of advertising and promotion. Finally, short-run and long-run benefit-cost ratios were estimated.

The California Avocado Industry

Avocados are an important, high-value fruit crop with annual sales revenue ranking among California's top ten fruit and nut crops. California produces 85 to 95 percent of the annual U.S. avocado crop; Florida accounts for the rest. Demand for avocados has grown over time as a result of rising consumer incomes, increasing population, and industry-sponsored advertising programs, and producers have responded by expanding planted acres and production. Bearing acreage, for example, did not surpass 25,000 acres until 1977 and total crop value did not exceed $25 million until the 1972/73 crop year. By 1994/95, bearing acres in California totaled 61,254 (after peaking at 76,307 acres in 1987/88), and the state's annual avocado production and value averaged 345.5 million pounds and $194.4 million, respectively, for the five-crop-year period from 1990/91 through 1994/95. Around this trend, avocado production and prices varied substantially from year to year as a result of variable yields and inelastic producer-level demand. With inelastic demand, a large crop returns less total revenue to producers than does a small crop, *ceteris paribus*. Avocado producers tend to exhibit extrapolative expectation behavior when making crop investment decisions—they respond to recent crop returns, expanding acreage and production after several years of favorable returns and decreasing acreage when recent returns have been low. Changes in bearing and nonbearing acreage for 1920 through 1995 are shown in Figure 7.1.

More than twenty varieties of avocados have been produced commercially in California since 1950 but four—Hass, Fuerte, Bacon, and Zutano—currently account for most bearing acres. The relative importance of particular varieties has changed significantly over time. The Fuerte variety's share of total acreage decreased steadily from almost 79 percent in 1950 to a little more than 10 percent in 1990. It was largely replaced by the Hass variety, which increased from 15.5 percent of total acreage in 1960 to more than 71 percent in 1990. In general, the Hass variety has two significant advantages over other varieties:

it typically has the highest average yields and the highest average price per pound.

Weather constraints concentrate California avocado production near the coast in Southern California and areas of the state where microclimates result in a low incidence of frost. The industry commonly divides California's commercial production areas into four districts: (1) North Counties–Santa Barbara, Ventura, and San Luis Obispo; (2) Mid-Counties–Los Angeles, Orange, Riverside, and San Bernardino; (3) San Diego County; and (4) San Joaquin Valley Counties–Fresno, Tulare, and Kern. For the 1995/96 crop year, the districts' estimated shares of bearing acres were San Diego County, 42.3 percent; North Counties, 41 percent; Mid-Counties, 14.4 percent; and San Joaquin Valley Counties, 1.7 percent (CAC annual reports).

The aggregate value of land and trees in California devoted to avocado production is relatively high even for a perennial fruit crop. Several factors are at work, including the value of the crop, land values, and development costs. The average gross revenue per acre from avocado production is typically high compared to many other crops, but land is expensive because the climate that is ideal for avocado production is also ideal for many other crops and for people. Budgeted costs for establishing a new avocado planting were estimated at more than $15,000 per acre in 1993. With land costing between $15,000 and $16,500 per acre, the total cost to establish a new avocado grove is more than

Figure 7.1. California Avocado Acreage by Category, 1920–1995

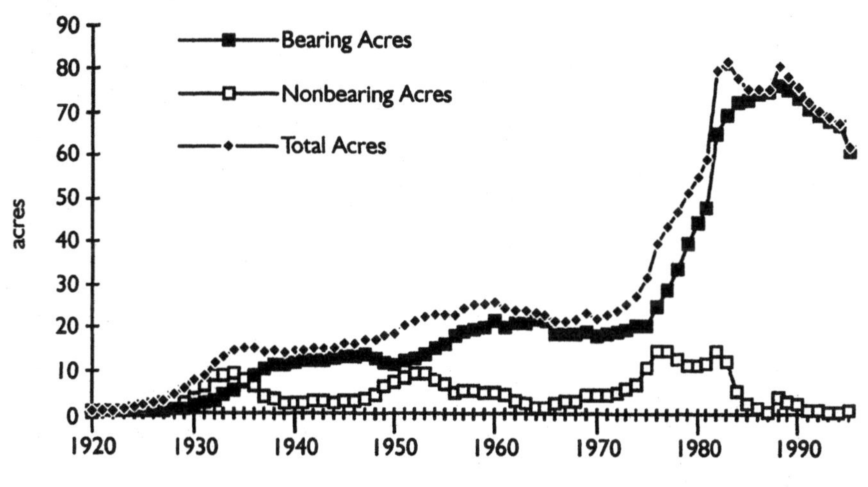

Source: California Agricultural Statistics Service, *California Fruit and Nut Acreage*, annual issues.

$30,000 per acre (Livingston et al. 1993). A recent survey of California land values[1] estimated the value of San Diego County's avocado groves in 1996 as ranging from $5,000 to $15,000 per acre (down from between $10,000 and $17,000 in 1993). Ventura County's groves were valued between $12,000 and $25,000, and Santa Barbara County's groves were valued between $13,000 and $35,000 per acre. Using a value of $20,000 per acre for groves in Ventura, Santa Barbara, San Luis Obispo, Orange, and Los Angeles counties and a value of $15,000 per acre for the remaining bearing acres (San Diego, Riverside, San Joaquin Valley, and others) conservatively estimates the value of California's 1994/95 bearing acres of avocados (61,254 acres) at more than $1 billion.

Analytical Framework

Specifying an empirical model to estimate the costs and returns from CAC's advertising programs required estimating the relationships for acreage responses (supply adjustments) and the demand facing producers. Details on the specification, estimation, and validation of annual acreage responses and demand equations are provided in Carman and Craft (1998). Following is a condensed description of the model's components.

Avocado Supply Response

Annual avocado production is the product of bearing area and average yield. While average yields vary significantly from year to year as a result of weather and other factors, bearing area trends up or down over time as a result of planting and tree-removal decisions that depend on profits and profit expectations. The theoretical framework for models of supply responses by producers of perennial crops has been developed and tested for several crops. A summary of empirical studies, modeling approaches, and estimated relationships is presented in Alston et al. (1995).

The bearing area of avocados at any point in time is based on previous plantings and removal decisions. This relationship can be expressed as:

$$(1)\quad BA_t = BA_{t-1} + NP_{t-k} - R_{t-1},$$

where BA is bearing area, the subscript t (time) designates the year, k is a lag of k years required from the time a tree is planted until it reaches bearing age, NP is area of new plantings, and R is area removed. Two approaches are commonly used to estimate bearing area in year t, and the choice of which one to use depends on the nature and quality of the data available. Detailed and accurate planting and removal data allow direct estimates of new plantings and removal relationships. When such accurate planting and removal data are not available, as is the case with avocados, it is possible to estimate the annual change in bearing area and indirectly account for plantings and removals by specifying lagged variables related to each. The change in bearing acreage relationship was derived by subtracting BA_{t-1} from both sides of Equation (1), which yielded

$$(2) \quad \Delta BA_t = NP_{t-k} - R_{t-1},$$

where the annual change in bearing area, $BA_t - BA_{t-1}$, was defined as ΔBA_t.

Estimation of Equation (2) involved separately specifying the variables related to plantings and removals and the time lags that determine their impact on changes in bearing area. We specified a plantings equation and a removals equation and then substituted the independent variables with time lags for *NP* and *R* in Equation (2).

New plantings of avocado trees during any year are based on the expected profitability of avocado production over the life of the trees. Since expectations cannot be observed, estimation of a plantings equation requires specification of a set of observable, proxy variables related to expectations. Based on previous work, the new-plantings equation for avocados was specified as:

$$(3) \quad NP_t = f(TRA_{t-1,m}, TAX),$$

where *TRA* was a lagged, moving average of farm-level total revenue per acre for avocados deflated by the index of prices paid by farmers for all commodities, services, interest, taxes, and wage rates (with the length of the moving average *m* based on the data) and *TAX* was a 0-1 variable that captured the effect of income tax rules in effect between 1970 and 1976 that reduced the after-tax cost of avocado orchard development and encouraged new plantings. Given that profit expectations are typically based on producers' recent experiences, we expected new plantings to increase as average real returns increase. Our attempts to specify other proxy variables related to the new-planting decision did not succeed.

Avocado trees are removed in response to expectations of poor profits, which may be due to disease, tree age, low prices and/or yields, increased costs, and opportunities to convert land to higher-value uses. Because estimates of annual removals have not proven reliable, most studies have had difficulty isolating statistically significant variables for removals and have instead used a constant percentage of acres or a measure of profitability at the time the removal decisions were made. In this study, the removals function was specified as:

$$(4) \quad R_t = f(TRA_{t-1,m}, WC, BA_t),$$

where *WC* was a 0-1 variable capturing the impact of sharply increased water costs with a value of 1 from 1990/91 through 1994/95 and a value of 0 otherwise. We expected removals to increase when returns were decreasing and to decrease when returns were increasing.

Substituting the explanatory variables for new plantings (*NP*) and removals (*R*) with appropriate time lags in Equation (2) yielded the change in bearing area relationship to be estimated:

$$(5) \quad \Delta BA_t = f(TRA_{t-5,m}, TAX, WC, TRA_{t-1,m}).$$

The major time lag is the time that passes between a producer making a planting decision and classification of the tree subsequently planted as bearing. Most of California's avocado trees (more than 85 percent) reach bearing age in three years. Since a one- to two-year delay between the decision and actual planting is not unusual (as a result of time required for land preparation and acquisition of nursery stock), the lag between the decision to plant an avocado tree and the time that tree reaches bearing age can be four to five years ($k = 4$ or 5).

Average annual yields and bearing area combine to determine total production during any given year. We examined the pattern of average yields over time but found no evident trend. Also, there is little producers can do to alter average yields during a crop year.

Avocado Demand

Since production of avocados during a marketing year is determined by previous decisions and events and because avocados generally are consumed fresh and not stored from one crop year to the next, the quantity of avocados placed on the market during a given year is essentially predetermined. Given these conditions, annual demand for avocados is best expressed in the price-dependent form (the inverse demand model) where the annual price of avocados is explained by available quantities and other factors. Along with quantity, key factors influencing demand include the price of related goods, the purchasing power of potential customers, the size of the market in terms of number of consumers, and consumer tastes and preferences. Here we present a broad overview of historical data that characterizes and explains annual demand for California avocados.

Several trends were evident in the variables expected to affect demand for California avocados. Both the U.S. population and the purchasing power of consumers as represented by real disposable income increased significantly during the period of analysis. With increasing production and crop values, the assessments collected from producers and available to fund industry advertising and promotion also increased.

One of the most prominent factors explaining the price of avocados for a given time period is the quantity offered on the market. Figure 7.2 illustrates total California avocado production, along with the average real price per pound, for the crop years 1961/62 through 1994/95. Supply fluctuated dramatically around an increasing trend for the period, with annual variability due largely to average yields. Beginning in 1981, however, there was little if any trend in production quantities and a substantial increase in annual variation. The "law of demand" holds that there is an inverse relationship between price and quantity, *ceteris paribus*. Thus, an increase in supply has a negative effect on

Figure 7.2. Annual California Avocado Production and Average Real Price (1995 = 100)

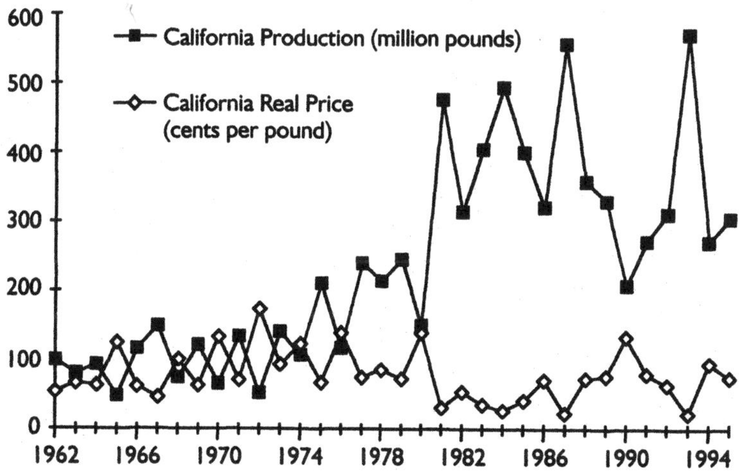

Source: California Agricultural Statistics Service, *California Fruit and Nut Statistics*.

price provided all other determinants of demand are fixed. This phenomenon is clearly seen in the avocado market, where real prices and quantities moved in opposite directions nearly every year (the simple correlation coefficient for the two series is −0.68). Thus, changes in production clearly explain a great deal of the variation in avocado prices over time.

The quantity demanded and price of California avocados are affected by supplies from other regions. California typically accounts for 85 to 90 percent of total U.S. production and Florida supplies the rest.[2] Historically, the quantity of avocados imported to the U.S. has been quite small. For instance, imports between 1962 and 1989 amounted to a little more than 1 percent of annual U.S. supply. Recent years, however, have seen substantial increases in foreign avocado purchases; imports have averaged more than 10 percent of the annual U.S. supply since 1990. While California clearly dominates the U.S. market, the avocado supply from Florida and foreign producers is significant and is expected to have a measurable negative impact on California prices.

Advertising is the final factor posited to explain consumer demand for avocados. Industry advertising efforts are expected to increase demand by providing information and changing preferences. A major objective of this study was to determine the extent to which generic advertising programs conducted by CAC have been successful in increasing the demand for California avocados.

The California avocado industry has conducted generic advertising programs since 1961. The current commission law allows a maximum assessment to producers of 6.5 percent of the gross dollar value of the year's avocado sales to support all CAC activities, including production research, industry affairs, marketing programs, and administration. The actual assessments during the 1990s were substantially below the allowable maximum, ranging from a low of 3 percent in 1990/91 to a high of 5.25 percent in 1992/93. The average annual assessment for crop years 1990/91 through 1994/95 was 4.05 percent.

Figure 7.3 describes CAC's marketing expenditures per crop year in real 1995 dollars. Since available funds were based on producer revenues, the marketing expenditures followed gross crop values fairly closely over the period. In fact, the correlation coefficient between marketing expenditures and the prior year's crop value was 0.83 in terms of levels and 0.50 in terms of annual rates of change, verifying the expected positive relationship.[3] Thus, program outlays trended upward throughout the period in both nominal and real terms. The overall rate of increase in expenditures, however, was smaller than the rate of increase in gross crop value. For instance, the real trend line

Figure 7.3. California Avocado Industry Annual Marketing Expenditures in Millions of Real Dollars (1995 = 100)

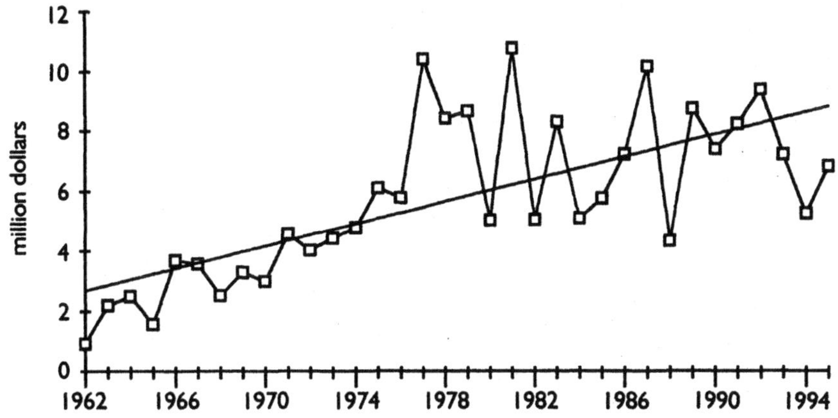

Source: California Avocado Commission.

in Figure 7.3 increases at an average annual rate of 3.7 percent, a percentage point below the trend line for real producer revenue. In addition, note that there appears to have been little in the way of a trend in real marketing expenditures after 1975/76.

Specific marketing efforts conducted on behalf of California avocado producers have taken a variety of forms. For instance, in its annual reports the CAC has divided total marketing expenditures into seven categories: consumer advertising, consumer promotion, trade advertising, food service, public relations, international promotion, and processed products. Figure 7.4 shows how funds have been allocated to the four major categories–consumer advertising, consumer promotion, trade advertising, and all other–from 1985/86 through 1994/95. Overall, consumer advertising received the greatest percent of funds (averaging 41.4 percent for the period), followed by trade advertising (25.1 percent), other (24.2 percent), and consumer promotion (9.3 percent). Year-to-year allocations varied somewhat but remained fairly constant for the most part, particularly when consumer advertising and promotion are considered together. Figure 7.5 illustrates the relative allocation of California avocado advertising dollars by media type from 1976 through 1994. Magazine advertising dominated the other media types, receiving an average of 78 percent of all advertising dollars for the period.[4] Magazines were the only advertising media used prior to 1976. The chart suggests that program administrators experimented with other media, varying the allocation of funds among television, newspapers, outdoor displays, and radio from year to year. Still, media

Figure 7.4. California Avocado Industry Annual Marketing Expenditures by Category, 1986–1995

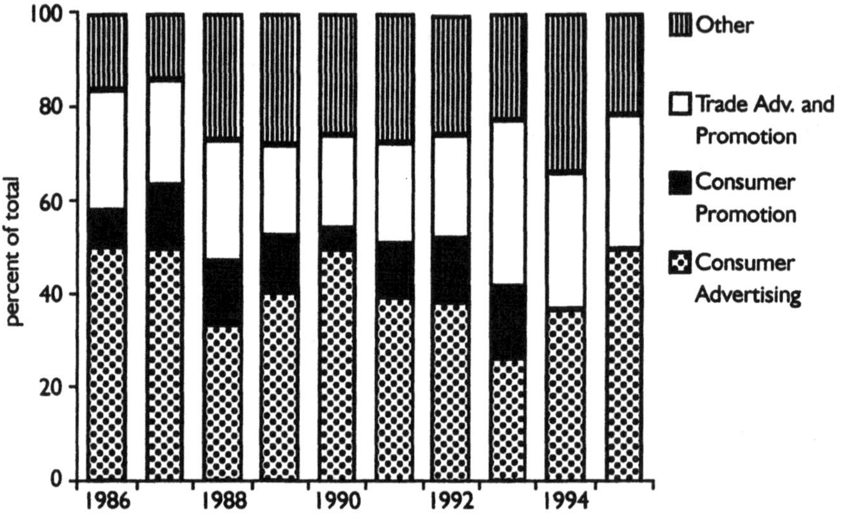

Source: California Avocado Commission.

Figure 7.5. California Avocado Industry Annual Advertising by Media Type, 1976–1994

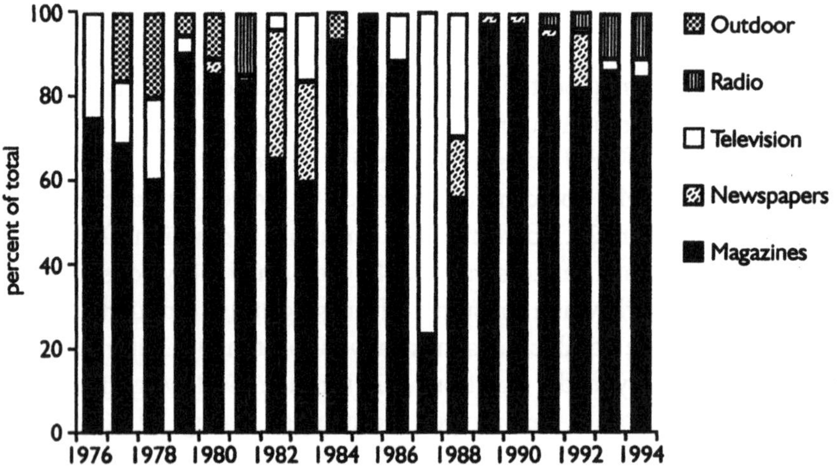

Source: California Avocado Commission.

other than magazines accounted for only small portions of the advertising effort for most years. A notable exception is a television campaign in 1987 that accounted for more than 75 percent of that year's advertising funds.

While the effectiveness of generic avocado marketing efforts may depend on the specific program implemented, a thorough analysis of that issue required more detailed data than was available for this study. Hence, we examined the effects of all marketing expenditures combined and refer to them simply as "advertising" or "advertising and promotion."

Estimation Results

Equation (5), which describes the change in bearing acres from new plantings and removals, was estimated using annual data for 1951/52 through 1994/95. The annual-price equation was estimated using data for 1961/62 through 1994/95. The latter period coincides with the existence of industry marketing efforts.

Avocado Supply Response

We estimated the change in bearing acreage in Equation (5) using lagged averages of deflated total revenue per acre that ranged from two to five years as the variable related to avocado plantings and removals. For new plantings, a four-year average of deflated total revenue per acre that was lagged five years provided the best statistical results. More recent total revenue per acre was not a good proxy for profit expectations affecting removals. Based on our statistical results, recent changes in avocado prices (*PR*) and costs measured by the index of prices paid by farmers (*PPI*), each lagged one year, were a better proxy for profit expectations affecting removals. It is not unusual in empirical work to find that removals respond to more recent economic factors than do plantings, even though one would expect profit expectations affecting plantings and removals to have similar time horizons. The equation estimating the change in bearing acreage is:

$$\text{(6)}\quad \Delta BA_t = \underset{(-5.30)}{-3{,}640.85} + \underset{(5.95)}{2.89}\, TRA_{t-5,4} + \underset{(2.89)}{1{,}753.63}\, TAX_{t-5} - \underset{(-3.35)}{2{,}544.72}\, DW$$

$$- \underset{(-1.69)}{7.17}\, PR_{t-1} + \underset{(2.65)}{28.21}\, PPI_{t-1} + \underset{(9.90)}{12{,}246.90}\, D81$$

$R^2 = 0.92$, Durbin Watson = 2.12

where the values in parentheses are the t-statistics for each coefficient. Each of the coefficients had the expected sign and all except the coefficient for the one-year lagged cost index were statistically significant at the 95 percent level or greater. The variables combined to explain 92 percent of the variation for annual changes in bearing acreage based on the R^2 statistic.

The proxy variable for expected profits that affect planting was a four-year average of total revenue per acre deflated by the index of prices paid by farmers for all production items (1977 = 1.00) lagged five years ($TRA_{t-5,4}$). The four-year average yielded better statistical results than did either a three- or five-year average. The five-year lag was also preferred to either four or six years based on standard statistical measures. The proxy variables used for expected profits in the removals equation were average prices and the cost index, each lagged one year.[5] Income tax laws had a significant impact on plantings during the six-year period from 1970/71 through 1975/76; the *DTax* variable had a value of 1 for each of those six years and 0 otherwise. The estimated coefficient on the *DTax* variable indicated that the shift of investor interest (due to restrictions on citrus and almonds) increased total avocado plantings by almost 1,753 acres each year during that period relative to the "no tax law" period. The impact of taxes on the change in bearing acreage was lagged five years because of the delays between planting and bearing noted previously. The

Figure 7.6. Annual Change in Bearing Acreage of California Avocados, Actual versus Predicted, 1953–1995

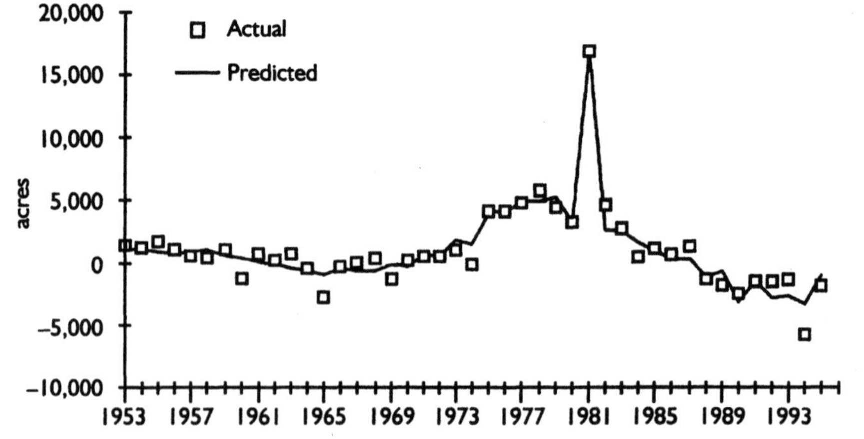

Source: 1953–1955, California Crop and Livestock Reporting Service; 1956–1987, California Agricultural Statistics Service (*California Fruit and Nut Acreage*); 1988–1995 California Avocado Commission.

estimated coefficient on the dummy variable for the increase in water prices indicated that bearing acreage decreased almost 2,545 acres each year during the five years from 1990/91 through 1994/95. Further decreases of this magnitude are unlikely; the industry has worked closely with local water agencies to receive some relief, and the most vulnerable acreage has been removed. The *D*81 dummy variable accounts for the discovery of previously uncounted acreage in the 1981 detailed acreage surveys for San Diego, Santa Barbara, and Ventura Counties (Carman and Craft 1998, pp. 10–11). Unexplained variation in the equation was relatively small.

The coefficients in Equation (6) were consistent with expectations. They explained a high proportion of the annual variation in bearing acreage and also were the easiest to work with in simulating acreage and supply response to various policies and programs. As shown in Figure 7.6, the equation did a reasonable job of predicting annual changes in bearing acreage given the values for each of the independent variables. When an initial value for bearing acreage was entered, the equation also provided estimates that tracked total bearing acreage quite well.[6] The comparison of actual and predicted bearing acreage shown in Figure 7.7 describes the model's tendency to overestimate bearing acreages slightly during the early years when acreage was comparatively stable and to underestimate the total when acreage was expanding rapidly. In overall terms, however, the estimated values were quite close to actual values during the years analyzed.

Avocado Demand

In this section we formalize the basic concepts previously discussed by developing and testing an econometric model of annual avocado demand. As shown, the variables consist of thirty-four annual observations from 1961/62 through 1994/95 corresponding to the factors already described.

Specification of the empirical demand model had to be based on theory, availability of data, and statistical feasibility. Given these considerations, we represented the annual demand for avocados by the following general function:

$$(7) \quad Pc_t = f(Qc_t, Qf_t, Qm_t, Y_t, A_t),$$

where, for a specific year t, Pc_t is the farm-gate price of California avocados, Qc_t is total sales of California avocados, Qf_t is total sales of Florida avocados, Qm_t is total avocado imports, Y_t is disposable income, and A_t is advertising

Figure 7.7. Bearing Acreage of California Avocados, Actual versus Predicted, 1953–1996

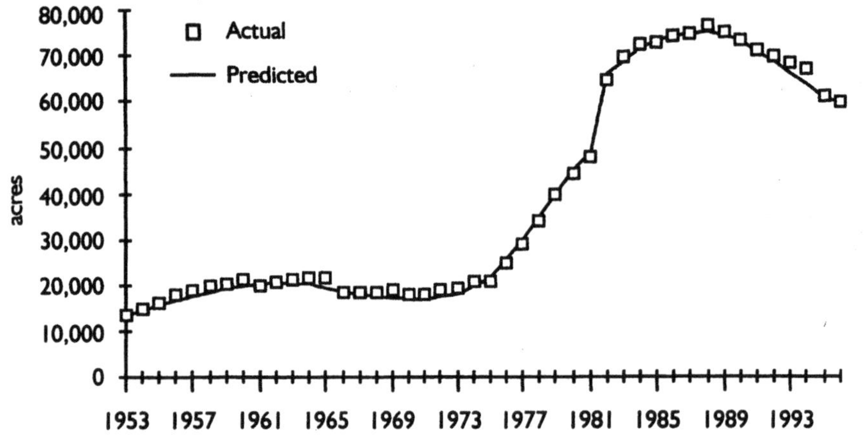

Source: 1953–1955, California Crop and Livestock Reporting Service; 1956–1987, California Agricultural Statistics Service (*California Fruit and Nut Acreage*); 1988–1995, California Avocado Commission.

Table 7.1. Definitions of Variables Used in the Annual Demand Model

Variable	Definition	Units	Mean Value	Standard Deviation
Pc_t	Average annual producer price of California avocados deflated by the CPI for all items (1982–84 = 100).	Real cents per pound.	51.2860	24.8340
Qc_t	Per capita sales of California avocados during the marketing year.	Pounds per capita.	1.0120	0.5888
Qf_t	Per capita sales of Florida avocados during the marketing year.	Pounds per capita.	0.1862	0.0705
Qm_t	Per capita imports of avocados during the marketing year.	Pounds per capita.	0.0327	0.0484
Y_t	U.S. per capita disposable income deflated by the CPI for all items (1982–84 = 100).	Thousands of real dollars.	10.4570	1.7392
A_t	Annual avocado advertising expenditures deflated by the CPI for all items (1982–84 = 100).	Millions of real dollars.	3.7405	1.7655

Note: Variables consisted of thirty-three annual observations, 1961/62 through 1993/94, based on a California crop year with timing adjustments as discussed in Carman and Craft (1998).

expenditures. Precise definitions and descriptive statistics for each variable are given in Table 7.1. We assumed that current-period California and Florida quantities were determined by prior-period production decisions and events independent of the current-period price. Thus, they were *predetermined* rather than endogenous variables.

We followed the accepted practice of assuming demand was homogeneous of degree zero in money income and prices; hence, all dollar-denominated variables were expressed in real terms using the Consumer Price Index (CPI) as a deflator. In addition, the quantity variables and income were given in per capita terms to account for the effect of population on demand. The advertising variable was expressed in real terms but was not adjusted for population growth. The choice of whether to include promotion in per capita terms or in total was discussed by Alston et al. (1997).

The quantity of imports may well depend on the current-period price. It was reasonable to suppose that, while foreign production is predetermined, a larger share of avocados enters the country when the domestic price is high and, hence, that Qm_t was endogenous. Nevertheless, it was possible that the statistical consequences of treating a theoretically endogenous variable as if it were an independent explanatory factor (exogenous) were negligible. Because the quantity of imports was very small relative to domestic avocado production, we chose to regard it as an exogenous variable and then perform tests to ascertain whether this simplification was likely to have important effects on the results.

Finally, since the funds available for advertising are based on producer revenue and since producer revenue depends on price, the A_t variable also could be endogenous. This did not appear to be a problem in our analysis. As noted, our investigation found that advertising expenditures bear some relationship to the *previous* year's revenues but very little to the *current* year's revenues–i.e., A_t was associated with Pc_{t-1} rather than Pc_t. In addition, the statistical relationship between advertising expenditures and producer revenues was diluted by the following circumstances: the avocado industry varied the percentage of the assessment over time, the proportion of the total assessment used for advertising varied annually, and there was carryover of funds from year to year. Thus, we assumed that advertising could be treated as an exogenous or predetermined explanatory factor.

Model Estimation and Testing

To estimate the annual demand model, a functional form that characterized the general relationship previously described had to be specified. Since choice of functional form could have important impacts on the results (e.g., Alston and Chalfant 1991), we began by following Carman and Green (1993) in specifying an extended Box-Cox model–which implied a quite general functional form. Thus, the empirical demand model initially assumed was given by

$$(8) \quad Pc_t^* = \beta_0 + \beta_1 Qc_t^* + \beta_2 Qf_t^* + \beta_3 Qm_t^* + \beta_4 Y_t^* + \beta_5 A_t^* + \varepsilon_t,$$

where for each variable X_t, X_t^* symbolized the Box-Cox transformation defined as $(X_t^{\lambda} - 1) / \lambda$ if $\lambda \neq 0$ and $X_t^* = \ln X_t$ if $\lambda = 0$. The ε_t term was assumed to be a normally distributed random error with zero mean and constant variance ($\varepsilon_t \sim N(0, \sigma^2)$). Note that the Box-Cox model encompasses the two most common specifications used in empirical applications: $\lambda = 1$ implies a linear functional form and $\lambda = 0$ implies a log-linear functional form.

Using annual data for crop years 1961/62 through 1993/94, the parameters β_i (i = 0–5) and λ were estimated using maximum likelihood, yielding the following results:

$$
\begin{array}{lllllll}
(9) \quad Pc_t^* = -15.68 & - 5.61\, Qc_t^* & + 0.80\, Qf_t^* & - 2.00\, Qm_t^* & + 5.38\, Y_t^* & + 0.37\, A_t^* & + \varepsilon_t. \\
\quad\quad (-2.79) & (-18.75) & (1.38) & (-2.08) & (6.04) & (1.65) & \\
 & [-1.21] & [0.09] & [-0.11] & [2.89] & [0.13] &
\end{array}
$$

$$\lambda = 0.39, \; R^2 = 0.94, \; \text{Durbin Watson} = 1.45$$

The numbers in parentheses are t-statistics and the numbers in brackets are the price flexibilities of demand with respect to the corresponding variables, evaluated at the data means.

The estimated coefficients in Box-Cox models were of little interest since they applied to the transformed rather than the original variables. We therefore report the estimated price flexibilities of demand defined as the percent change in price resulting from a 1 percent change in a given explanatory factor.[7] For example, the estimated model implied that a 1 percent increase in advertising expenditures would result in an approximately 0.13 percent increase in the price of California avocados when evaluated at average levels of price and advertising in the sample data.

The coefficients on Qc_t, Qm_t, and Y_t had the expected signs and were statistically significant at the 95 percent confidence level. The coefficient on Qf_t had the "wrong" sign–it was expected to be negative–but was not statistically significant. The coefficient on A_t had the expected sign but was not statistically significant.

The R^2 statistic implied that the estimated equation generally fit the data well. However, the Durbin Watson statistic suggested that the error term might be autocorrelated; if so, it would not satisfy the assumptions under which the equation was estimated and would suggest a problem with the model. Since the value of the Durbin Watson statistic was in the inconclusive range, we performed a second test, which gave strong evidence that the regression errors did exhibit a high degree of autocorrelation (Savin and White 1978).[8] Autocorrelated errors can result from model misspecification (incorrect functional form or omitted explanatory variables) or may actually follow an autoregressive process. The former calls the entire model into question while the latter makes the conventionally calculated t-statistics invalid. Thus, we sought a remedy.

Our examination of alternative functional forms found that a variation on the originally estimated Box-Cox model performed well. Rather than all of the variables being transformed by the λ parameter as in the first regression (called the *extended* Box-Cox model), the alternative model transformed only the dependent variable (called the *classical* Box-Cox model). The results from estimating the classical Box-Cox model were:

$$(10)\quad Pc_t^* = \underset{(18.37)}{1.91} - \underset{\substack{(-20.22)\\ [-1.33]}}{0.53\, Qc_t} + \underset{\substack{(0.70)\\ [0.06]}}{0.14\, Qf_t} - \underset{\substack{(-2.96)\\ [-0.10]}}{1.23\, Qm_t} + \underset{\substack{(7.51)\\ [2.77]}}{0.11\, Y_t} + \underset{\substack{(1.53)\\ [0.13]}}{0.01\, A_t} + \varepsilon_t.$$

$\lambda = -0.23$, $R^2 = 0.95$, Durbin Watson = 2.11

As before, the numbers in parentheses are t-statistics and the numbers in brackets are the price flexibilities of demand with respect to the corresponding variables evaluated at the data means.

The first thing to note about these results is that the estimated flexibilities and t-statistics were very close to those estimated with the extended Box-Cox model; thus, our conclusions about both the economic and statistical significance of each explanatory variable were essentially the same. Moreover, the R^2

statistic indicated that the classical Box-Cox model fit the data to basically the same degree as the extended model. In a statistical sense, the key difference between the two models was that autocorrelation did not seem to be a problem with the classical Box-Cox model. The Durbin Watson statistic implied that the errors were not autocorrelated, and the test of Savin and White (1978) confirmed that finding.[9] These results suggested that an incorrect functional form was the cause of autocorrelation in the extended Box-Cox regression, that the classical Box-Cox model corrected the problem, and that therefore the latter was a better model.

There was a possible difficulty with this conclusion, however (which is why we report both models). Contrary to the results from the extended Box-Cox model, the estimated classical Box-Cox function implied that price rises at an increasing rate as the explanatory variables increase. Thus, as the level of advertising rose, each unit increase would result in a greater change in price than did the previous unit increase. Clearly, this functional form could not prevail for all levels of advertising. Economic theory and common sense prescribed that there had to be diminishing marginal returns to advertising beyond some point and, moreover, that the optimal level of advertising was in the range of diminishing returns. If we assumed that the avocado industry was in equilibrium and behaving rationally, the extended Box-Cox regression model that exhibited decreasing returns would be preferred. It was quite possible, however, that CAC had been operating in the area of increasing returns because of previously limited information on the impact of its advertising on avocado prices. In other words, it could be extremely difficult for a commodity organization such as CAC to determine the nature of returns to advertising without detailed empirical analysis.

A common hypothesis in the marketing research literature is that market response to advertising is s-shaped (e.g., Little 1979; Lilien, Kotler, and Moorthy 1992). The concept is illustrated by the hypothetical market response function shown in Figure 7.8. Initially, the market responds (e.g., price changes) slowly when the levels of advertising applied are small. At less than fifty units of advertising, the market responds at an increasing rate; at more than fifty units of advertising, the market responds at a decreasing rate. By one hundred units, advertising has reached a saturation point at which more effort has negligible effects. Thus, the theory asserts, price can exhibit both increasing and decreasing marginal returns over different ranges of advertising.

Considering our estimation results from this perspective suggested the possibility that advertising efforts in the California avocado industry between 1961/62 and 1993/94 were in the range of increasing returns (i.e., corresponding to advertising levels below fifty, as shown in Figure 7.8). If this was the case, then the appropriate model was the classical Box-Cox. However, that model, while perhaps appropriate for past levels of advertising, would not apply to greater and greater levels of advertising. In addition, if we believed that observed advertising expenditures were at or near optimal levels, then a regression model exhibiting constant or increasing returns would not be satisfactory since the optimal advertising level must be in the region of diminishing returns.

It was reasonable to assume that the impacts of advertising extended over some period of time and, hence, that this dynamic response should be accounted for in the regression model. However, determination of the nature and duration of the lagged effect is difficult (Nerlove and Waugh 1961; Clarke 1976; Reynolds, McFaul, and Goddard 1991; Kinnucan and Forker 1986; Ward and Dixon 1989). Since the estimated model utilized annual data, we expected little carryover of advertising effects from one year to the next. As partial verification, we estimated a model that included lagged effects of advertising. The estimated coefficients for one- and two-year lags were not statistically different from zero and, thus, variables for lagged effects of advertising were not included in the final model. We did, however, expect to find lagged effects from advertising expenditures when moving to a monthly period of analysis.

Figure 7.8. Hypothetical Market Response to Advertising

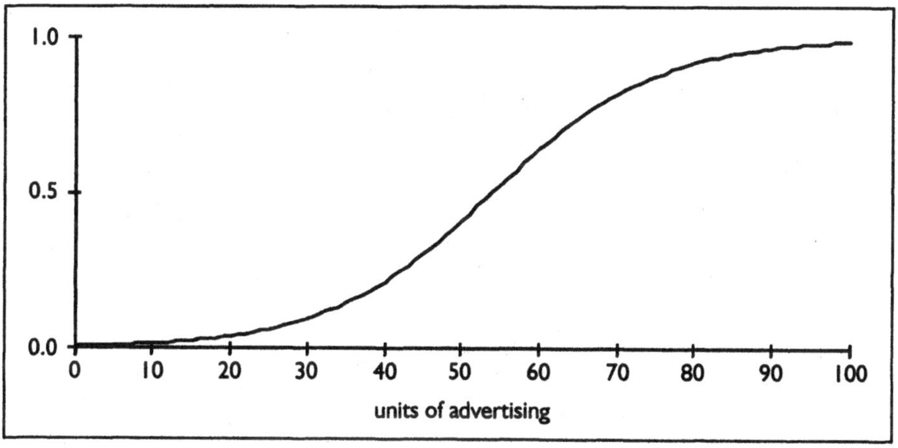

Evaluation of the Estimated Demand for California Avocados

While there were some theoretical concerns about the implications of increasing returns to advertising, the classical Box-Cox regression ultimately was our preferred model. In most respects, the model was statistically sound and logically consistent. We therefore used the results from the classical Box-Cox regression model to draw inferences about annual demand for California avocados during the study period (1961/62 through 1994/95).

As expected, the model indicated that the quantity of avocados offered on the market was an important explanatory factor as it had a precise, negative impact on price. The estimated price flexibility of demand of −1.33 (at the data means) indicated that the price elasticity of demand was approximately −0.75, implying avocado demand is inelastic, as predicted.

We expected Florida avocados to be competitive with California avocados and that increased sales of Florida avocados would have a negative impact on the price of California avocados. The coefficient on the quantity of Florida avocados, however, had a positive sign, but it was not statistically different from zero. There were at least two possible explanations for this outcome. First, it could have resulted from the data inconsistencies previously discussed in some detail. Second, it could be that the timing of Florida shipments made them competitive with California avocados at some times and complementary with them at other times *within* the same year. For instance, if Florida avocados tended to be in greater supply when California avocados were in relatively low supply, their availability could help keep consumers in the habit of purchasing avocados and hence have a complementary effect on California demand. More detailed data would be necessary to resolve this question.

Imports were found to have a statistically significant, negative impact on California avocado prices. This indicated that foreign avocados compete directly with California avocados for consumer dollars. Disposable income was found to have a very large and significantly positive impact on avocado prices, indicating that avocados are a normal good.

Finally, the classical Box-Cox regression model indicated that advertising has had a positive impact on California avocado prices. While the value of the t-statistic for the estimated advertising coefficient was relatively close to the critical value, it was not statistically significant at the usual confidence levels (95 percent or better). This result did not lead to the conclusion that advertising was ineffective. Rather, it implied that, with the available annual data and assumed model, our estimates were not precise enough to conclude with 95 percent certainty that the advertising coefficient could not in fact equal zero.

Improved data likely would increase the precision of our estimates of the effects of advertising.[10]

Analysis of Advertising Benefits and Costs

Several approaches were available for measuring estimated benefits and costs of the California avocado industry's advertising programs given the demand- and acreage-response equations developed in this study. Using the estimated demand equation, we developed short-run (within-year) estimates of the benefits of increased demand due to advertising. These estimated annual benefits were then divided by actual annual costs to calculate average benefit-cost ratios. A ratio greater than 1 indicated that total returns from advertising were greater than the costs; the higher the ratio, the higher the returns from advertising. A positive benefit-cost ratio less than 1 indicate that advertising increased revenues but that the increase was less than the costs.

The short-run benefit-cost ratios based solely on estimated demand did not account for the lagged supply response to short-run price improvements and, thus, tended to overstate the benefits from an advertising program conducted over a long period of time. We combined the acreage-response and annual-demand equations to form a recursive model of supply response and used that model to simulate annual total revenues over time both with and without the advertising program. The differences in total crop revenues measured advertising benefits, which we then compared with annual program costs.

The short-run and longer-run benefit-cost ratios outlined previously were averages of benefits and costs. The marginal response of revenues to additional advertising was also of interest since it indicated whether the industry had overallocated or underallocated funds to advertising. We examined marginal benefit-cost ratios by increasing annual or monthly advertising expenditures by 1 percent and calculating the ratio of the change in benefits to the change in expenditures.

Analytical Model of Supply and Demand

The measurement of returns from advertising can be illustrated with the hypothetical supply and demand relationships shown in Figure 7.9. The annual-demand curve for avocados without advertising is shown by the demand curve *D*. With an effective advertising program, the demand curve *D* increases (shifts to the right) to D^A. Since avocado supply is essentially fixed for a given marketing year, as represented by the vertical supply curve *S*, average prices

Figure 7.9. An Economic Model of Avocado Supply and Demand

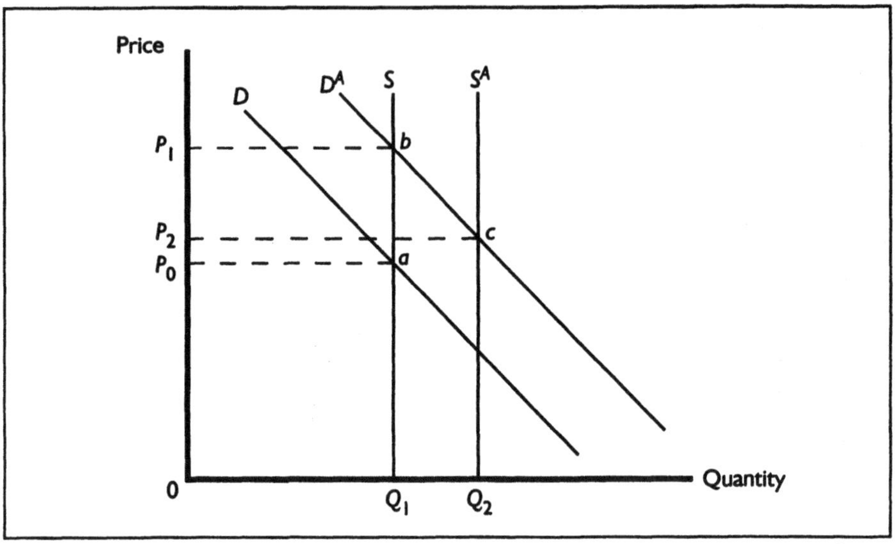

increase from P_0 to P_1. Increased revenues (and profits) encourage producers to expand acreage and, after a lag of several years, production. The lagged increase in production is shown by the outward shift of a fixed annual supply from S to S^A. Because of significant delays between the time a decision to expand production is made and actual output is available, higher prices to producers persist during the early years of an advertising program. Then, as new trees reach bearing age, expanded production shifts the vertical supply curve to the right and prices decrease from P_1 to P_2, as shown in the diagram. As illustrated, total revenue with advertising (the rectangle $0P_2cQ_2$) is greater than total revenue without advertising (the rectangle $0P_0aQ_1$).

We measured the short-run returns from advertising by the rectangle P_0P_1ba. The longer-run returns from advertising, which accounted for the effects of supply response, were measured by the difference between total revenue with advertising and total revenue without advertising. The returns from advertising had to be compared to the costs of advertising to evaluate the profitability of the program. There were two possible measures of costs. The obvious measure was the total dollars spent on advertising, which implicitly assumed that avocado producers bore all of the costs of the program. However, when the advertising cost is financed by a percentage assessment on total crop revenues at the producer level, some of the incidence of the assessment, over time, falls on buyers through the operation of supply and demand. A second measure of costs allowed some of the costs to be borne by buyers, resulting in the producers' share of costs being less than actual expenditures.

The amount of the price increase to buyers depends on the slopes of the supply and demand curves. When supply is fixed and unresponsive to price, as it is in the short run, there is no increase in the price to buyers and all of the costs are borne by producers. The more responsive quantity supplied is to price (the more price elastic the supply is), the smaller is the proportion of the assessment borne by producers.

Avocado Supply Response

We combined the price and change-in-bearing-acreage equations estimated in previous sections into a recursive model of supply response. First, we entered initial values for lagged, real total revenue and actual values for per capita disposable income, the Consumer Price Index, the index of prices paid by farmers, population, average yield per acre, Florida avocado production, quantity of avocado imports, and advertising expenditures. Starting with these values, the model generated values for annual bearing acreage, average price per pound, and total revenue per acre for California avocados. Bearing acreage multiplied by average yield determined California production. Total production was combined with demand factors to calculate the average price of avocados. The year-to-year change in bearing acreage was a function of profit expectations, which were based on lagged per-acre total revenue (price multiplied by average yield) adjusted by the index of prices paid by farmers and on avocado prices and the producer cost index lagged one year.

A comparison of actual and simulated values for bearing acreage is presented in Figure 7.10. The actual peak bearing acreage was 76,307 in 1986/87 while the simulated peak was almost identical at 76,289 acres but occurred two years earlier (1984/85). While the actual and simulated peak acreages were very close, the simulation model typically underestimated or overestimated acreage during much of the period. The largest overestimated acreage was 5,260 acres for 1979/80 and the largest underestimated acreage was 1,903 acres for 1992/93. The average annual absolute difference between actual and simulated acreage was 1,390 acres (3.6 percent). Overall, the model did a very good job of tracing the total bearing acreage adjustments that occurred between 1961/62 and 1994/95; the difference in actual and simulated 1995 acreage was only 187 acres.

To derive long-run estimates of the impact of advertising on production and prices of California avocados over time, we ran the simulation model

Figure 7.10. California Avocado Bearing Acreage, Actual and Simulated with and without Advertising, 1962–1995

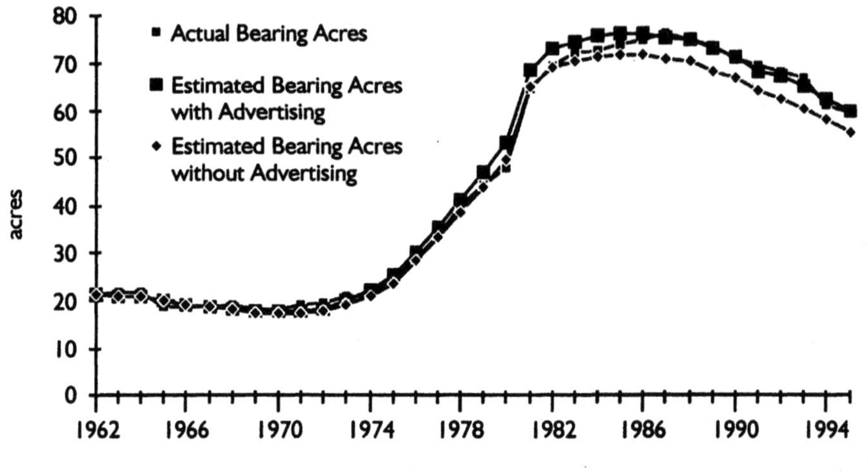

Source: 1962–1987, California Agricultural Statistics Service (*California Fruit and Nut Acreage*); 1988–1995 California Avocado Commission.

without advertising. Comparison of the simulated advertising and no-advertising scenarios showed that advertising increased prices and per-acre returns and that those improved returns led to expansion of bearing acreage and production. Simulated bearing acreage reached a peak of 76,289 acres in 1984/85 with advertising and 71,819 acres during the same year without advertising (Carman and Craft 1998, p. 60). As shown in Figure 7.10, the last observation of bearing acreage without advertising was 55,196 acres, which was 4,568 acres (almost 8 percent) less than with advertising.

Annual Short-Run Benefit-Cost Ratios for Advertising

The estimated annual demand relationship was used to estimate benefits from advertising in a year-to-year (short-run or fixed supply) framework by calculating price and total revenue for actual production each year with and without advertising. Given the positive coefficient for the advertising variable in the estimated demand equation, increased advertising would result in higher average prices during a given crop year if all other factors were equal. The short-run comparison of annual estimated prices with and without advertising indicated that advertising yielded positive net returns (the benefit-cost ratio was greater than 1) for all crop years beginning with 1965/66. The short-run total benefit-cost ratio for advertising ranged from 0.76 to 0.93 during the first four years of the program (1961/62 through 1964/65) and then ranged from 1.32 to 14.37 during the remaining period of analysis. The weighted annual average was 7.09 for the total period (Figure 7.11). Thus, each dollar spent on advertising increased average total sales revenue in the same year by $7.09 and, after subtracting the cost of advertising, yielded a net return of $6.09 (the benefit-cost ratios for each year were reported in Carman and Craft (1998) in Appendix Table 11a).

Readers will note that some of the largest returns were observed during the most recent eight years of CAC's advertising program. Since the costs and benefits of the advertising program change from year to year and accrue over time, it is more meaningful when calculating returns to account for changing price relationships and discount the stream of costs and benefits. Thus, the current costs and returns from advertising were restated in 1994/95 dollars and were used to calculate the present value of the program at discount rates of 0 and 3 percent (Carman and Craft 1998, p. 61). Note that the annual benefit-cost ratio was the same in current and in 1994/95 dollars. When advertising costs and returns were discounted at 0 and 3 percent, the cumulative benefit-cost ratios were 6.01 and 5.33, respectively.

This fixed supply analysis did not fully account for the supply response that *could* occur over time, which could limit the applicability of the weighted

Figure 7.11. Estimated Short-Run and Long-Run Benefit-Cost Ratios for Crop Years 1961/62 through 1994/95

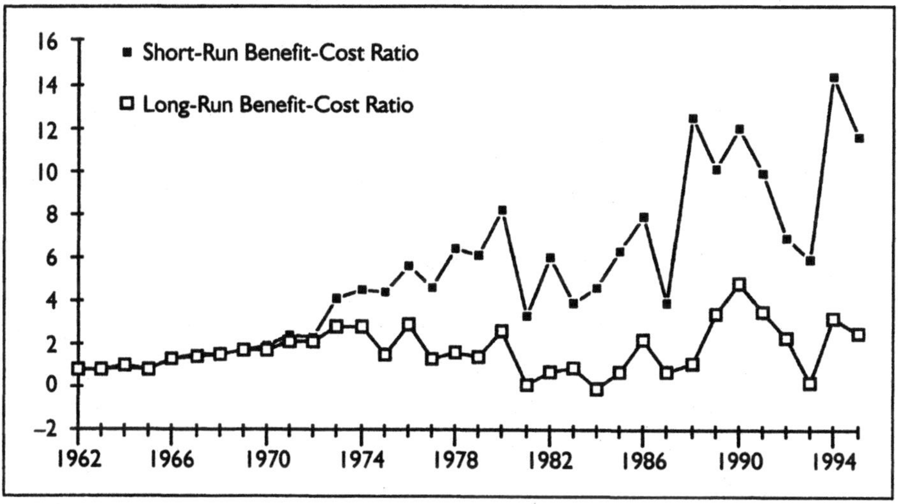

average based on summation across the years previously presented. The acreages used were those actually occurring with advertising. If advertising had been eliminated in 1975, 1985, or any other year, subsequent quantities with and without advertising would have changed. We avoided the problem of adding up over time by using a simple average of the annual fixed-supply benefit-cost ratios. This average, 5.25, was close to the estimated aggregate short-term returns from advertising discounted at 3 percent ($B / C = 5.33$).[11]

Long-Run Benefit-Cost Ratios for Advertising

As documented in the acreage-response analysis, producers expand acreage and production when returns are favorable, and improved returns from advertising expand the acreage and supply of avocados. The short-run net returns from advertising already described were eroded over time by increased supplies stimulated by earlier increased returns. Given flexible prices (inelastic demand), increased production from more acres of avocado trees partially offsets the demand-enhancing effects of advertising. As illustrated in Figure 7.11, the long-run benefit-cost ratio was again less than 1 during the first four years of the program, exceeded 1 from 1965/66 through 1979/80, and then dropped below 1 for the next five years (it was even negative one year). The estimated benefit-cost ratio was greater than 1 in 1985/86 and for seven of nine years thereafter. The average benefit-cost ratio for the thirty-four years of the analysis was 1.89; the estimated increase in total industry revenue was $218.8 million and CAC marketing expenditures were $116 million. The long-run benefit-cost ratios, when costs and returns were discounted at 0 and 3 percent, were 1.78 and 1.71, respectively (Carman and Craft 1998, p. 62). A simple average of marginal benefit-cost ratios for the same period, derived by increasing advertising 1 percent during each year, was equal to 1.48. The marginal ratios tended to be less than 1 at the beginning of the period and greater than 1 at the end. For example, the average marginal benefit-cost ratio for the first ten years of the analysis was 0.47 while the average for the last ten years was 3.23.

Just as increased demand stimulated increased production of avocados over time, industry assessments tended to decrease supply. The long-run impact of the adjustment to assessments for advertising was to shift a portion of the costs to buyers. Thus, the long-run benefit-cost ratios previously estimated tended to overstate the true costs of the program to producers. As noted earlier, we computed an estimate of the producers' share of costs by subtracting total producer revenue after the assessment from total producer revenue before the assessment. The annual average benefit-cost ratio for the thirty-four years of the analysis was 2.84. The long-run benefit-cost ratios, when costs and returns were discounted at 0 and 3 percent, were 2.48 and 2.26, respectively (Carman and Craft 1998, p. 63).

Concluding Comments

This chapter presents the results of research on the impacts of CAC's advertising and promotion programs on the demand (and price) for California avocados between 1961/62 and 1994/95. Annual demand and supply-response relationships were estimated with generally good results as measured by standard statistical tests and concurrence with theoretical expectations. The results of a monthly demand analysis (discussed in detail in Carman and Craft (1998)) generally confirmed the results of the annual analysis but with improved statistical measures and tests. The similarity of estimated annual and monthly price flexibilities of demand makes us confident that we have accurately measured the important determinants of California avocado demand and, in particular, the effects of advertising and promotion expenditures.

Returns for CAC advertising and promotion programs have been very attractive regardless of one's perspective. On a short-run, fixed-supply, month-to-month and year-to-year basis, returns have typically averaged \$5 to \$6 for every dollar expended. These are the relevant returns to consider when making short-run decisions about CAC advertising and promotion program expenditures. Over time, the supply response resulting from increased returns reduces prices and returns. This is the nature of short-run versus long-run returns to advertising when an industry does not control supply and there is ease of entry and exit. But even in the long run producers earned more than \$1.70 for every dollar spent on advertising and promotion since 1961/62. This ratio increased to about \$2.26 when it was based only on the producers' share of advertising expenditures.

Future research could attempt to isolate the separate impacts of various types of advertising and promotion expenditures on the demand for and price of California avocados. We believe that carefully designed market experiments are required to best assess the comparative impacts of various programs. Future data collection efforts must be mindful of the need to match measures of advertising effort with the timing of program execution to derive improved estimates of the dynamic effects of advertising programs. Finally, we remind readers that the study results are for a specified past time period and that, while most of the estimated supply and demand relationships can reasonably be expected to continue in the near future, any projections using these relationships must be regarded with caution.

Notes

Research funding was provided by the California Avocado Commission.

1 These values are from the California Chapter of the American Society of Farm Managers and Rural Appraisers, "Trends in Agricultural Land and Lease Values," Spring Ag Outlook Forum, March 1996, pp. 14, 16.

2 The state of Hawaii also has a small avocado industry that, on average, produces about 0.25 percent of the total U.S. crop.

3 If the common trend is accounted for, there does not appear to be a strong relationship between marketing expenditures and the *current* crop value. For instance, the correlation coefficient is 0.60 in levels and −0.32 in annual rates of change. This is an important point for the econometric estimation that follows.

4 Information on advertising expenditures by media type was obtained from an independent marketing research firm (LNA-Mediawatch). The data were based on a calendar year, while data reported by the California marketing order and commission program administrators were based on a California crop year (November through October).

5 The lagged price and cost index were used after the four-year average of real total revenue lagged one year was not statistically different from zero.

6 The estimated change in bearing acreage for each year t was added to the estimated bearing acreage for the previous year, $t-1$, to yield a cumulative summation of bearing acreage.

7 For the general Box-Cox model, it can be shown that the price flexibility of demand with respect to one of the explanatory variables, say x, is given by

$$\eta_x = \beta x \frac{x^{\lambda_x}}{y^{\lambda}},$$

where λ is the estimated Box-Cox parameter for the dependent variable y and λ_x is the estimated Box-Cox parameter associated with explanatory variable x.

8 Following Savin and White (1978), the joint hypothesis of λ unrestricted and $\rho = 0$ (the first-order autocorrelation coefficient) was tested against the alternative of λ and ρ both unrestricted, yielding $\rho = 0.84$ and $\chi^2 = 19.13$.

9 The joint hypothesis of λ unrestricted and $\rho = 0$ was tested against the alternative of λ and ρ both unrestricted, yielding $\rho = -0.33$ and $\chi^2 = 1.04$.

10 Possible problems with the data series on marketing expenditures included the changing composition and categories of expenditures included over time. For example, administration and marketing research were necessary expenditures, but neither was expected to affect demand directly. The relative importance of these and other similar categories of expenditures changed over time, resulting in an advertising variable for which measurement was subject to possibly large unexplained variability. A linear, monthly inverse-demand equation estimated for the nine years from 1986/87 through 1994/95 had a statistically significant advertising coefficient and an estimated price flexibility of advertising at mean values of 0.137. This result was very close to the annual estimated price flexibility of advertising of 0.13.

11 Carman and Craft (1998) also estimated short-run benefits for nine crop years (1986/87 through 1994/95) using an estimated monthly demand relationship–see footnote 10. Aggregated benefits and costs over the nine years yielded a benefit-cost ratio of 5.71 for all CAC marketing expenditures (1998, p. 45).

References

Alston, J.M., H.F. Carman, J.A. Chalfant, J.M. Crespi, R.J. Sexton, and R.S. Venner. *The California Prune Board's Promotion Program: An Evaluation.* Berkeley CA: University of California Division of Agriculture and Natural Resources, Giannini Foundation Research Report 344, March 1998.

Alston, J.M., H.F. Carman, J.E. Christian, J.H. Dorfman, J.R. Murua, and R.J. Sexton. *Optimal Reserve and Export Policies for the California Almond Industry: Theory, Econometrics and Simulations.* Berkeley CA: University of California Division of Agriculture and Natural Resources, Giannini Foundation Monograph 42, February 1995.

Alston, J.M., and J.A. Chalfant. "Unstable Models from Incorrect Forms." *American Journal of Agricultural Economics* 73 (1991):1172–1181.

Alston, J.M., J.A. Chalfant, J.E. Christian, E. Meng, and N.E. Piggott. *The California Table Grape Commission's Promotion Program: An Evaluation.* Berkeley CA: University of California Division of Agriculture and Natural Resources, Giannini Foundation Monograph 43, November 1997.

California Agricultural Statistics Service. *California Fruit and Nut Acreage.* Sacramento CA, annual issues.

——. *California Fruit and Nut Statistics.* Sacramento CA, annual issues.

California Avocado Commission. Annual reports. Irvine CA, various years.

California Chapter of the American Society of Farm Managers and Rural Appraisers. "Trends in Agricultural Land and Lease Values." University of California, Davis, Agricultural Issues Center Spring Ag Outlook Forum, Davis CA, 1996.

California Crop and Livestock Reporting Service. *California Fruit and Nut Crops, 1909–1955.* Sacramento CA: California Department of Agriculture Special Publication 261, July 1956.

Carman, H.F., and R.K. Craft. *An Economic Evaluation of California Avocado Industry Marketing Programs, 1961–1995.* Berkeley CA: University of California Division of Agriculture and Natural Resources, Giannini Foundation Research Report 345, July 1998.

Carman, H.F., and R.D. Green. "Commodity Supply Response to a Producer Financed Advertising Program: The California Avocado Industry." *Agribusiness: An International Journal* 9 (1993):605–621.

Clarke, D. "Econometric Measurement of the Duration of Advertising Effect of Sales." *Journal of Marketing Research* 13 (1976):345–357.

Kinnucan, H., and O.D. Forker. "Seasonality in the Consumer Response to Milk Advertising with Implications for Milk Promotion Policy." *American Journal of Agricultural Economics* 68 (1986):562–571.

Lilien, G.L., P. Kotler, and K.S. Moorthy. *Marketing Models.* Englewood Cliffs NJ: Prentice Hall, 1992.

Little, J.D.C. "Aggregate Advertising Models: The State of the Art." *Operations Research* 27 (1979): 629–667.

Livingston, P., G. Bender, B. Faber, K. Klonsky, E. Takele, and G. Whitney. "Sample Costs to Establish and Produce Avocados in California in the 90's." Oakland CA: University of California Cooperative Extension and California Avocado Society, Inc. Circular CAS-93/1, September 1993.

Nerlove, M., and F.V. Waugh. "Advertising without Supply Control: Some Implications of a Study of the Advertising of Oranges." *Journal of Farm Economics* 43 (1961):813–837.

Reynolds, A., A. McFaul, and E. Goddard. "Testing for Causality between Advertising Expenditures and Canadian Demand for Cheese and Butter." *Agribusiness: An International Journal* 7 (1991):197–202.

Savin, N.E., and K.J. White. "Estimation and Testing for Functional Form and Autocorrelation." *Journal of Econometrics* 8 (1978):13–21.

Ward, R.W., and B.L. Dixon. "Effectiveness of Fluid Milk Advertising Since the Dairy and Tobacco Adjustment Act of 1983." *American Journal of Agricultural Economics* 71 (1989):730–740.

8

Evaluating the Effectiveness of California Almond Promotion: How Much Did the Litigation Cost Producers?

John M. Crespi & Richard J. Sexton

The Almond Board of California was established in 1950 by a federal marketing order. The almond order provides the industry with various tools to influence the demand and supply of almonds with the goal of increasing grower returns. Among its provisions, the order authorizes the industry to undertake advertising and promotion with funds collected through an assessment on almond handlers. Provisions of the industry's advertising and promotion program have varied over time, but the order allows handlers to receive full or partial credit against their assessments for advertising their own products, and specific programs have generally included provisions for full or partial credit. The industry also has conducted generic advertising. For three crop years, 1994/95 through 1996/97, the entire advertising program was suspended due to litigation (see Chapter 3 for details), providing a natural setting for an evaluation of the advertising program's effectiveness.

The central questions addressed by this study are: (1) Have the advertising programs conducted under the almond order been effective in increasing demand for almonds in the United States? (2) If so, have the expenditures been cost effective in the sense of yielding net benefits to growers? This focus on grower returns is pursuant to the mandate of the order that its programs be conducted for the benefit of growers. For example, §981.41 paragraph (d) of the order reads in part that "No promotion or advertising shall be undertaken without reason to believe that returns to producers will be improved by such activity."

The study proceeded first by reviewing relevant prior work on the economic evaluation of promotion and on the California almond industry. We then estimated the effect of almond advertising and promotion on demand for almonds in the United States using data for the 1962/63 through 1997/98 crop years. The analysis indicated that promotion has been an effective tool in expanding almond demand. The results of the demand analysis were next incorporated into an equilibrium model of the U.S. almond market. This model was used to simulate the effect on the market of a small (10 percent) increase in almond promotion expenditures for the 1990/91 through 1993/94 crop years. A four-year period was chosen because in that time frame new acreage could not be planted and reach a bearing state, so production could be treated as fixed during the period considered. Because the actual program was suspended from 1994/95 through 1996/97, the 1990/91 through 1993/94 period represented the most recent four years that represented "normal" operations for the industry.

Results are reported under a variety of scenarios pertaining to demand for almonds in the United States and the price elasticity of the almond supply in the U.S. market. Across the variety of scenarios considered, a consistent set of conclusions emerged. Almond advertising and promotion has worked, yielding marginal benefits in the range of $3 to $10 per dollar expended. Average benefits, although not computed explicitly in this study, have been even higher. These results suggest that the industry could have profited from investing more money in advertising and promotion during the time period studied.

The almond industry in California suffered substantially in terms of profits due to suspension of the industry advertising program imposed by litigation. Our analysis suggests that the three-year suspension of the advertising program cost the industry as much as $200 million.

Previous Work on Almond Demand and Almond Promotion

Results from previous studies of the U.S. almond market provided the modeling foundation for our analysis. Bushnell and King (1986) examined the world almond market, including demand, production, market structure, trade, and government intervention. They specified supply and demand relationships for the U.S. and several international markets. They estimated a linear demand model of the following form using annual data from 1953 to 1980:

$$Q_t = b_0 + b_P RP_t + b_{EXP} REXP_t + b_{EUR} QEUR_t + b_{Qlag} Q_{t-1} + e_t.$$

In an econometric model of this type, the *b*'s are coefficients to be estimated. The variable to be explained is *Q*, U.S. per capita consumption of almonds. The explanatory variables are *RP*, the real price per pound of almonds; *REXP*, a measure of real per capita income; and *QEUR*, U.S. per capita consumption of European almonds. The subscript *t* is the time period for each observation (thus Q_{t-1} is U.S. per capita consumption lagged one period) and e_t is an error term measuring the difference between the actual quantity of almonds consumed and the quantity predicted by the model.

Using their estimated coefficients and the mean value for price and quantity, Bushnell and King calculated a price elasticity of demand for almonds of −0.39, which represented the percent change in quantity demanded due to a 1 percent increase in price. This elasticity is in the "inelastic" range, as it suggests that a 10 percent increase in the price of almonds results in only about a 4 percent decrease in U.S. per capita consumption.

Alston et al. (1995) conducted the most in-depth study of the almond industry to date. They developed models of almond demand in the U.S. and major importing countries and both short- and long-term models of almond supply in this country and in Spain, the U.S.'s major international competitor. The Alston et al. model of U.S. demand was a key foundation for our analysis; they tested the reliability and statistical significance of several model specifications before arriving at a preferred functional form. Although their choice of explanatory variables was similar to that of Bushnell and King, Alston et al. chose to model U.S. domestic demand for almonds in a double-logarithmic form:

$$\ln(Q_t) = b_0 + b_P \ln(RP_t) + b_{EXP} \ln(REXP_t) + e_t.$$

Their estimations used observations from 1961 to 1991. For much of that period, U.S. almond production had increased to the point that imports of European almonds were inconsequential, causing Alston et al. to omit the European almond import variable contained in Bushnell and King's model. Alston et al. found a price elasticity of demand of −1.08 in the U.S. market, much larger than that found by Bushnell and King. This elasticity implied that a 10 percent increase in the price of almonds results in roughly a 10 percent decrease in quantity demanded in the U.S. Alston et al. found, however, that demands for most major importers of almonds tended to be inelastic.

Christian (1994) was the first to analyze the effectiveness of advertising in the almond industry.[1] Christian focused specifically on the U.S. market and advertising expenditures by Blue Diamond Growers, a marketing cooperative that was the single largest handler of almonds throughout the time period under study. The demand equation was based on the work of Alston et al. and covered the same sample period (1961 to 1991):

$$\ln(Q_t) = b_0 + b_P \ln(RP_t) + b_{EXP} \ln(REXP_t) + b_{PRO} \ln(RBD_t) + e_t,$$

where *RBD* is the real per capita advertising and promotion expenditure by Blue Diamond (*BD*). Christian found a price elasticity of demand in line with that of Alston et al. (−1.16) and an advertising elasticity of 0.14, implying that a 10 percent increase in Blue Diamond's advertising would have resulted in a 1.4 percent increase in per capita almond consumption.

Crespi and Chacón-Cascante (2004) examined the demand for U.S. almonds in both the domestic and export markets from 1962 to 1997. The focus of their study was the extent to which the Almond Board's setting of a reserve (a feature of the almond marketing order that allows the board to divert almonds from the consumer market) affected world almond prices. In the course of their analysis, however, they also estimated price and advertising elasticities. Using a system of equations representing world demand for almonds, the supply of almonds from U.S. competitors in the world market, and a hypothetical profit maximization condition facing the Almond Board, Crespi and Chacón-Cascante found that the estimated price elasticity of demand was slightly larger (in absolute value, −1.33) and that the estimated advertising elasticity was slightly lower (0.09) in the U.S. domestic market than those found by Christian.

Accounting for Changes in U.S. Almond Consumption

In this section, we specify demand models for California almonds and report the results from estimating these models using crop-year (July–June) data covering the years from 1962/63 through 1997/98, a total of thirty-six observations. We specifically studied per capita quantities of almonds consumed in the U.S.

Let Q_t represent the quantity of almonds demanded by a representative consumer during a particular year, *t*. The theory of consumer demand suggests a model in which Q_t depends on the corresponding real (inflation-adjusted) price of almonds, RP_t, and the consumer's total real money income or expenditure on all goods, $REXP_t$. This model can be expressed as

$$(1) \quad Q_t = f(RP_t, REXP_t).$$

In addition, the prices of goods that are close substitutes for the commodity under analysis should be included in the demand equation. However, Alston et al. (1995) concluded that substitute goods were not important factors in determining almond consumption in the U.S., although filberts were an important substitute commodity in some European markets.

We expected *RP* to have a negative effect on *Q* (i.e., a higher real price for almonds will decrease per capita consumption, implying a negative coefficient on RP_t) and the income effect (the coefficient on $REXP_t$) to be positive. An increase in total money income should lead to an increase in almond consumption. Moreover, the effect of *REXP* should be in the range for a normal good, corresponding to an income elasticity of demand for almonds between 0 and 1.

Equation (1) implicitly assumes constant tastes and preferences for almonds. To accommodate changes in preferences arising from promotion or anything else that may affect demand, the model can be augmented with other demand-shift variables. Promotion is one such variable. To obtain reliable estimates of the influence of the factors that are most important for the present study, particularly the responsiveness of demand to price and promotion, it was necessary to take into account the influence of additional demand-shift variables. Otherwise, there was a risk that the effects of omitted shifters would be attributed falsely to the variables included in the model.

In a model of consumer demand for almonds, demand-shift variables could include factors such as: (1) consumer health consciousness and interest in natural foods, (2) demographic factors such as the age structure and ethnic composition of the population and the rate of labor-force participation by women, (3) generic promotion by the Almond Board and brand promotion by Blue Diamond Growers and others, and (4) measures of the quality of California almonds.

To deal with all of these various effects was impossible given our limited data set and the difficulty of separating the influence of the various effects;

many of the factors change smoothly together over time. Instead, we focused on the shift variables we believed were likely to be the most important. However, because of the extensive prior work on demand for almonds, there was little reason to examine a broad set of added variables. Thus, we included only two shift variables in our study: (1) the quantity of promotion, represented by the combined promotion expenditures made by the Almond Board and Blue Diamond, $RPROMO_t$ (expressed in real terms by dividing expenditures by the Consumer Price Index (CPI));[2] and (2) a linear time-trend variable, $TIME_t$, included to represent the effects of other factors that were not explicitly modeled.

(2) $Q_t = f(RP_t, REXP_t, RPROMO_t, TIME_t)$.

The effects of the demand-shift variables were not clear *a priori*. Successful promotion increases demand, but unsuccessful promotion has little or no effect on demand. The effect of the time trend can be either positive or negative, reflecting a general shift of consumer demand toward or away from almonds over time.

We aggregated Almond Board and Blue Diamond promotion for two reasons. First, for most of the time period studied, almond handlers like Blue Diamond had been allowed to satisfy at least a portion of their promotion assessment by advertising their own products. Thus, promotion funded under the auspices of the almond marketing order appropriately included the amount of assessment credits to Blue Diamond and other handlers that advertised their own products.[3] Second, variation in the data is crucial in the estimation of any econometric model. A problem when working with annual data, as in this study, is that variations that occur during a year are eliminated. If all variables in a model change smoothly across years, it is difficult to discern their individual effects, a problem known as multicollinearity. The multicollinearity problem was mitigated in our data set by combining the advertising expenditures of the Almond Board and Blue Diamond.

Empirical Approach

This section summarizes the econometric models used to estimate the effects of promotion on annual per capita almond consumption in the United States. Further detail on data gathering, variable construction, market-sector linkages, and other demand specification choices is provided in Alston et al. (1995).

The supply of California almonds (measured in pounds (lbs), kernel weight) was calculated as harvests plus stocks carried in from the previous year less stocks carried out. These data came from the Almond Board's monthly position reports. The quantities were converted into pounds per capita using an estimate of the U.S. midyear population provided by the International Monetary Fund (IMF). The mean for per capita quantity demanded was 0.54 pounds per year with a standard deviation of 0.22 pounds per year.

The price used in the demand models was the average grower return in California per pound (kernel weight) of almond production from the U.S. Department of Agriculture's *Fruit and Tree Nuts Situation and Outlook Yearbook*, an annual bulletin. This price was converted into 1998 dollars using the CPI. Deflation by the CPI has the effect of treating an aggregate of all other goods as a general substitute for almonds. Average real grower return was \$2.20 per pound with a standard deviation of \$0.78 per pound.

Promotion expenditures, $RPROMO_t$, included the sum of generic promotion expenditures by the Almond Board and promotion expenditures by Blue Diamond. Branded promotion expenditures by other handlers were insignificant. The promotion data were obtained directly from the Almond Board and Blue Diamond. These expenditures were also expressed in real terms by deflating the nominal expenditures by the CPI.

A total expenditure variable was also included as a measure of consumer income. The expenditure variable used was total private domestic consumption expenditures (billions of dollars) in the United States (IMF). This figure was then deflated by the CPI to put it in real terms and divided by the estimate of the U.S. population to put it in per capita terms. The annual average of this per capita income variable was \$15,743, and its standard deviation was \$2,901. The time-trend variable rose from 62 in 1962/63 to 97 in 1997/98. Table 8.1 provides definitions of the variables used in the demand models. Summary statistics for the variables are included in Table 8.2.

Estimation

To make the model in Equation (2) operational, we had to specify a functional form for $f(\cdot)$. Based on the prior work conducted on the almond industry, we limited our analysis to the linear form used by Bushnell and King (1986) and the double-logarithmic form used by Alston et al. (1995) and Christian (1994). A key feature of the double-logarithmic form is that it yields constant elasticities with respect to each of the explanatory variables. This feature

Table 8.1. Description of Variables in the Demand Models

Variable	Definition	Units	Data Source
Q_t	Annual per capita consumption of almonds in the U.S.	Pounds, kernel weight.	Almond Board, *Position Reports*.
RP_t	Average grower return in California.	Real (1998 = 1) dollars per pound per year, kernel weight.	U.S. Department of Agriculture, *Fruit and Tree Nuts Situation and Outlook Yearbook*.
$RPROMO_t$	Aggregate domestic promotion expenditures by the Almond Board and Blue Diamond Growers.	Millions of real 1998 dollars per year.	Almond Board and Blue Diamond Growers.
$REXP_t$	U.S. per capita private domestic consumption expenditures.	Real 1998 dollars per year.	International Monetary Fund, *International Financial Statistics*.
$TIME_t$	"Year."	62 = 1962/63 to 97 = 1997/98.	

Table 8.2. Summary Statistics for the Variables in the Demand Models

Variable	N	Mean	Standard Deviation	Minimum	Maximum
Q_t	36	0.540	0.221	0.057	1.031
RP_t	36	2.198	0.779	1.160	4.743
$RPROMO_t$	36	7.999	5.271	1.160	21.004
$REXP_t$	36	15,743	2,900.60	10,559	20,819
$TIME_t$	36	79.500	10.536	62.00	97.00

makes the double-logarithmic model popular in many econometric studies. However, addition of the promotion variable created statistical problems that were not encountered by Alston et al. (1995), whose work did not incorporate this variable. These problems led us to reject the double-logarithmic model in favor of a model that was linear in all the variables except the promotion variable, which was entered in square root form. This specification imposed diminishing marginal returns to promotion.[4] Models of this square root form were used to study demand and promotion of table grapes and prunes by

Alston et al. (1997, 1998). We refer to this model as the linear square root (LSR) form.

For the LSR specifications, we examined models with and without a time-trend variable. Thus, the estimations were as follows:

$$(3)\quad Q_t = b_0 + b_P RP_t + b_{EXP} REXP_t + b_{PRO}(RPROMO_t)^{1/2} + e_t.$$

$$(4)\quad Q_t = b_0 + b_P RP_t + b_{EXP} REXP_t + b_{PRO}(RPROMO_t)^{1/2} + b_T TIME_t + e_t.$$

In these models, the b coefficients are parameters to be estimated econometrically based on the data described previously. The coefficients on RP_t and $REXP_t$ are marginal effects–the predicted change in Q_t in response to a one-unit increase in RP_t or $REXP_t$. In the case of $RPROMO_t$, the marginal effect due to the square root specification is $b_{PRO} / 2(RPROMO_t)^{1/2}$. The "error" term e_t in Equations (3) and (4) represents residual changes in per capita consumption that are not accounted for by changes in the explanatory variables.

Estimation Results and Model Evaluation

We estimated the models in Equations (3) and (4) using the ordinary least squares (OLS) estimation procedure. The results are shown in Table 8.3. The next step in the analysis was to submit the models to a battery of tests of their performance. A first test was how well each model explained movements in per capita almond consumption. For this test, we use the adjusted R^2 statistic shown at the bottom of Table 8.3. The adjusted R^2 measures the proportion of the variation in per capita almond consumption explained by movements in the explanatory variables contained in the model. The LSR models explained about 85 percent of variation in Q_t. As an example, the close correspondence between actual per capita consumption and the estimated (fitted) values for Equation (4) over the sample period is shown in Figure 8.1a. Figure 8.1b includes a plot for Equation (4) showing the fraction of the fitted values accounted for by all of the variables other than promotion. In other words, it is the fitted values net the estimated effects of promotion (calculated by subtracting 0.051 × $(RPROMO_t)^{1/2}$ from the fitted value in each year). The importance of promotion in the overall demand for almonds is thus illustrated in Figure 8.1b.[5]

A second set of tests involved examining the estimated coefficients and determining their consistency with expectations based on the economic theory of consumer demand. The estimates indicated that demand is inelastic with respect to price. The elasticities evaluated at the data means are contained in brackets in Table 8.3. The price elasticity of demand for both versions of the LSR model was approximately −0.7. This elasticity implies that a 10 percent increase in the price of almonds results in a 7 percent decrease in almond consumption. Both of these elasticities were statistically significantly different from zero at the 1 percent level of significance. Likewise, the promotion elasticities were roughly the same in the two models–about 0.13 (in line with Christian's findings (1994))–indicating that a 10 percent increase in annual promotion expenditures results (on average) in a 1.3 percent increase in almond consumption. The elasticities with respect to income, evaluated at the data means, were about 0.7 in Equation (3) and 0.6 in Equation (4), both in the predicted range of what is considered a normal good. The effect, however, is not statistically significant in Equation (4), most likely due to collinearity between the time-trend and the income variable.

Table 8.3. Ordinary Least Squares Estimates of Equations (3) and (4)

Independent Variables	Equation (3)	Equation (4)
Constant	0.387 (3.330)	0.345 (1.190)
RP	−0.168 (−7.900) [−0.682]	−0.166 (−6.990) [−0.676]
RPROMO	0.052 (2.896) [0.128]	0.051 (2.710) [0.126]
REXP	0.00002 (3.894) [0.709]	0.00002 (0.580) [0.559]
TIME		0.002 (0.159)
Adjusted R^2	0.848	0.843
Durbin Watson	1.839	1.837
Log-Likelihood	39.240	39.255

Notes: Dependent variable is per capita consumption of almonds (pounds). The sample period is 1962/63 through 1997/98 (thirty-six observations). The t-statistics are in parentheses and elasticities evaluated at the variable means are in brackets. The elasticity of the promotion variable is the elasticity with respect to the level of promotion, not its square root.

Figure 8.1. U.S. Per Capita Annual Almond Consumption – Actual versus Fitted Values, 1962/63 through 1997/98

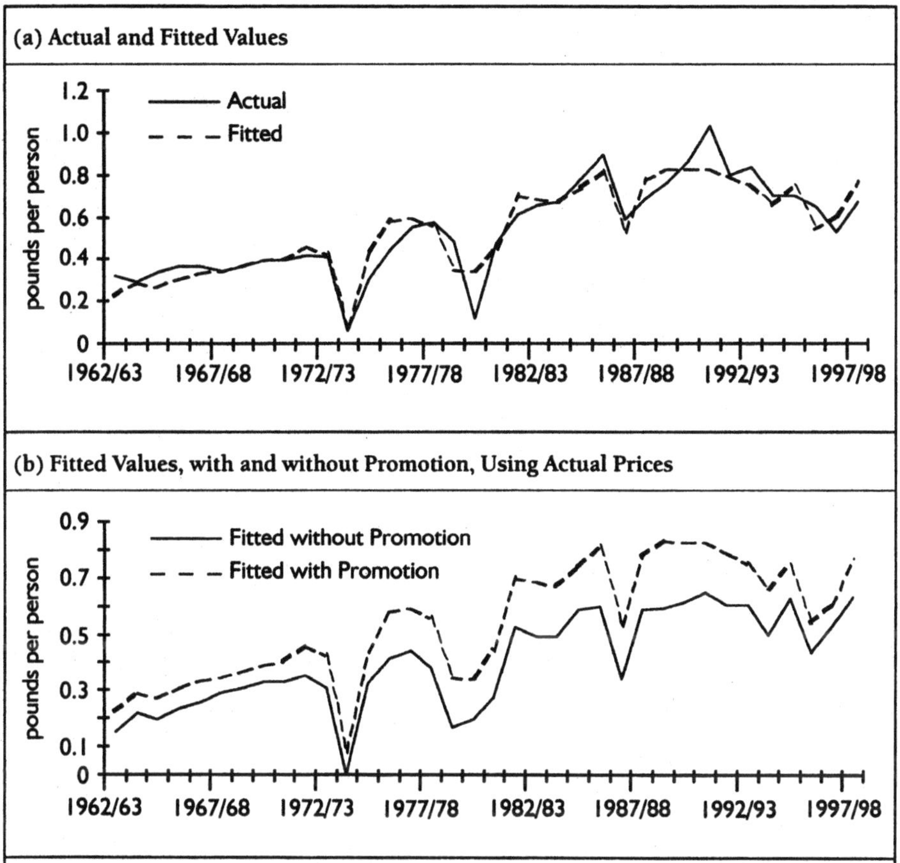

Source: Actual values from Almond Board of California. Fitted values from authors' calculations.

Our OLS estimation procedure assumed that the variance in the error term was constant across observations. If this were not true (i.e., if heteroskedasticity were present), then estimates of the standard errors for the coefficients were biased and conventional hypothesis tests were invalid. Several tests for heteroskedasticity were performed on the models shown in Table 8.3. For each test and each model, we failed to reject the hypothesis of no heteroskedasticity. Similarly, we found no evidence of autocorrelation among the residuals based upon tests performed using the Durbin Watson statistic reported at the bottom of Table 8.3.[6]

Simulation Model and Benefit-Cost Analysis

The estimated annual demand parameters described in the previous sections were used to estimate the gross and net benefits to the California almond industry from its expenditures on promotion. Measuring the welfare impacts of promotion funded by check-off programs requires (1) a conceptual structural model of the industry market equilibrium, (2) estimates of supply and demand parameters that can be used to define the values for the parameters in the structural model, (3) estimates of the demand response to promotion expenditures, and (4) information to allow a transformation of the effects of promotion (through demand shifts) and assessments or check-offs (through commodity supply shifts) into measures of benefits and costs.

The econometric models discussed in the previous section allowed us to estimate the change in quantities of almonds sold in response to a given increase in promotion expenditures, holding prices and other variables constant. This did not, however, tell us how much sales would actually increase when promotion changed because prices were assumed to remain constant. Indeed, the increase in prices following a promotion-induced shift in demand is an important source of the benefits from promotion. To evaluate the industry's promotion activities properly, we therefore had to account for both demand effects and the response of supply to increased price.

The diagram in Figure 8.2 illustrates the conceptual supply and demand relationships for a typical year *t*. In the figure, the curve labeled S_t represents the residual supply curve for almonds to the domestic (U.S.) market. It shows the quantities available to domestic consumers at various prices; at higher prices, more almonds are available domestically; at lower prices, larger quantities of almonds are diverted to other uses such as the export market or left unharvested or stored across crop years. The curve labeled $D_t(RPROMO_t)$ represents the demand curve: consumers purchase a smaller quantity of almonds at higher prices than they do at lower prices, holding other factors constant. In particular, the promotion expenditure was held constant at its actual value, $RPROMO_t$, along this curve. The market equilibrium occurs at point *E*. The market price adjusts until the quantity demanded and the quantity supplied are equated at price P_t.

In this example, the effect of a 10 percent increase in promotion is illustrated by the outward shift in the demand curve. The new curve is labeled $D_t(RPROMO_t \times 1.1)$. The econometric model allowed us to estimate the horizontal distance of the demand shift in the U.S. market, identified by Δ in the diagram. In the LSR model given in Equation (4) of Table 8.3, for example, the estimated coefficient on $(RPROMO_t)^{1/2}$ is 0.051. Suppose the actual monthly promotion expenditure was \$1 million (in 1998 dollars). This means that a \$0.1 million (i.e., 10 percent) increase in total promotion expenditures would be expected to lead to an increase in U.S. per capita almond consumption of $0.051 \times (1.1^{1/2} - 1.0^{1/2}) = 0.0025$ pounds per year if there were no change in price.

Multiplying by the population (POP_t = 267.9 million in 1997/98) yielded the total horizontal demand shift in the U.S. from a 10 percent increase in promotion expenditures–about a 0.5 percent increase in average consumption at constant prices. However, this was greater than the actual increase in consumption that would result. An increase in price was needed to bring forth the additional quantities to satisfy the increased demand in the domestic market. The new equilibrium is represented by the point E', where the new demand curve intersects the supply curve. Price and quantity both increase to the new equilibrium values P_t' and Q_t'.

The evaluation of promotion expenditures required an estimate of the increments in both prices and quantities due to the expenditures and a measure of the costs of supplying the additional quantities to the market. Over time, increased prices call forth additional supplies, but these supplies come at a

Figure 8.2. Conceptual Supply and Demand Model

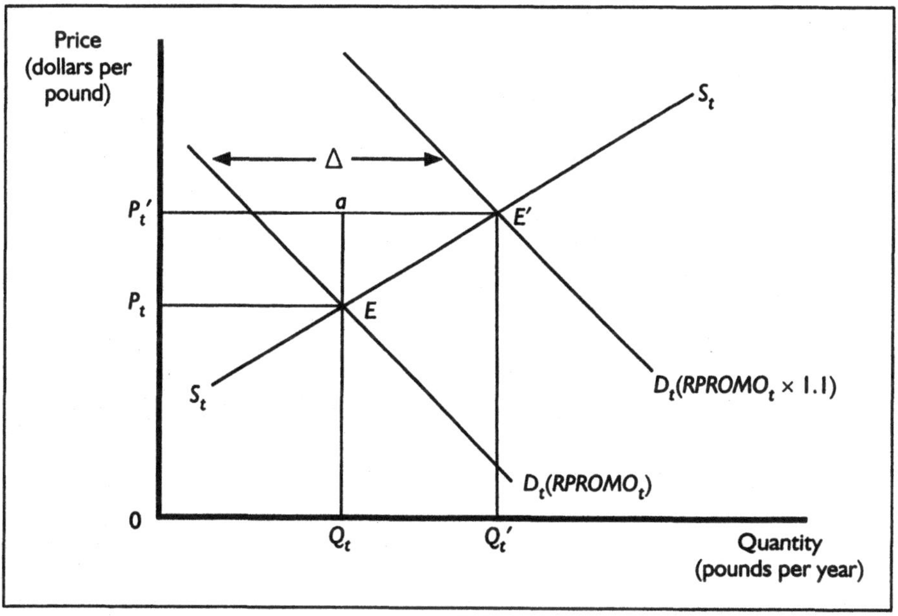

cost. In the case of a perennial crop such as almonds, the cost may be largely earnings foregone from other uses of existing production (e.g., exports) rather than costs of new production. Further, one expects an effective promotion program to induce greater supplies in the future as growers react to higher almond prices by increasing plantings.

Since newly planted almond trees do not bear for three to four years, we examined the promotion program over a recent four-year period. In this way, we could treat bearing acreage as fixed. Harvests were thus determined by the product of the bearing acreage and the yield, with yield determined primarily by weather conditions and largely unaffected by the current year's price. In other words, in the short run (a time window of about four years for almonds), price has little effect on total supply. To simplify our analysis, we assumed that the total supply was perfectly inelastic in the short run (i.e., was unaffected by price). Given this assumption, the increase in producer surplus (grower profits) was measured simply as the change in the producer price as a result of the promotion multiplied by the amount of the harvest. Further, because supplies were fixed, producers bore the full incidence of the increase in the promotion expenditure.[7] Given our assumptions about supply, the main challenge in the

Figure 8.3. Producer Surplus, Export Effects, and Assessments

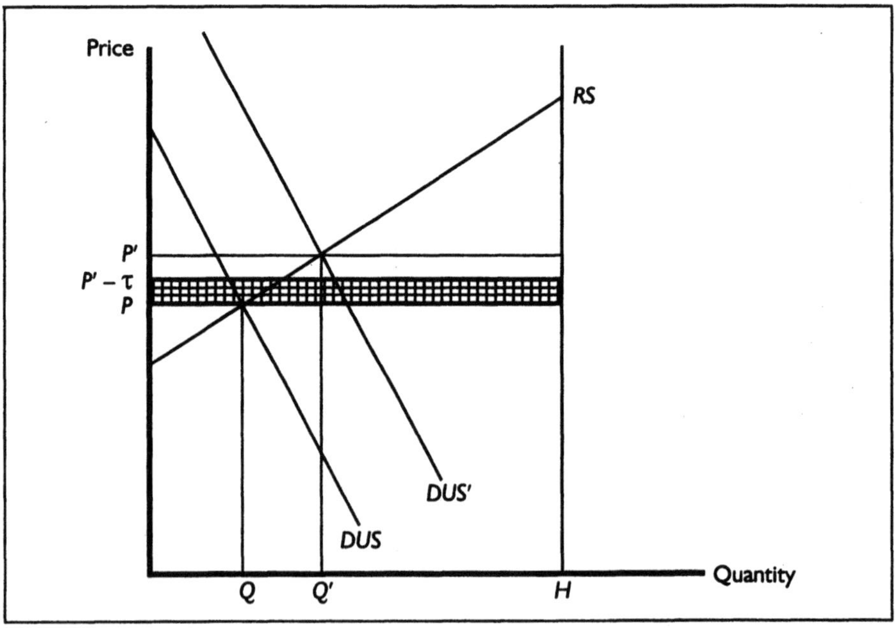

simulation modeling lay in determining the change in producer price due to promotion activities.

Figure 8.3 shows a short-run supply-and-demand analysis that takes the export market and the assessment for promotion into account. The world price of almonds was originally at price P and the U.S. domestic demand at that price was given by quantity Q. This price and quantity are found where the U.S. residual supply curve (RS) intersects the U.S. domestic demand curve (DUS). The residual supply curve is constructed by subtracting export demand (not shown on the graph for simplicity) at each price from the total harvests, H. An increase in promotion shifts out the domestic demand curve to DUS', leading to the new market price P'. However, this increased promotion comes at a cost of increased assessments, τ, that apply to the entire harvest. Thus, producers' net increase in price from the promotion program is $P' - \tau - P$. The increase in producer surplus from the promotion is measured by the shaded rectangle–the change in the producer price due to the promotion multiplied by the entire harvest quantity.

Simulation Model

In the simulation analysis, we compared the pattern of consumption and prices predicted by the model, given actual historical promotion activities, with corresponding values predicted by the model following a counterfactual 10 percent increase in promotion expenditures each year over the four-year period from 1990/91 through 1993/94.[8] We also simulated changes in the assessments jointly with corresponding changes in promotion. The differences between these actual and counterfactual scenarios were then used to calculate measures of the marginal gross and net benefits to producers from the joint increase in promotion expenditures and assessments.

To conduct the benefit-cost analysis, the preferred demand model was combined with an assumed residual supply function. First, from the demand side, predicted domestic quantities were calculated by substituting the actual values for each of the explanatory variables in the estimated equation. For instance, for Equation (3) in Table 8.3,

$$(5) \quad \hat{Q}_t = 0.387 - 0.168\, RP_t + 0.00002\, REXP_t + 0.052\, (RPROMO_t)^{1/2},$$

where the hat (^) on Q_t denotes the predicted value given the estimated coefficients and actual values of the exogenous variables. We replicated this step for each of the LSR models.

Next, the U.S. residual supply function was defined to be of the constant elasticity form and to pass through the points defined by the predicted quantities from the demand model. That is, the residual supply function was of the form

$$(6)\quad Q_t = A_t R_t^{\varepsilon},$$

where $A_t \equiv \hat{Q}_t / R_t^{\varepsilon}$ and ε is the price elasticity of supply. R_t is the producer return per pound in year t, defined as $R_t = (1 - \tau_t)P_t$ where τ_t is the actual promotion expenditure per pound consumed in year t expressed as a fraction of the market price in year t (i.e., the rate of assessment required to finance the actual promotion expenditure). A_t was varied from year to year to ensure that, given the actual values of prices and the other exogenous variables each year, the supply equation passed through the point defined by the predicted quantity from the demand model and the actual price. To account for export effects as in Figure 8.3, the residual supply was shifted by the additional assessment when an increase in promotion was simulated. We combined the calibrated supply model and the estimated demand model to simulate past actual prices and predicted U.S. quantities.

As noted, the residual supply of almonds to the U.S. market consists of the total (fixed) supply minus the amount that would be exported at various prices. Given an inelastic total supply, the elasticity, ε, of the residual supply was found by the formula $\varepsilon = -\eta_E(Q_E / Q_{US})$, where η_E is the price elasticity of export demand and (Q_E / Q_{US}) is the ratio of export sales to sales in the U.S. market. We did not estimate export demand as part of this study, but export demands were analyzed by Alston et al. (1995) and, later, in Christian and Crespi (1996). In particular, Alston et al. (1995, Table 3.2, p. 59) reported values of η_E for Germany, France, the Netherlands, Great Britain, Japan, Canada, Italy, and the rest of the world. The elasticity estimates ranged from a low of −0.43 for Japan to a high of −1.28 for Canada.[9] Assuming a ratio of $(Q_E / Q_{US}) = 2$ (i.e., about two-thirds of a typical crop is exported), we obtained a lower-bound estimate on the residual supply elasticity of $\varepsilon_L = -(-0.43) \times 2 = 0.86$ and an upper-bound estimate of $\varepsilon_H = -(-(1.28) \times 2 = 2.56$. Residual supply functions were calibrated using ε_L, ε_H, and intermediate values of 1.0, 1.5, and 2.0. Using this range of values for the residual supply elasticity enabled us to examine the degree to which our results were sensitive to choices of this parameter, where our knowledge was admittedly less than precise.

Using these definitions of the supply and demand equations, we first replicated the past. By equating the equations for supply and demand and solving for market equilibrium, we obtained values of actual prices and predicted quantities (from the demand model) given the actual values for the explanatory variables. We then simulated counterfactual scenarios using a hypothetical marginal increase in the amount of promotion each year for 1990/91 through 1993/94 of 1.1 times the actual amount of promotion. We also increased the assessments to pay for that promotion by 1.1 times the actual assessment rate. We defined the actual assessment rate as the ratio of total promotion expenditures to the total value of production.

An important issue was how much confidence could be placed in the particular values of the benefit-cost measures. For example, how confident could we be that the net benefits were greater than $1 million given a "best" estimate of, say, $2 million? The precision of the estimates of the benefits depended on the precision of the estimates of the underlying demand parameters but in ways that were not easy to see clearly. To evaluate the precision of our measures of benefits and costs, we conducted Monte Carlo simulations following the approach developed in Alston et al. (1997). These simulations yielded confidence intervals on our welfare measures, permitting us to make probability statements, such as that a 95 percent confidence interval for the benefit-cost ratio was formed by the interval from *a*:1 to *b*:1, where *a* is a lower confidence limit on benefits per dollar expended and *b* is an upper confidence limit.[10]

Table 8.4 reports the marginal benefit-cost ratios for almond promotion from 1990/91 through 1993/94 implied by each of the LSR demand models, along with the mean and lower and upper bounds from the Monte Carlo simulations, using the five alternative values for the residual supply elasticity. A real discount rate of 3 percent per annum was used to compound the costs and benefits. In each case, we report producer benefits relative to the total costs of the change in promotion expenditures.[11] All monetary calculations were deflated and are expressed in 1998 dollars.

The information across the columns in Table 8.4 shows the impact of changes in the hypothetical supply elasticity. For example, the first row in Table 8.4 refers to Equation (3) and the first column represents our lower-bound, residual supply elasticity of 0.86. In this instance, the benefit-cost ratio for a 10 percent increase in promotion expenditures was estimated to be 6.88; in other words, a marginal dollar expended on promotion would yield an estimated return to growers of $6.88. Looking across the columns in this row, we

Table 8.4. Marginal Benefit-Cost Ratios for Almond Promotion from 1990/91 through 1993/94: Point Estimates and Simulation Statistics

Producer Benefits / Total Expenses for Each Series	U.S. Residual Supply Elasticity 0.86	1.00	1.50	2.00	2.56
Benefit-Cost Ratios with a 3 Percent Annual Interest Rate					
Point Estimates					
Equation (3)	6.88	6.17	4.51	3.55	2.87
Equation (4)	6.79	6.09	4.44	3.50	2.83
Monte Carlo Means					
Equation (3)	6.97	6.25	4.56	3.59	2.90
Equation (4)	6.81	6.10	4.45	3.50	2.83
Monte Carlo: Lower 95 Percent Boundary					
Equation (3)	2.32	2.08	1.53	1.21	0.98
Equation (4)	2.03	1.82	1.33	1.05	0.85
Monte Carlo: Upper 95 Percent Boundary					
Equation (3)	11.71	10.47	7.60	5.97	4.80
Equation (4)	11.90	10.64	7.72	6.06	4.89

can see the effects of the increase in the residual supply elasticity. As the supply elasticity rises, producers receive progressively smaller benefits from a given demand increase because price rises less due to the demand shift. Hence, the benefit-cost ratio falls with increases in the supply elasticity, declining from 6.88 to 2.87 when the upper bound of the residual supply elasticity is reached (2.56). This pattern was repeated for both demand models, as well as for the statistics from the Monte Carlo simulations.

Choosing a particular supply elasticity and reading down Table 8.4 allows a comparison of the ratio of producer benefits to costs generated from the point estimates with the corresponding results (the mean and the lower and upper bounds of the 95 percent confidence interval) from the Monte Carlo simulation for a given residual supply elasticity. For example, looking at the results from Equation (3) for a supply elasticity of 1.0 shows that the point estimate (6.17) and mean from the Monte Carlo simulation (6.25) were roughly equivalent and that a 95 percent confidence interval for the producer marginal benefit-cost ratio was given by a lower bound of 2.08 and an upper bound of 10.47. Both the point estimates and the means from the Monte

Carlo simulations indicated that the marginal benefit-cost ratio based on the demand models fell within a range from approximately 7 to 1 at a supply elasticity of 0.86 to 3 to 1 for a supply elasticity of 2.56.

Absence of Promotion in Crop Years 1994/95 through 1996/97

Suspension of the industry's promotion program from 1994/95 through 1996/97 provided a natural setting for evaluation of the impact of the absence of advertising on the almond industry. The average promotion expenditure by the industry in the three years preceding the suspension, expressed in 1998 dollars, was $10.5 million per year; during the suspension, it was only $3.9 million per year. Although prices initially dropped in 1994/95 and then rose in 1995/96 and 1996/97, harvests were particularly small in the two later crop years. Because of this, price alone does not allow one to discern the effect of the suspension. However, our econometric modeling and subsequent simulation did allow us to model the effect of the suspension by controlling for all of the relevant factors affecting demand.

Using annual harvest data, we could estimate the effective assessment rate necessary to pay for promotion. For the three years prior to the suspension, the average effective assessment was $0.02091 per pound (kernel weight). Using this assessment rate with harvests for the 1994/95 through 1996/97 period computed the following expenditures on promotion: $15.3 million in 1994/95, $7.7 million in 1995/96, and $10.6 million in 1996/97. Actual promotion expenditures were $5.3 million, $4.2 million, and $2.2 million for 1994/95, 1995/96, and 1996/97, respectively. Using the counterfactual estimates of promotion and the corresponding assessment rate for the years 1994/95 through 1996/97 in the market equilibrium model allowed us to compare the producer surplus estimated using the actual promotion expenditures and assessments with that estimated under the counterfactual expenditures. For this analysis, we report results for the LSR model that included the time-trend variable (Equation (4)). Given the similarity in estimates between the two LSR models, the results are insensitive to the choice of demand model.

The presumption described in this and the previous section was that industry promotion during the suspension period was below optimum. To test this proposition, we repeated the benefit-cost analysis for the period 1994/95 through 1996/97, first using a small, 10 percent increase in promotion and then using the levels of promotion from the counterfactual scenario previously discussed. Table 8.5 compares the benefit-cost estimates from these two simulations with the estimates that were given in the second row of Table 8.4.

Table 8.5. Marginal Benefit-Cost Ratios before and during the Promotion Suspension: A Comparison of Point Estimates from Equation (4)

Producer Benefits / Total Expenses for Each Series	U.S. Residual Supply Elasticity 0.86	1.00	1.50	2.00	2.56
Benefit-Cost Ratios with a 3 Percent Annual Interest Rate					
Marginal Effects 1990/91–1993/94	6.79	6.09	4.44	3.50	2.83
Marginal Effects 1994/95–1996/97	14.60	13.25	9.98	8.02	6.57
Counterfactual Effects 1994/95–1996/97	7.53	6.80	5.05	4.01	3.26

The first row in Table 8.5 shows the benefit-cost estimates using the point estimates from Equation (4) over the period 1990/91 through 1993/94 (i.e., the second row from Table 8.4). The second row gives the same marginal benefit-cost estimates during the years of the suspension, 1994/95 through 1996/97. Note that the marginal benefits for the suspension period are more than twice the level of the period prior to the suspension. This result indicates that the industry's promotion expenditures during the suspension period were less optimal than those in previous years. The counterfactual estimations shown in the third row are closer in magnitude to the estimations in the first row. Thus, if promotion expenditures had been at levels consistent with those in the years prior to the suspension, the marginal benefit-cost ratios would likewise have been consistent with those of previous years.[12] This consistency between the presuspension and the counterfactual benefit-cost estimations gives us confidence in the producer surplus estimations that follow.

Table 8.6 shows the estimates of price and producer surplus using actual levels of promotion and promotion assessments compared with counterfactual levels for 1994/95 through 1996/97. The accumulated loss from suspension of the promotion program was estimated to be between $87.93 million and $230.89 million, depending on the value chosen for the residual supply elasticity. These estimates are of profit (not just revenue) losses since the costs of the increased promotion are already accounted for in the model.

Table 8.6. Actual and Counterfactual Prices and Producer Surpluses for 1994/95 through 1996/97 (1998 Dollars)

Crop Years	Actual Prices ($/pound)	Counterfactual Prices ($/pound)	Actual Surplus (million $)	Counterfactual Surplus (million $)	Surplus Loss (million $)
U.S. Residual Supply Elasticity = 0.86					
1994/95	1.47	1.61	1,075.01	1,165.25	90.24
1995/96	2.65	2.76	966.81	1,002.84	36.04
1996/97	2.14	2.36	1,083.41	1,188.02	104.61
		Total	*3,125.23*	*3,356.11*	*230.89*
U.S. Residual Supply Elasticity = 1.00					
1994/95	1.47	1.60	1,075.01	1,154.30	79.29
1995/96	2.65	2.75	966.81	999.74	32.94
1996/97	2.14	2.34	1,083.41	1,177.63	94.22
		Total	*3,125.23*	*3,331.67*	*206.45*
U.S. Residual Supply Elasticity = 1.50					
1994/95	1.47	1.56	1,075.01	1,129.17	54.16
1995/96	2.65	2.73	966.81	991.75	24.94
1996/97	2.14	2.29	1,083.41	1,152.09	68.69
		Total	*3,125.23*	*3,273.01*	*147.79*
U.S. Residual Supply Elasticity = 2.00					
1994/95	1.47	1.54	1,075.01	1,115.05	40.04
1995/96	2.65	2.71	966.81	986.62	19.81
1996/97	2.14	2.26	1,083.41	1,152.09	53.24
		Total	*3,125.23*	*3,253.76*	*113.09*
U.S. Residual Supply Elasticity = 2.56					
1994/95	1.47	1.53	1,075.01	1,105.15	30.14
1995/96	2.65	2.71	966.81	982.70	15.89
1996/97	2.14	2.24	1,083.41	1,125.31	41.90
		Total	*3,125.23*	*3,213.16*	*87.93*

Notes: Estimates in this table use the LSR model based on Equation (4). Producer prices for the surplus calculations are equal to Price × (1 – Assessment Rate). Columns may not sum due to rounding.

Conclusion

The focus of this study was to evaluate the effectiveness of almond promotions conducted under the almond marketing order and, in particular, the impacts of suspension of the promotion program due to litigation during the period from 1994/95 through 1996/97. The results indicate that almond promotion has been a highly effective tool in stimulating almond demand and increasing producer profits. We presented results under a variety of scenarios that, in turn, yielded a variety of estimates of the benefit-cost ratio to producers from a marginal increase in promotion expenditures. Among these various scenarios, a best guess was that marginal dollars expended promoting almonds yielded a return to producers in the range of from 3 to 1 to 10 to 1. Of course, these rates of return are very favorable when compared to returns available to other investments. Given the LSR specifications used in this study, the *average* return on investments in almond promotion is necessarily higher than the marginal returns noted in Tables 8.4 and 8.5.

Suppose, for example, that a 10 percent rate of return on investment is normal. Then any producer benefit-cost ratio in excess of 1.1 to 1.0 indicates a profitable expenditure of funds at the margin. In fact, to maximize its return from investment in almond promotion, the industry should expand promotion efforts to the point where the marginal expenditure on promotion just yields a return comparable to that available on investments elsewhere. Thus, the evidence suggests quite strongly that the industry spent too little on promotion over the time period analyzed. Given that the evidence suggests a pattern of spending too little on promotion, it is unfortunate from the industry's perspective that expenditures were curtailed from 1994/95 through 1996/97 because of litigation. Our analysis suggests that suspension of the advertising program during this period cost the industry accumulated profits ranging from \$84 million to \$231 million.

Notes

The authors are grateful to the staff of the Almond Board of California for contributing data to the study. Blue Diamond Growers also contributed key data to the study, and the cooperation of Blue Diamond's Director of International Sales, Rex Lake, is much appreciated.

1 Subsequent work by Halliburton and Henneberry (1995) examined the effectiveness of promoting almonds abroad through the U.S. Department of Agriculture's Foreign Market Development Program and Market Access Promotion Program. See also the commentary on Halliburton and Henneberry's analysis by Kinnucan and Christian (1997).

2 The choice of whether to include promotion in per capita terms or in total was discussed by Alston et al. (1997).

3 Suspension of the industry's promotion program for crop years 1994/95 through 1996/97 provided evidence of the effect of the industrywide program on Blue Diamond's advertising. Advertising and promotion expenditures by Blue Diamond declined in nominal terms by 61.3 percent during those three years from their average in the prior three years. This result lent strong support to the proposition that advertising by Blue Diamond is done largely in response to the presence of industry programs.

4 The marginal return to promotion refers to the incremental benefit from increasing promotion expenditures by a small amount, such as one dollar. Diminishing marginal returns mean that each successive dollar spent yields a smaller incremental benefit than its predecessor. Diminishing returns must hold eventually or it would in principle be optimal to spend an infinite amount on promotion. Thus, the square root specification of promotion expenditures is preferred over a model that is linear in all variables because the latter is characterized by constant marginal returns.

5 There is a caveat to this figure. Subtracting $0.051 \times (RPROMO)^{1/2}$ from the fitted equation is equivalent to setting *RPROMO* equal to 0. However, the farther one sets the level of a variable from its actual value, the less certain one can be of the model's prediction.

6 As a test of robustness, we also estimated a shorter time series covering 1980/81 through 1997/98. The estimated coefficients on price were very similar across the two time periods. However, the price elasticity of demand at the data means for the shorter time series was considerably lower–about −0.35 compared to −0.68. Given that the estimated b_p coefficients were similar, the difference in elasticity estimates was due to the much larger average consumption for 1980/81 through 1997/98. In other words, the greater production and consumption of more recent years apparently moved the almond industry down along a fairly stable linear demand curve into the more inelastic ranges of the curve.

7 As noted, assessments in the almond industry are imposed nominally on handlers. However, per-unit taxes can be shifted in whole or part to other participants in the market. It is straightforward to demonstrate that producers bear the entire burden or incidence of the assessment when their supply is perfectly inelastic. See Alston et al. (1998) for further discussion of shifting of commodity promotion assessments.

8 Beginning with the 1994/95 crop year, the industry's promotion program was suspended due to litigation. Thus, 1990/91 through 1993/94 comprised the most recent four-year period of promotion activity that was in some sense "normal" from the industry's perspective. Thus, those four years were chosen as the base years for the simulation analysis. We provide a separate analysis of the impacts to the industry from suspension of the promotion program.

9 The elasticity estimates for the other countries were as follows: Germany −0.70, France −0.69, Netherlands −1.10, Great Britain −0.64, Italy −0.72, and rest of the world −0.85. These estimates were consistent with the results obtained by Christian and Crespi (1996), whose analysis was based on an alternative data set and included more recent observations than were available to Alston et al.

10 The Alston et al. approach is an iterative process where first a particular set of values of the demand parameters is drawn at random from the estimated joint statistical distribution. This set of parameters is substituted into the demand equation and used to generate predicted quantities that are, in turn, used to set parameters for the supply equation for each choice of supply elasticity. Then these supply equations are used with the demand equation to conduct the counterfactual scenarios and then to evaluate the scenarios in terms of producer benefits and costs. This process is repeated 10,000 times, thereby generating a

distribution of producer benefits and costs. A 95 percent confidence interval is then generated by eliminating the upper and lower 2.5 percent tails of the distribution.

11 Producers bear the entire incidence of the promotion program because supply is perfectly inelastic in this analysis. Thus, the magnitude of producer benefits relative to total program costs is the correct basis for comparison.

12 Means from the Monte-Carlo simulations were very similar to the point estimates. For the second row of Table 8.5, the means for the simulations were 14.65, 13.29, 10.00, 8.02, and 6.57 given residual supply elasticities of 0.86, 1.0, 1.5, 2.0, and 2.56, respectively. Given these same supply elasticities, the simulation means corresponding to the third row of Table 8.5 were, respectively, 7.56, 6.82, 5.05, 4.01, and 3.25.

References

Alston, J.M., H.F. Carman, J.A. Chalfant, J.M. Crespi, R.J. Sexton, and R.J. Venner. *The California Prune Board's Promotion Program: An Evaluation.* Berkeley CA: University of California Division of Agriculture and Natural Resources, Giannini Foundation Research Report 344, March 1998.

Alston, J.M., H.F. Carman, J.E. Christian, J.H. Dorfman, J.R. Murua, and R.J. Sexton. *Optimal Reserve and Export Policies for the California Almond Industry: Theory, Econometrics and Simulations.* Berkeley CA: University of California Division of Agriculture and Natural Resources, Giannini Foundation Monograph 42, February 1995.

Alston, J.M., J.A. Chalfant, J.E. Christian, E. Meng, and N.E. Piggott. *The California Table Grape Commission's Promotion Program: An Evaluation.* Berkeley CA: University of California Division of Agriculture and Natural Resources, Giannini Foundation Monograph 43, November 1997.

Bushnell, P.G., and G. King. *The Domestic and Export Markets for California Almonds.* Berkeley CA: University of California Division of Agriculture and Natural Resources, Giannini Foundation Research Report 334, 1986.

Christian, J. "The Economic Effects of Advertising in the California Almond Industry." Working Paper, Department of Agricultural Economics, University of California, Davis, 1994.

Christian, J., and J.M. Crespi. "Export Demand Estimations for California Almonds." Working Paper, Department of Agricultural Economics, University of California, Davis, 1996.

Crespi, J.M., and A. Chacón-Cascante. "Do U.S. Marketing Orders Have Much Market Power? An Examination of the Almond Board of California." *Agribusiness* 20(1) (2004):1–15.

Halliburton, K., and S.R. Henneberry. "The Effectiveness of U.S. Nonprice Promotion of Almonds in the Pacific Rim." *Journal of Agricultural and Resource Economics* 20 (1995):108–121.

International Monetary Fund. *International Financial Statistics.* Washington DC, various issues.

Kinnucan, H.W., and J.E. Christian. "A Method for Measuring Returns to Nonprice Export Promotion with Application to Almonds." *Journal of Agricultural and Resource Economics* 22 (1997):120–132.

U.S. Department of Agriculture, Economic Research Service. *Fruit and Tree Nuts Situation and Outlook Yearbook.* Washington DC, various years.

9

An Economic Analysis of the Walnut Marketing Board's Promotion Programs

Harry M. Kaiser

The California walnut industry promotes walnuts through several related marketing programs run by the Walnut Marketing Board (WMB) and the Walnut Commission. The Walnut Commission, which is sanctioned as a state marketing order, runs promotion programs exclusively for the export market, a crucial and growing market for walnuts. On the other hand, WMB offers public relations and promotion (nonadvertising) programs servicing the domestic market. WMB, which was established in 1948 as a federal marketing order representing walnut growers and handlers, promotes utilization of walnuts through publicity, product promotions, research, and education programs. Under the 1996 Federal Agricultural Improvement and Reform (FAIR) Act, all federal check-off programs are required to have economic evaluations conducted to determine their market impacts on prices and demand[1] and their benefits relative to costs for stakeholders. While the Walnut Commission and WMB work in conjunction, only the federally sanctioned WMB is required to be evaluated under the FAIR Act.

This chapter examines the domestic market impacts of WMB's marketing programs and focuses on three important issues.[2] The first issue is the degree of demand responsiveness, if any, to WMB's promotion activities. To examine this issue, a domestic per capita walnut-demand equation was estimated to compute the corresponding marketing elasticity–the percentage change in domestic per capita walnut demand given a 1 percent increase in WMB's marketing budget while holding constant all other factors affecting domestic walnut demand. As such, it provides a net measure of the impact of marketing on walnut demand. Statistical tests were performed to determine whether or not the marketing elasticity was positive and statistically different from zero.

The second objective of this study was to measure any reduction in total domestic walnut utilization that would have occurred in the absence of WMB's promotion activities. To address this objective, the estimated demand equation was used to simulate two scenarios for the period from 1984 through 1999: (1) a baseline scenario based on historical WMB marketing levels, and (2) a hypothetical "no WMB" scenario that assumed there were no WMB promotion activities. A comparison of these two scenarios provided a measure of the total impact of WMB's marketing programs on walnut demand.

A positive impact on domestic utilization does not guarantee that the monetary benefits of WMB's promotion activities are greater than the costs. Hence, the third and most important objective of this study was to estimate the gain in industry welfare due to WMB's promotion activities and compare those benefits to the associated costs. To examine this issue, an average benefit-cost ratio for WMB promotion activities was computed. The benefits to the walnut industry were measured as the increase in producer surplus (a measure of producer welfare) over the study period due to promotion. Costs were measured by the actual promotion expenditures during the study period. The average benefit-cost ratios, also known as average returns on investment, provided a measure of the returns (in dollars) to the walnut industry for every dollar invested in marketing.

Conceptual Overview of Economic Evaluation

In an economic evaluation of generic (not brand-specific) marketing programs, three basic questions must be answered. First, does the marketing program result in increased demand? To be effective, a marketing program must increase demand for the product. Second, does generic marketing result in a higher price? It is possible for a program to increase market demand but not price if the increased demand is completely offset by an increase in the quantity supplied by growers (domestically and/or internationally). Finally, do the industrywide benefits exceed the total cost of the marketing program? This is the bottom-line, most important criterion for effectiveness to the industry funding the program.

To evaluate the economic impacts of promotion programs on quantity, price, and profits, economists use a market supply and demand framework. Obviously, other factors besides price affect market demand, and those factors affect the position of the demand curve. Consequently, to accurately isolate the impact of promotion activities from other factors affecting demand (i.e., the demand determinants–see the methodology section for detail), all of those factors must be accounted for in any quantitative analysis of market demand.

Price determination in a market is based on the interaction of market demand and supply. The market supply curve measures how quantity supplied in the market responds to increases and decreases in price. The first possibility is that supply is fixed regardless of price. Such a situation most often occurs only in the very short run, when producers do not have time to adjust production in response to a price change. The second possibility is a positive supply response, reflecting the "law of supply," where quantity supplied increases as price increases. The last possibility is called an "infinite supply response"–any increase in demand is completely offset by an increase in quantity supplied. This possibility is more likely in markets that have few barriers to entry or exit by producers.

Generic marketing programs are intended to increase the market demand for the commodity. If the program succeeds in increasing demand, the market effects that result depend in large part on the nature of the supply response. Figure 9.1 illustrates the situation in which supply is fixed. Initial market equilibrium without a marketing program occurs where supply and demand are equal, resulting in a market price and quantity of P_1 and Q_1, respectively.

Figure 9.1. Impact of Marketing When There Is No Supply Response

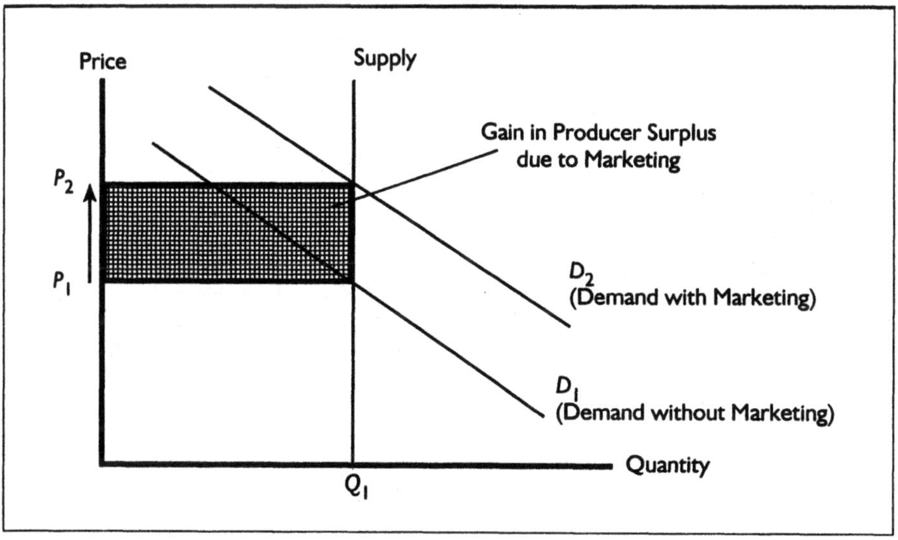

Suppose that a successful marketing program causes the demand curve to increase from D_1 to D_2.[3] Since supply is fixed, the only way to satisfy the increase in demand is for the market price to increase from P_1 to P_2. The benefit to producers in this market from the promotion program is the gain in industrywide producer surplus given by the shaded area in the figure. This gain in producer surplus measures the promotion program's gross benefits to producers and must be compared to total marketing costs to determine the net benefits of the program.

Figure 9.2 illustrates a similar case for a positive supply response. Here the successful marketing program again causes demand to increase from D_1 to D_2. Since supply is no longer fixed, there is now a quantity as well as a price response to the increase in demand. Price increases from P_1 to P_2 and quantity supplied increases from Q_1 to Q_2. The benefit to producers is the gain in producer surplus depicted by the shaded area, which must be compared to the total cost of the marketing program.

The final case, of infinite supply response, is illustrated in Figure 9.3. In this case, the increase in demand due to the marketing program is accompanied by an equal increase in quantity supplied. The net result is no change in price and therefore no gain in producer surplus. In markets characterized by an infinite supply response, increasing demand is not a good investment since there are no benefits to doing so. An example of this is a small importing country in which most of the demand is satisfied by imports from other countries.

Figure 9.2. Impact of Marketing When There Is Positive Supply Response

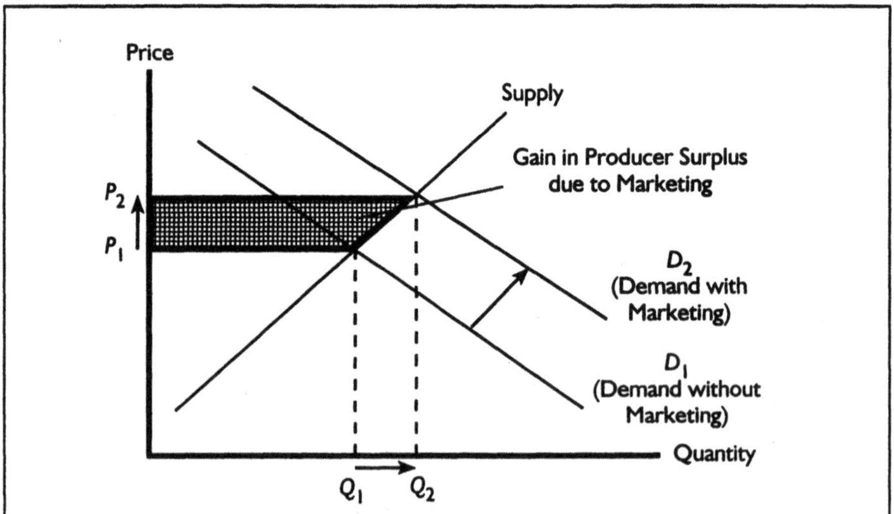

Figure 9.3. Impact of Marketing When Supply Response Is Infinite

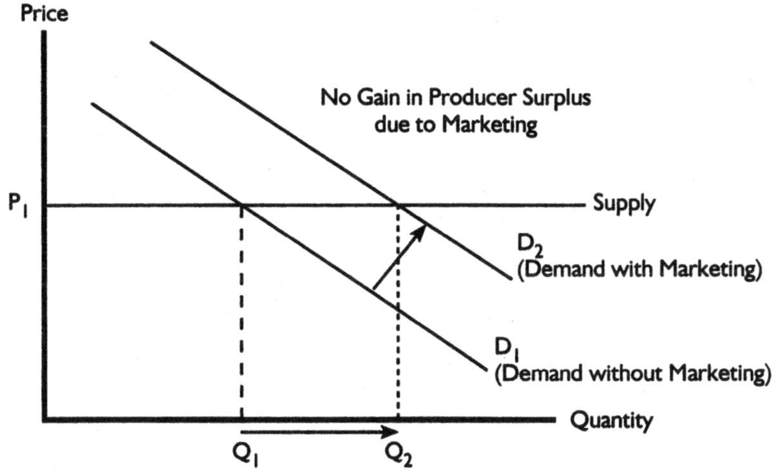

Economic Methodology

This study quantified the relationship between WMB's marketing efforts and U.S. domestic demand for walnuts. The export market was ignored since the focus was solely on WMB, which serves only the domestic walnut market. The model was based on the economic theory of consumer demand, under which promotion activities are expected to be beneficial to walnut growers and handlers because they increase the demand for walnuts and therefore increase sales and revenue. However, there are also other factors that affect domestic walnut demand. To distinguish the impact of WMB's promotion activities from the impacts of other factors influencing demand, an econometric framework was adopted.

The econometric approach quantifies economic relationships using economic theory and statistical procedures with data. It enables one to account for the impact of a variety of factors simultaneously. These demand-determining factors or "determinants" include the price of walnuts, prices of walnut substitutes and complements, population, proportion of women in the workforce, consumer income, consumer tastes and preferences, and WMB's marketing expenditures. By casting the economic evaluation in this type of framework, one can filter out the effect of other factors and, hence, quantify directly the net impact of WMB's promotion activities on domestic walnut demand.

The walnut-demand model developed in this study used annual time-series data for 1980 through 1999.[4] The model can be used to assess how strongly various walnut-demand determinants are correlated with domestic walnut utilization (i.e., demand). For example, with the model one can determine how important a change in the price of walnuts is relative to a change in WMB's marketing expenditures regarding their impacts on domestic walnut demand. The following factors were included in the walnut-demand equation to ascertain the extent, if any, of their impact on annual domestic per capita utilization for walnuts since 1980.[5]

1. **Walnut Price.** Either retail- or handler-level prices would have been the ideal data with which to determine the relative magnitude of price effects on domestic walnut utilization, but those data were not available. Instead, the annual grower price received for all walnuts was used as a proxy for the relevant demand price. Changes in walnut price should be negatively associated with walnut utilization; e.g., an increase in price should be associated with a decrease in utilization.
2. **WMB Marketing Program Expenditures.** Program expenditures were the key factor to be statistically tested for a significant and positive impact on walnut shipments, which would show that WMB's promotion activities increase domestic walnut demand.
3. **Price of Pecans and Almonds.** Again, ideally it would be desirable to have either retail- or handler-level price data, but the grower price for pecans and almonds was used to keep this data consistent with the data for walnuts. Almonds and pecans may be either substitutes or complements to walnuts. One would find a positive cross-price elasticity for a substitute product–an increase in the price of the substitute results in an increase in walnut demand because it makes walnuts more price-attractive. For a complementary product, the cross-price elasticity is negative–an increase in the price of the complementary product results in a decrease in walnut demand since the two are consumed jointly.
4. **Consumer Disposable Income.** Disposable income is expected to be positively related to walnut utilization; i.e., as consumers' disposable income increases, walnut demand should also increase.
5. **Proportion of Women in Workforce.** It has been hypothesized that an increase in the number of women entering the workforce is

associated with a reduction in domestic walnut utilization, presumably because fewer women have time to bake.

6. **Trend Term.** A linear trend term is included to capture demand factors that have been omitted from the model, such as changes in consumer tastes and preferences.

To compare the relative importance of each factor on shipments, the results from the statistical (econometric) model were converted into demand elasticities. Recall that a demand elasticity measures the percentage change in per capita domestic walnut utilization given a 1 percent change in a specific demand factor when all other factors are held constant. For example, the computed price elasticity measures the percent change in per capita domestic walnut utilization given a 1 percent change in price. The computed marketing elasticity measures the percent change in per capita domestic walnut demand given a 1 percent change in marketing. Since demand elasticities were calculated for each demand factor listed, they were compared to determine the factors that had the largest impact on walnut demand.

Econometric Results

The estimated per capita domestic-walnut-demand equation is reported in Table 9.1. The equation was specified on a per capita basis and used a logarithmic specification. A convenient feature of the logarithmic specification is that each of the estimated coefficients has the interpretation of being the demand elasticity for the variable in question. A first-order autoregressive error structure was appended to the equation to correct for serial correlation in the residual series and the Durbin Watson statistic reported in the table suggested that the resulting estimated equations were free from serial correlation problems. Further, the equation fit the data well; the adjusted R^2 statistic indicated that 80 percent of the variation in per capita walnut demand was explained by the independent variables in the equation. The equation had elasticity signs that were consistent with economic theory, and at least half of the estimated coefficients were statistically significant at conventional confidence levels. All elasticities reported were based on mean values for the period from 1980 through 1999. Finally, several econometric diagnostic tests were conducted on the residuals in the regression, and no serial correlation or heteroskedasticity problems were detected. Hence, the estimated demand model was deemed suitable for this analysis.

Table 9.1. Estimated Elasticities in the Domestic per Capita Walnut Demand Equation

Demand Determinant	Elasticity[a]	t-Ratio
Constant	-10.514	-1.30
Walnut Grower Price	-0.082	-0.83
Pecan Grower Price	0.140	3.59
Almond Grower Price	-0.078	-1.25
Disposable Income	1.123	1.25
Proportion of Women in Workforce	-1.922	-2.01
Trend Term	-0.521	-1.51
WMB Marketing Expenditures	0.005	1.64
Durbin Watson: 2.46[b]		
Adjusted R^2: 0.80[c]		

[a] Elasticity measures the percentage change in domestic per capita walnut demand given a 1 percent change in any demand determinant, holding constant all other determinants.

[b] A first-order autoregressive error process is appended to the equation to correct for serial correlation in the residual series. Further econometric diagnostic tests suggest that the resulting estimated equation is free from serial correlation problems.

[c] The adjusted R^2 indicates that the estimated equation explains 80 percent of the variations in domestic per capita walnut demand.

The estimated demand equation suggested that the price of walnuts at the grower level is not as important in explaining annual variations in per capita domestic demand as expected. The estimated own-price elasticity was −0.08, which implied that a 1 percent increase in the grower price would result in only a 0.08 percent decrease in domestic utilization. The estimated coefficient was not statistically different from zero as measured by conventional confidence levels. This result suggests that walnut utilization is not very sensitive to changes in grower price, which is intuitively plausible since walnuts represent a relatively small portion of U.S. consumers' total budgets. This result may also be due to the data being based on grower prices rather than on retail or processor prices.

The price of pecans was found to be an important demand determinant for walnuts. The cross-price elasticity of per capita walnut demand with respect to pecans was estimated to be 0.14, which is statistically significant at the 1 percent confidence level. That is, a 1 percent increase in the pecan price would result in a 0.14 percent increase in per capita demand for walnuts. This indicates that walnuts and pecans are substitute products since the demand for walnuts is enhanced by an increase in the price of pecans.

The cross-price elasticity of per capita walnut demand with respect to almonds was estimated at −0.08. That is, a 1 percent increase in the almond price would result in a 0.08 percent decrease in per capita walnut demand. This result suggests that walnuts and almonds are complementary products since the demand for walnuts is enhanced when the price of almonds decreases.

Income is one of the most important determinants of per capita walnut demand in the United States. The disposable income elasticity in the United States was estimated at 1.12, indicating that a 1 percent rise in U.S. disposable income results in a 1.12 percent increase in per capita demand for walnuts. This implies that walnuts are a "normal good" since demand rises with income.

The proportion of women in the workforce is the most important factor affecting domestic walnut demand. The results indicate that a 1 percent increase in the proportion of women in the workforce would result in a 1.92 percent decrease in per capita walnut demand. This result confirmed our earlier hypothesis that an increase in women in the workforce is associated with a decrease in walnut demand, likely from a decline in baking overall in households.

A trend term was also included in the model to capture changes in consumer preferences for walnuts over time. The coefficient associated with this variable (−0.52) is negative and statistically significant at a 10 percent level of significance. This finding indicates a mild decline in consumers' preferences for walnuts since 1980.

The coefficient associated with WMB's marketing programs is positive and statistically different from zero. Thus, the statistical evidence supports the notion that WMB's promotion activities have the effect of increasing demand for walnuts in the United States. The estimated marketing elasticity of 0.005 means that a 1 percent increase in walnut marketing expenditures would result in a 0.005 percent increase in per capita domestic walnut demand. While this elasticity is not large relative to the other demand determinants, it is statistically different from zero and positive. Moreover, two points should be considered. First, a large marketing elasticity is not expected since the annual marketing budget of WMB is quite small. Second, since WMB's expenses are relatively small, it does not take a large marketing elasticity to lead to positive net benefits from the program. This leads us to the next step in the economic analysis—simulating the effect of walnut promotion on total domestic quantity using the estimated demand equation.

Simulation Analysis

The estimations just described indicate that WMB's marketing program has had a positive and statistically significant effect on domestic walnut demand. The estimated demand equation was simulated to address the remaining three objectives of this study. Two scenarios were considered in the simulation:

1. **Baseline Scenario** – WMB's marketing program in effect from 1984 through 1999.
2. **No-WMB-Marketing Scenario** – WMB's marketing program not in effect from 1984 through 1999.

In the no-WMB scenario simulation, all demand determinants except WMB marketing expenditures were set equal to their historical levels. The WMB-marketing variable was set to 0 and the corresponding market demand was simulated over time.[6] The difference between the two scenarios gave the total impact of WMB's promotion efforts on domestic walnut utilization.

Figure 9.4 displays the results for the two scenario simulations. The figure clearly shows the positive impact of WMB's promotion programs on domestic walnut utilization. From 1984 through 1999, WMB promotion activities increased total walnut utilization by 342.4 million pounds in total, or 21.4 million pounds per year. Hence, WMB's promotion efforts have had a positive effect on domestic walnut demand.

A remaining key concern was the impact WMB's marketing has had on industry producer surplus (i.e., profit) compared with the costs of the program. The increase in walnut utilization determined previously assumed that all other demand determinants, including price, remained constant. However, an increase in demand generally causes price to increase as well, provided that the demand increase is not perfectly offset by an increase in quantity supplied (an infinite supply response). Hence, to evaluate the full effect of WMB's marketing program on quantity and price, one must incorporate the supply of walnuts into the model by estimating the supply response by walnut growers.

Previous econometric studies of tree nuts have indicated that it is problematic to obtain a reliable estimate of supply response to price because of the fairly lengthy time required to bring newly planted trees to harvest. Consequently, the harvest in any particular year is generally a function of yield, which is influenced by weather conditions but is largely unaffected by

Figure 9.4. U.S. Walnut Shipments with and without WMB Marketing Programs, 1984–1999

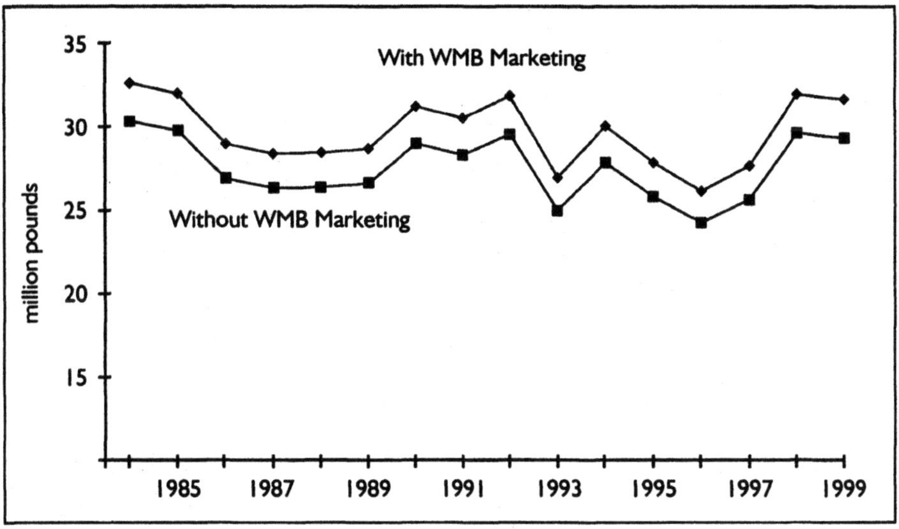

price. This makes it difficult statistically to determine any positive correlation between nut production and price. Consequently, this study did not develop an econometric supply model. Instead, an approach developed by Alston et al. (1996) and used by others (e.g., Crespi and Sexton 2001; Schmit and Kaiser 1998) was followed. The supply response was incorporated using a constant elasticity form, and sensitivity analysis was conducted on a range of assumed own-price elasticities of supply.

The simulation procedure began on the demand side, where predicted quantities of walnut demand (Q_t^D) were estimated from the demand equation. Then, using Alston et al.'s (1996) procedure, supply was defined in constant elasticity form and equated with predicted demand quantities. Changes in demand due to WMB's marketing efforts then affected the level of production and resulting grower price. Specifically, the supply function was defined as:

(1) $Q_t^S = A_t R_t$,

where $A_t = (Q_t^D + CS_t + NX_t) / R_t^{\varepsilon}$, and

(2) $R_t = P_t - \delta_t$,

where R_t is the net grower return per pound in year t, ε is the supply elasticity, and δ_t is the assessment rate required to finance WMB's efforts. The change in stocks (CS_t) and net exports (NX_t) are included as exogenous variables to close the model. The defined value A_t varies by year and ensures that, given the actual values of prices and other variables, the supply equation passed through the quantity defined by Q_t^D. This allowed us to combine the supply response and estimated demand model to simulate past prices and quantities. To estimate a supply response, an estimate of the own-price elasticity of supply was necessary. Given the lack of previous estimates of own-price elasticity of walnut supply, ε was varied over a wide range of possible values–from 0.5 to 3.0.

Average Benefit-Cost Analysis

Given the simulation procedures described in the previous section, the change in net economic benefits due to WMB promotion efforts was computed for each year from 1984 through 1999 as the difference in producer surplus (ΔPS) between the two scenarios, which mathematically is equal to the following:

(3) $\Delta PS_t = (R'_t Q'_t - R_t Q_t) / (1 + \varepsilon)$,

where $R'_t Q'_t$ represents the scenario with WMB marketing and $R_t Q_t$ represents the scenario without WMB marketing.

Table 9.2 presents the average annual impacts and benefit-cost ratios (1984–1999) for WMB's marketing efforts for the various assumed own-price elasticities of supply. WMB had a positive impact on the walnut growers' price over the period under all supply-response scenarios. The average increase in price ranged from just under 3 cents per pound in the case of the most inelastic supply response ($\varepsilon = 0.5$) to just under 0.5 cents per pound in the case of

Table 9.2. Average Annual Market Impacts and Benefit-Cost Ratios due to WMB's Marketing Programs, 1984–1999

	Own-Price Elasticity of Supply					
	0.5	1.0	1.5	2.0	2.5	3.0
Change in Grower Price (cents/pound)	2.71	1.37	0.92	0.69	0.55	0.46
Change in Producer Surplus (million $)	7.87	3.98	2.67	2.00	1.61	1.34
Change in Marketing Cost (million $)	0.81	0.81	0.81	0.81	0.81	0.81
Benefit-Cost Ratio	9.72	4.91	3.30	2.47	1.99	1.65

the most elastic supply response ($\varepsilon = 3$). Price impacts decrease as the supply response gets larger because the additional quantities producers place on the market negatively affect the price gains. The estimated average impact over all of the supply responses was 1.12 cents per pound. In other words, had there not been a WMB marketing program, the average grower price would have been 1.12 cents per pound lower (2 percent) than it actually was between 1984 and 1999.

WMB's marketing efforts had a positive impact on producer surplus over the period as well. The average increase in producer surplus due to the marketing programs ranged from $7.87 million per year in the case of the most inelastic supply response ($\varepsilon = 0.5$) to $1.34 million per year in the case of the most elastic supply response ($\varepsilon = 3$). The rationale for the negative relationship between supply elasticities and producer surplus is identical to that previously described for price and supply elasticities. The average increase in producer surplus over all of the supply responses was $3.25 million per year. Hence, it is clear that WMB's domestic marketing efforts have had a positive impact on grower profits since 1984.

How does the gain in producer surplus compare with the costs of WMB's program, which averaged about $810,000 per year from 1984 through 1999? As mentioned before, this was the most important question—was the program profitable to the industry? To obtain the answer, an average benefit-cost ratio was computed and is presented in the bottom row of Table 9.2. A benefit-cost ratio greater than 1.0 implies that the benefits of WMB's program exceeded its costs. The average benefit-cost ratio for WMB's program exceeded 1.0 for every supply response considered in the simulation. For the most inelastic supply response ($\varepsilon = 0.5$), the average ratio was 9.72. This implies that, on average over the period, the benefits of WMB marketing programs were almost ten times greater than the costs. At the opposite end of the spectrum in supply response ($\varepsilon = 3$), the average benefit-cost ratio was computed to be 1.65, implying that the benefits of WMB's marketing programs were 1.65 times greater than the costs. The wide range of supply responses considered in this analysis and the fact that the benefit-cost ratio exceeded 1.0 in all cases provides significant evidence that WMB's marketing programs have been profitable for the domestic walnut industry. The average benefit-cost ratio over all of the supply responses was 4.01; in other words, the benefits of WMB's marketing activities exceeded the costs fourfold. Also note that the actual supply response is likely to be in the lower elasticity range since walnut trees require at least five years to reach economic bearing age. Given the relatively short period under study, then, any supply response on the part of growers would be quite limited. Thus, higher benefit-cost ratios based on the lower price elasticities of supply are most likely.

To allow for error inherent in any statistical estimation, a 90 percent confidence interval was calculated for the average benefit-cost ratios described. The confidence interval provides a lower and an upper bound for the average ratio. Between these bounds is a range of possible average ratios and one can be "confident" 90 percent of the time that the true average benefit-cost ratio lies within that range. Table 9.3 presents the lower and upper bounds for the 90 percent confidence interval. For the lowest assumed supply response, the estimated lower and upper bounds were 5.25 and 15.54, respectively. For the highest assumed supply response, the estimated lower and upper bounds were 0.91 and 2.64, respectively. It is important to point out that the average benefit-cost ratio is greater than 1.0 for all assumed supply responses except the highest one (ε = 3.0). Hence, it was reasonable to conclude that the benefits of WMB's marketing programs have been considerably greater than the cost of the programs.

Conclusions

The objective of this research was to: (1) determine the domestic market impacts of WMB's marketing programs, and (2) compute a benefit-cost ratio for WMB's promotion activities. Specifically, this research examined whether WMB's domestic marketing activities between 1984 and 1995 had a positive and statistically significant impact on domestic shipments of California walnuts. The impact of all factors affecting domestic walnut demand (where data were available) was measured statistically. In this way, the impacts of other important factors affecting domestic demand were accounted for over time.

The empirical walnut-demand model developed in this study used annual time-series data for the years 1980 through 1999. To distinguish the impact of WMB's marketing activities on demand for walnuts from the impacts of

Table 9.3. Lower- and Upper-Bound 90 Percent Confidence Interval for Benefit-Cost Ratio due to WMB's Marketing Programs, 1984–1999

	Own-Price Elasticity of Supply					
	0.5	1.0	1.5	2.0	2.5	3.0
Benefit-Cost Ratio – Lower Bound	5.25	2.65	1.78	1.34	1.06	0.91
Benefit-Cost Ratio – Upper Bound	15.54	7.87	5.26	3.96	3.18	2.63

other factors influencing demand, an econometric framework was adopted. The econometric approach quantified economic relationships using economic theory and statistical procedures with data. It enables one to account for the impact of a variety of factors affecting walnut demand simultaneously. These demand-determining factors included the price of walnuts, prices of walnut substitutes and complements, the proportion of women in the workforce, consumer income, consumer tastes and preferences, and WMB's marketing expenditures.

The results indicate that WMB's marketing programs have had a positive and statistically significant impact on domestic per capita walnut demand since 1980. The estimated marketing elasticity was 0.005, which means that a 1 percent increase in walnut marketing expenditures would result in a 0.005 percent increase in per capita domestic walnut demand. While this elasticity is not large relative to other demand determinants, it is statistically different from zero and positive. Given WMB's relatively small program budget, a large marketing elasticity was not expected and a small elasticity can still lead to positive net benefits from the program.

The estimated demand equation was simulated to determine WMB's market impacts. In the baseline scenario, marketing expenditures were set equal to actual levels spent by WMB from 1984 through 1999. In the no-WMB scenario, marketing expenditures were set to zero for the same period. The difference between the two scenarios gave the total impact of WMB marketing efforts on domestic walnut utilization. The results indicate that WMB had a major impact on annual walnut utilization in the U.S. In the years between 1984 and 1999, WMB marketing activities increased total walnut utilization by 124,665 tons, or 7,333 tons per year.

The results also indicate that WMB had a positive impact on the walnut grower price over the period. The average increase in price ranged from just under 3 cents per pound in the case of an inelastic supply response to just under 0.5 cents per pound in the case of a very elastic supply response. The average impact over all of the supply responses was 1.12 cents per pound. In other words, had there not been a marketing program by WMB, the average grower price would have been 1.12 cents per pound, or 2 percent, lower than it actually was during the study period.

WMB's marketing effort had a positive impact on producer surplus (i.e., producer profits) over the period as well. The average increase in producer surplus due to WMB's marketing programs ranged from $7.87 million per year in the case of an inelastic supply response to $1.34 million per year in the case of a very elastic supply response. The average increase in producer surplus over all of the supply responses was $3.25 million per year. Hence, it is clear that WMB's domestic marketing efforts have had a positive impact on grower profits since 1984.

An average benefit-cost ratio was computed for WMB's marketing activities. The average ratio exceeded 1.0 for every supply response considered in the simulation. For the most inelastic supply response, the average benefit-cost ratio was 9.72. This implied that, on average over the period, the benefits of WMB marketing programs were almost ten times greater than the costs. At the opposite end of the spectrum in supply response, the average benefit-cost ratio was computed to be 1.65, implying that the benefits were 1.65 times greater than the costs. Given the wide range of supply responses considered in this analysis and the fact that the benefit-cost ratio exceeded 1.0 in all cases, there is significant evidence that WMB's marketing programs have been profitable for the domestic walnut industry. The average benefit-cost ratio over all of the supply responses was 4.01; in other words, the benefits of WMB's marketing activities exceeded the costs fourfold.

To allow for error inherent in any statistical estimation, a 90 percent confidence interval was calculated for the average benefit-cost ratios. For the lowest considered supply response, the estimated lower and upper bounds were 5.25 and 15.54, respectively. For the highest supply-response scenario, the lower and upper 90 percent confidence bounds were 0.91 and 2.64, respectively. The average benefit-cost ratio was greater than 1.0 for all assumed supply responses except the highest one. Hence, the confidence intervals give credence to the previous finding that the benefits of WMB's marketing programs have been considerably greater than the cost of the programs.

Notes

1 The terms demand, utilization, and shipments are used interchangeably throughout this chapter.

2 In this study, "marketing activities" refers to all promotion and public relations campaigns conducted by WMB in the United States. As previously noted, WMB does not advertise walnuts in the domestic market. In the econometric model, the marketing effort was measured by the total budget for promotion and public relations by WMB.

3 As pointed out in Chapter 1, generic marketing programs also have an impact on market supply that is similar to the impact of a tax. The mandatory assessment would therefore cause the supply curve to decrease (i.e., shift back to the left). For simplicity, this shift is not drawn here.

4 For the statistical analysis, more observations would have been desirable. However, the walnut market was substantially different before 1980 and there were no marketing

expenditures by WMB prior to 1983. Consequently, the shorter time period was used to reflect a more stable market environment.

5 Much of the data used in the estimation were compiled by Ms. Jodi Newman of WMB. All prices, income, and promotion expenditures were deflated by the general Consumer Price Index to remove the impact of inflation on all monetary series.

6 Technically, WMB's marketing expenditures in the second scenario were set to a very small number rather than zero since the logarithm of zero is undefined.

References

Alston, J.M., J.A. Chalfant, J.E. Christian, E. Meng, and N.E. Piggott. "The California Table Grape Commission's Promotion Program: An Evaluation." *Economic Evaluation of Commodity Promotion Programs in the Current Legal and Political Environment.* Proceedings of the NEC-63 Fall Meeting, October 8, 1996, Monterey CA, pp. 33–62.

Council of Economic Advisors. *Economic Report of the President.* Washington DC, various years.

Crespi, J.M., and R.J. Sexton. "Have Expenditures to Advertise California Almonds Been Effective?" *NICPRE Quarterly* 6 (2001):1–5.

Federal Agricultural Improvement and Reform Act. U.S. Code. Vol. 7, sec. 7401 (1996).

Schmit, T., and H.M. Kaiser. "Egg Advertising, Dietary Cholesterol Concerns, and U.S. Consumer Demand." *Agricultural and Resource Economics Review* 27 (1998):43–52.

U.S. Department of Agriculture, Economic Research Service. *Fruit and Tree Nuts Situation and Outlook Yearbook.* Washington DC, various years.

10

An Economic Analysis of California Raisin Export Promotion

Harry M. Kaiser, Donald J. Liu & Ted Consignado

One of the missions of the Raisin Administrative Committee (RAC) is to conduct export promotion programs in various countries with the principal aim of increasing exports of California raisins. RAC is part of a federal raisin marketing order that was authorized under the 1937 Agricultural Marketing Agreement Act and was established to foster orderly marketing conditions for the raisin industry. Under the 1996 Federal Agricultural Improvement and Reform (FAIR) Act, all federal marketing orders that operate promotion programs are required to have economic evaluations conducted to ascertain the extent of their impact on the market. This chapter presents the results of an economic analysis of the market impacts of RAC's raisin export promotion programs. The two largest markets for California raisin export promotion are Japan and the United Kingdom, so this study focused on the effectiveness of export promotion in those two countries.

Specifically, the study addressed four important questions about the effectiveness of RAC's export promotion programs between 1965 and 1998. First, what was the responsiveness of demand for California raisins in Japan and the United Kingdom with respect to the RAC's export promotion? To answer this question, we estimated per capita import demand equations for California raisins in the two importing countries and computed the corresponding export promotion elasticity.

Second, what would exports of California raisins to Japan and the United Kingdom have been if there had not been any RAC export promotion there? To answer this question, we used the estimated import demand equations to simulate two scenarios: (1) a baseline scenario based on historical program levels, and (2) a hypothetical scenario assuming that there had been no RAC export promotion programs. A comparison of these two scenarios provided a measure of the total impact of RAC's export promotion on California raisin exports to the two importing countries.

Third, how did the gain in export revenues due to RAC's export promotion in Japan and the United Kingdom compare to the costs of the export promotion? To answer this question, we computed average benefit-cost ratios for export promotion in Japan, in the United Kingdom, and in the two countries combined. Benefits to the California raisin industry from export promotion were measured as the increase in export revenue over the study period due to the promotion, which was calculated as the simulated gain in export quantities multiplied by deflated export prices. The costs in the benefit-cost ratio computation were measured by actual program expenditures over the study period, again deflated. The average benefit-cost ratios were useful since they provided a measure of the returns (in dollars) to the California raisin industry for every dollar invested in export promotion.

To explore the "optimality" of RAC's export promotion investment in Japan and the United Kingdom, we also conducted a marginal simulation analysis to address the fourth question of the study: What was the marginal return of RAC's export promotion programs in Japan and the United Kingdom? Specifically, what would be the gain in export revenues due to a 10 percent increase in export promotion expenditures? We computed a marginal benefit-cost ratio for export promotion where the marginal benefits were the increase in export revenue due to a 10 percent increase in export promotion expenditures and marginal costs were equal to 10 percent of historic export promotion expenditures, both deflated. As noted, the marginal benefit-cost ratio was useful in assessing whether or not current RAC export promotion expenditure levels in each of the two importing countries were optimal.

California Export Promotion Programs

RAC operates several different export promotion programs, all with the fundamental objective of increasing the demand for imports of California raisins in foreign markets. The following section describes the export promotion programs and activities implemented by RAC in Japan and the United Kingdom.

Japan has been the most intensively promoted market by the California raisin industry. During the study period, export promotion was funded from four sources: (1) the Merchandise Incentive Program (MIP), (2) the Industry Market Promotion Funds (IMPF), (3) the U.S. Department of Agriculture's

(USDA's) Foreign Agricultural Service (FAS), and (4) the California Raisin Advisory Board. Each is briefly discussed.

The objective of the Merchandise Incentive Program, MIP, is to enhance exports through volume discounts to importers. The California raisin industry offers merchandise incentives to Japanese importers and pays all of the incentives directly to the California Raisin Division of the Japan Dried Fruit Importers Association. Since its inception in Japan in 1962, the program has been funded in various ways. For the 1999/00 marketing year, the incentive program required that importers purchase a minimum of 22,000 packed short tons and provide a minimum market share for California raisins of 80 percent. The base incentive was $12.50 per packed ton, with an additional $2.50 for each 1,000 tons over the 22,000 minimum and a maximum payment of $25 per packed ton.

The objective of the Industry Market Promotion Funds (IMPF) program is to generate funds to promote California raisins in foreign markets. Funds are earned for each ton imported by participating countries.[1] Rules and procedures are in place to monitor use of the funds, and importers must first gain approval from RAC before making any actual expenditures. The program year is from February 1 through January 31 and importers have one year after the close of each program year to spend the funds, making it in effect a two-year program.

FAS is another source of funds for promotion. Since 1984, financial assistance has been made available through FAS's Targeted Export Assistance and Market Access programs. Promotion funds are generated both by members of the Japan Dried Fruit Importers Association and by nonmembers and are used by the local RAC office in Japan to supplement other FAS programs already in place. Currently, the Market Access program is the only FAS program operating in Japan. Its activities there are generic rather than branded, and the amount of FAS funds has involved around $20 million since 1984.

The final program, support by the California Raisin Advisory Board, began contributing funds for export promotion in Japan in 1987. However, in 1994, the California Raisin Advisory Board was abolished and the program ended.

Two specific examples of RAC's promotion activities in Japan are the New Product Development Contest and the California Raisin Sticker program. Each year RAC conducts a new-product-development contest for the bakery and confectionery trade, and both have become very popular in Japan, as demonstrated by the increasing number of entries RAC receives each year. The goal is to roll out as many new raisin products for consumers as possible.

More than thirty products that are currently on store shelves in Japan have come from these contests. RAC's California Raisin Sticker program is also a promotional contest for Japanese consumers, who are encouraged to purchase raisin products with a special California Raisin sticker that they can pull off and send in for a chance to win a variety of prizes. The tie-in between the trade and consumers is that the trade must first sign on to the promotion and put the sticker on its raisin products. Increases in both number of products and company participation are deemed essential for increasing consumption of California raisins among Japanese consumers.

The United Kingdom is RAC's second largest export market; however, there was no RAC export promotion there prior to 1985. The four sources of funds for programs implemented since 1985, in descending size, are: (1) USDA FAS support, which was implemented in the United Kingdom in 1985 and peaked in 1993; (2) programs sponsored by the California Raisin Advisory Board that ended in 1994; (3) FAS branded matching-fund sponsored programs; and (4) recent bakery programs sponsored by RAC. All four programs operated at significantly lower levels in the United Kingdom than in Japan. Currently, only the FAS programs for marketing and branded matching funds are in effect in the United Kingdom.

RAC's primary marketing approach to United Kingdom consumers is print advertising via ad placements in ten to twelve of the most popular ladies' magazines. In addition to consumer advertising, RAC has produced a school pack, *Sun Fruit/Fun Fruit*, that is offered to teachers in the primary school system. Since the program's inception in 1995, RAC has distributed the pack in more than 13,000 of the country's 25,000 schools. The school packs present information about California raisins as part of the school curriculum, and they are popular among teachers and students.

Methodology

To answer the four questions posed, this study quantified the relationship between export promotion efforts by the California raisin industry and imports of California raisins by Japan and the United Kingdom. Single-equation raisin import demand models were developed using annual time-series data for Japan and the United Kingdom for 1965 through 1998. A single-equation approach was used rather than a demand-systems approach primarily because of difficulty obtaining all the relevant data on related commodities required by the demand-systems approach.

The following factors were included in the Japanese import demand equation to ascertain the extent, if any, of each factor's impact on annual per capita import demand for California raisins by Japanese consumers:[2] the deflated price of California raisins in Japan,[3] deflated prices of Australian and South African raisins in Japan, the deflated retail price index of bakery products in Japan, the deflated, per capita gross domestic product in Japan, and deflated U.S.-raisin-export-promotion expenditures in Japan. The prices of Australian and South African raisins were included since these two countries are the chief competitors to California raisins in the Japanese market. The relationship between the bakery price index and demand for California raisins can be negative or positive, depending on whether bakery products are used by Japanese consumers as a substitute for or complement to California raisins. If the two goods are used as substitutes, one would expect the coefficient associated with the variable to be positive; as the price of one good rises, demand for its substitute increases. On the other hand, if the two goods are complements, the coefficient of the variable should be negative. The export promotion effort was measured as total expenditures on the four promotion programs in Japan. Since previous research (e.g., Kinnucan 1982, 1983, 1985; Pritchett, Liu, and Kaiser 1998; Schmit and Kaiser 1998; Le, Kaiser, and Tomek 1998) has indicated that the carryover effect of promotion is generally less than one year, our annual model did not include lagged export promotion. Furthermore, preliminary regression with one- and two-year lags resulted in highly insignificant coefficients for the lagged variables.

Similarly, the following factors were included in the import demand equation for the United Kingdom to ascertain the extent, if any, of each factor's impact on annual per capita import demand for California raisins by British consumers: deflated prices of California raisins in the United Kingdom, the deflated price of Greek and Turkish raisins in the United Kingdom, the deflated retail price index of bakery products in the United Kingdom, the deflated per capita gross domestic product in the United Kingdom, and deflated U.S. raisin export promotion expenditures in the United Kingdom. While Australia and South Africa are California's major competitors for raisin exports to Japan, Greek and Turkish raisins compete directly with California raisins in the United Kingdom. The export promotion effort was measured as total expenditures on the four promotion programs in the United Kingdom.

Table 10.1. Estimated Elasticities in the Japan Equation

Import Demand Determinant	Elasticity	t-Ratio
Constant	−0.029	−3.0
Price of California Raisins in Japan	−1.117	−4.5
Price of Australian Raisins in Japan	0.442	1.4
Price of South African Raisins in Japan	0.610	2.3
Retail Bakery Product Price Index in Japan	1.547	1.8
Japanese Per Capita Gross Domestic Product	−0.079	−0.2
California Raisin Industry Export-Promotion Expenditures in Japan	0.029	3.0
Durbin Watson: 2.01[a]		
Adjusted R^2: 0.72[b]		

[a] A first-order autoregressive error process was appended to the equation to correct for serial correlation in the residual series. The Durbin Watson statistic of 2.01 suggests that the resulting estimated equation is free from serial correlation problems.

[b] The adjusted R^2 indicates that the estimated equation explains about 72 percent of the variations in Japan's per capita import demand for California raisins.

Econometric Results

The estimated per capita import demand equations are reported in Tables 10.1 (Japan) and 10.2 (United Kingdom). The equations are specified on a per capita basis and in double-logarithmic form. Because of the logarithmic specification, each of the estimated parameters has the interpretation of elasticity, which measures the percent change in per capita import demand for California raisins given a 1 percent change in the demand determinant in question, holding constant all other variables.[4] A first-order autoregressive error structure was appended to each of the two equations to correct for serial correlation in the residual series and the Durbin Watson statistics reported in the tables suggested that the resulting estimated equations were free from serial correlation problems. Further, both equations fit the data well; the adjusted R^2s indicated that the Japan equation explained about 72 percent of the variations in Japan's per capita import demand for California raisins and the United Kingdom equation explained about 87 percent of the variation there. Both equations have elasticity signs that are generally consistent with economic theory, and the estimated coefficients are usually statistically significant at conventional confidence levels. No multicollinearity was detected in either equation, based on Pindyck and Rubinfeld's suggestion (1991, p. 84) of examining whether there is a relatively high (in absolute value) covariance between estimated parameters, which is a symptom of multicollinearity.

Table 10.2. Estimated Elasticities in the United Kingdom Equation

Import Demand Determinant	Elasticity	t-Ratio
Constant	8.207	4.2
Price of California Raisins in United Kingdom	−1.280	−5.0
Price of Turkish Raisins in United Kingdom	0.905	1.6
Price of Greek Raisins in United Kingdom	−0.095	−0.2
Retail Bakery Product Price Index in United Kingdom	−0.278	−1.0
United Kingdom Per Capita Gross Domestic Product	0.295	2.2
California Raisin Industry Export-Promotion Expenditures in United Kingdom	0.133	5.4
Durbin Watson: 2.19[a]		
Adjusted R^2: 0.87[b]		

[a] A first-order autoregressive error process was appended to the equation to correct for serial correlation in the residual series. The Durbin Watson statistic of 2.19 suggests that the resulting estimated equation is free from serial correlation problems.

[b] The adjusted R^2 indicates that the estimated equation explains about 87 percent of the variations in the United Kingdom's per capita import demand for California raisins.

The estimated equations suggest that the price of California raisins was an important factor in explaining annual variations in per capita import demand in Japan and the United Kingdom. In the Japan equation, the estimated own-price elasticity was −1.117. The own-price elasticity for the United Kingdom was estimated at an even higher level of −1.28.

The price of raisins from other countries was found to be important in both equations. For Japan, the cross-price elasticity of per capita demand for California raisins with respect to the Australian and South African prices was estimated at 0.442 and 0.610, respectively. These results suggested that, while both countries compete with California, South Africa is slightly more important in terms of its ability to affect demand for California raisins in Japan. For the United Kingdom, the cross-price elasticity of per capita demand for California raisins with respect to the Turkish price was estimated at 0.905. Thus, Turkey is a major competitor with California in the United Kingdom raisin market. The cross-price elasticity with respect to the Greek raisin price was estimated at −0.095; however, the associated t-statistic indicated that this coefficient was not statistically different from zero.[5]

The retail bakery price index was included in both equations since it was expected to be a determinant of raisin demand. Interestingly, this variable turned out to have opposite signs in the model for the two countries. For

Japan, the cross-price elasticity of per capita demand for California raisins with respect to the retail bakery price was estimated at 1.547. One explanation for the estimate of a positive cross-price elasticity is that Japanese consumers regard bakery products and California raisins as substitutes and, hence, as the price of one product rises, the demand for the other increases. On the other hand, the cross-price elasticity for the United Kingdom was estimated at a negative value of −0.278, confirming that bakery products and California raisins are complements there.[6]

While income was not an important determinant of per capita raisin demand in Japan, it was an important determinant in the United Kingdom. The coefficient associated with gross domestic product in Japan was actually negative but highly insignificant from a statistical point of view, which means that changes in Japanese income have no impact on demand for California raisin imports. This result is consistent with a 1988 study by Nuckton, French, and King, which also found a negative relationship between gross domestic product and California raisin imports to Japan. However, the coefficient for gross domestic product in the United Kingdom was positive (0.295) and was significantly different from zero.

In both equations the coefficient associated with the export promotion variable was positive and statistically different from zero. This means that the statistical evidence supports the notion that California export promotion programs have the effect of increasing demand for its raisins in both countries. In the Japan equation, the estimated export promotion elasticity was 0.029. In the United Kingdom equation, the estimated export promotion elasticity was 0.133.

Simulation Results

Based on the estimated import demand equations, we concluded that the answer to the first question in this study is affirmative: the California raisin industry's export promotions have a positive and significant effect on its exports to Japan and the United Kingdom. The estimated equations were next simulated under two scenarios to address the remaining three questions posed in this study:

1. **Baseline Scenario** – All export promotion programs have been in effect.
2. **No-Export-Promotion Scenario** – There have been no California export promotion programs in the two importing countries in question.

In the second scenario, all demand determinants except export promotion expenditures were set equal to historic levels. However, the export promotion variable was set to 0 and the corresponding import demand was simulated over time for each country.[7] The difference between the two scenarios gave the total impact of RAC's export promotion on imports of California raisins in the foreign markets in question.

Figures 10.1 and 10.2 illustrate the simulation results on the quantity of imports by Japan and the United Kingdom. The two figures strikingly show the impact of RAC's export promotion programs on California raisin imports by the two countries. From 1965 to 1998, raisin export promotion programs by RAC and the California Raisin Advisory Board in the Japanese market increased Japanese imports of California raisins by 201,539 metric tons, or 6,107 metric tons per year (to an annual average of about 18,928 metric tons for the period). The results were even more profound for the United Kingdom, which did not have a raisin export promotion program until 1984. From 1985 to 1998, raisin export promotion programs by RAC and the California Raisin Advisory Board in the United Kingdom increased raisin exports to the country by 253,624 metric tons, or 18,116 metric tons per year (to an annual average

Figure 10.1. Simulated California Raisin Exports to Japan with and without Export Promotion Programs in Metric Tons

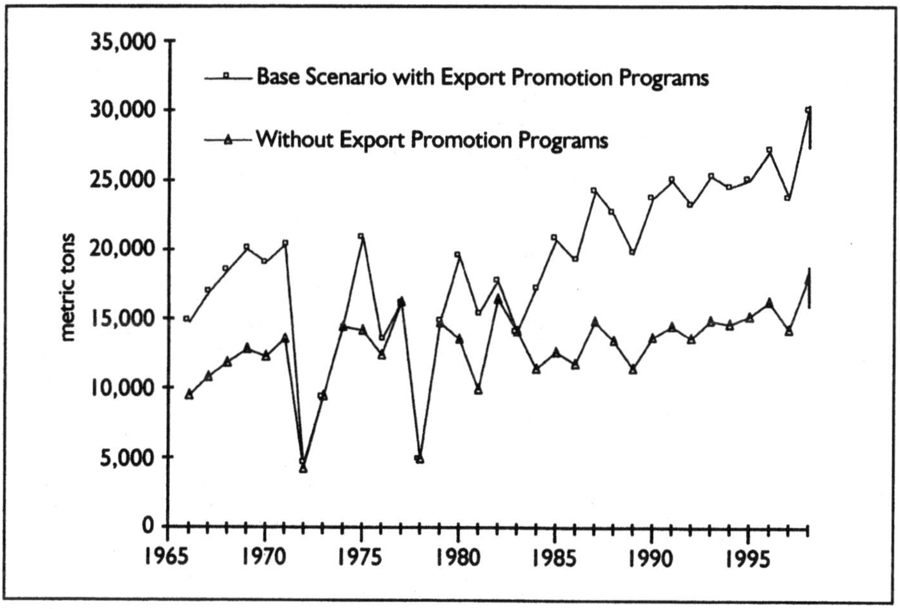

Figure 10.2. Simulated California Raisin Exports to the United Kingdom with and without Export Promotion Programs in Metric Tons

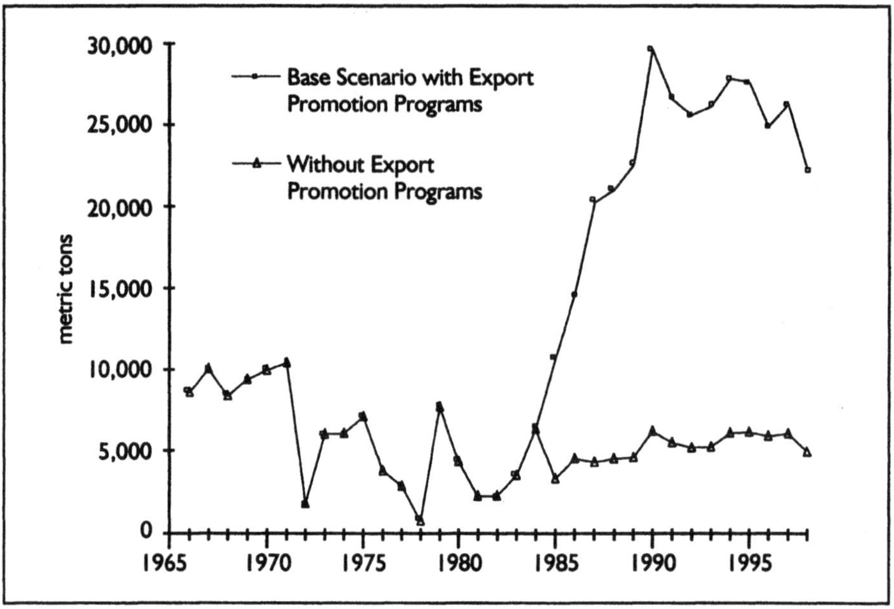

of about 23,263 metric tons for the period).[8] Hence, the answer to the second question posed in the study is that RAC's export promotion programs clearly had a large positive effect on the level of exports to both Japan and the United Kingdom.

While it is clear that export promotion of California raisins had a major impact on boosting exports, the third question posed in this study is more bottom-line in nature: the comparison of benefits with costs. To answer this question, we computed an average benefit-cost ratio for export promotion in Japan, in the United Kingdom, and in the two countries combined. The average benefit-cost ratios, also known as average rates of return on investment, were useful since they provided a measure of returns (in dollars) to the California raisin industry for every dollar invested in export promotion.

The following procedure was used to compute the benefit-cost ratio in each country. To compute the benefits of export promotion, the gain in export quantity due to export promotion estimated from the simulation previously described was multiplied by the deflated export price for each year of the simulation.[9] The average benefit-cost ratio was then computed by dividing the resulting monetary benefit value by the deflated combined cost of all export promotion activities in each country.[10] The resulting average benefit-cost ratios for each year in the simulation were then averaged to get an overall average benefit-cost ratio for Japan, the United Kingdom, and the two countries combined.

The resulting benefit-cost ratios for Japan and the United Kingdom were considerably larger than 1, indicating that the benefit of export promotion in terms of expanding export revenue was greater than the cost of the programs. For Japan, the average benefit-cost ratio for 1965–1998 was 6.64 and the average benefit-cost ratio for the more recent period of 1985–1998 was 5.13. This means that every dollar invested in export promotion in Japan returned, on average, $6.64 or $5.13 in export revenue to the California raisin industry over the two time periods examined. The average benefit-cost ratio was even higher in the United Kingdom, which averaged 15.29 for 1985–1998 (there was no export promotion in the United Kingdom prior to 1985). Hence, every dollar invested in export promotion in the United Kingdom returned $15.29 in export revenue to the California raisin industry. The average benefit-cost ratio for the two countries combined for 1985–1998 was 7.32, indicating a return of $7.32 for every dollar invested in export promotion in Japan and the United Kingdom.

To make allowances for the error inherent in any statistical estimation, a 90 percent confidence interval was calculated for the average benefit-cost ratios previously calculated. The lower and upper bounds for the average benefit-cost ratio were obtained by re-simulating the two scenarios with the export promotion coefficient in the estimated import demand equations set to their lower and upper bounds of the 90 percent confidence interval.

The estimated lower and upper bounds for the average benefit-cost ratio in Japan for 1985–1998 were 4.60 and 5.62, respectively. The lower and upper 90 percent confidence bounds for the average benefit-cost ratio in the United Kingdom for the same period were 14.63 and 15.87, respectively. These confidence intervals give credence to the previous finding that the revenue expansion benefit of the raisin industry's export promotion programs was considerably greater than the cost of the programs.

How do the average benefit-cost ratios computed previously compare with other export promotion programs conducted by various U.S. commodity promotion organizations? Table 10.3 provides a summary of comparable average benefit-cost ratios for several selected export promotion programs. The benefit-cost ratios computed in this study for California raisin export promotion compare favorably with those reported in Table 10.3, which range from 1.29

Table 10.3. Selected Benefit-Cost Ratios for Other Export Promotion Programs

Study		Ratio
Pecan Export Promotion – Asia	Onunkwo and Epperson 2000	6.45
Pecan Export Promotion – EU	Onunkwo and Epperson 2000	6.75
Almond Export Promotion – Japan	Halliburton and Henneberry 1995	4.95
Almond Export Promotion – Taiwan	Halliburton and Henneberry 1995	5.94
Almond Export Promotion – Hong Kong	Halliburton and Henneberry 1995	3.69
Potato Export Promotion – Japan	Lanclos, Devadoss, and Guenther 1997	1.29
Potato Export Promotion – Philippines	Lanclos, Devadoss, and Guenther 1997	11.77
Potato Export Promotion – Thailand	Lanclos, Devadoss, and Guenther 1997	16.36
Red Meat Export Promotion – Pacific Rim Excluding Japan	Le, Kaiser, and Tomek 1998	15.62

for potato export promotion in Japan to 16.36 for potato export promotion in Thailand.

Marginal Benefit-Cost Analysis

To explore the optimality of the raisin industry's export promotion investment in Japan and the United Kingdom, we also conducted a marginal simulation analysis, which answered the fourth question of the study regarding the marginal return of export promotion programs. The estimated import demand equations were used to simulate the outcome of an additional scenario (Scenario 3) for each country and the results were then compared with the baseline scenario (Scenario 1):

Scenario 1: Baseline Scenario – All export promotion programs have been in effect.
Scenario 3: 10-Percent-Plus Scenario – A 10 percent increase in historical export promotion expenditures.

The difference between the two scenarios measured the marginal impact of a 10 percent increase in export promotion expenditures on California raisin export quantities. This gain in exports was multiplied by deflated export prices to arrive at a dollar measure of the gain in export revenue. A marginal benefit-cost ratio for export promotion programs was then computed in which the marginal benefits were the increase in export revenue due to the 10 percent increase in export promotion expenditures, and marginal costs were equal to 10 percent of deflated historical export-promotion expenditures.

For Japan, the marginal benefit-cost ratio was 0.56 for 1965–1998 and 0.42 for the more recent period, 1985–1998. That is, had the industry spent 10 percent more in export promotion in the Japanese market, the rate of return for that additional investment would have been only about $0.50 for every dollar invested. Together with the previous result that the average benefit-cost ratio for Japan was about $5 to $6 per dollar invested, the marginal benefit-cost ratio of about 0.5 suggests that increasing export promotion expenditures in Japan beyond the present spending level would not be worthwhile. On the other hand, if promotion expenditures (between 1985 and 1998) in the Japanese market had been reduced by 10 percent, the ratio of savings (from reduced expenditures) to costs (of the reduced export revenue) would have been 0.46. Thus, it also would not have been worthwhile cutting promotion expenditures in Japan, which suggests that the current spending level there has been optimal. In the United Kingdom, the marginal benefit-cost ratio for 1985–1998 was 3.19, indicating that the industry should explore investing more money in export promotion activities in the United Kingdom.

Conclusions and Policy Implications

The purpose of the study was to assess the effectiveness of the California raisin industry's export promotion programs in Japan and the United Kingdom. An econometric import demand equation was estimated for each of these two foreign markets. Based on the estimated promotion elasticity, the statistical evidence supports the notion that the programs have a positive impact on demand for California raisins in Japan and the United Kingdom. Furthermore, the simulation results indicate that the magnitudes of the impact are not trivial. Between 1965 and 1998, RAC's export promotion programs accounted for an increase in Japanese imports of California raisins of 6,107 metric tons per year. From 1985 to 1998, RAC's export promotion program accounted for an increase in the United Kingdom's imports of California raisins of 18,116 metric tons per year. Hence, the answer to the first policy question is that RAC has been effective in its mission to enhance foreign demand for California raisins.

To address the second policy issue, the simulated benefits of export promotion were compared to the cost of the programs and the associated benefit-cost ratios were computed. The resulting benefit-cost ratios for the Japanese and United Kingdom markets were computed at about 5:1 and 15:1, respectively. Thus, the conclusion regarding the second policy issue is that the benefit of export promotion in terms of expanding export revenue has been much greater than the cost of the programs.

Finally, the marginal simulation analysis indicates that RAC's export promotion spending level in Japan has been optimal but that the industry should explore investing more money in its export promotion activities in the United Kingdom. Indeed, RAC recently adopted this policy recommendation by significantly increasing its promotion expenditures in the United Kingdom.

Notes

This chapter originally appeared as an article entitled "An Economic Analysis of California Export Promotion" by Kaiser, Liu, and Consignado in *Agribusiness: An International Journal* (© 2003 John Wiley and Sons). This material is used by permission of John Wiley and Sons, Inc.

1 In all participating countries except Japan, importers that earn the funds may use them to promote their own brands.

2 The data used in the estimation were compiled by RAC staff members. All prices, income, and promotion expenditures were deflated by the general Consumer Price Index in Japan (and the United Kingdom) to remove the impact of inflation on all monetary series. In addition, all monetary values were in yen for Japan and in pounds for the United Kingdom. Data were obtained from the following sources: promotion expenditures were obtained from RAC; gross domestic product and exchange rates were from the International Monetary Funds' *International Financial Statistics*; data on bakery prices for Japan were obtained from the Japanese Ministry of Finance and for the United Kingdom from Business and Trade Statistics Limited; and data on volumes and values of raisin exports to Japan and the United Kingdom and raisin prices in competing countries were obtained from USDA's Foreign Agricultural Service.

3 To ascertain whether there was a simultaneity problem in treating the price of California raisins as exogenous, a version of the Hausman endogeneity test was conducted (Pindyck and Rubinfeld 1991, pp. 303–304). Specifically, a price instrumental equation was first estimated by regressing the California raisin price against its one-period lag and other exogenous variables originally appearing in the demand equation. The estimated residuals from the instrumental equation were then appended to the original demand equation and a t-test performed on its estimated coefficient. The coefficient was found to be statistically not different from zero at the 10 percent level, failing to reject the null hypothesis that there is no simultaneity between the price and quantity variables in the Japan import-demand equation. A similar result of no price endogeneity was found when the same test was applied to the United Kingdom import demand equation.

4 The double-logarithmic model was selected for four reasons. First, it provides a convenient nonlinear approximation of the function in question without requiring introduction of numerous additional parameters into the model. Second, the double-logarithmic specification is popular in the advertising and promotion literature because the functional form allows for the desired property of diminishing returns to promotion. Third, the estimated promotion coefficient has the convenient interpretation of being the promotion elasticity. Finally, two alternative functional forms (linear and square root) were explored in the estimation but yielded less satisfactory results. Specifically, the goodness-of-fit was similar among the three functional forms. However, the significance level of the estimated parameters

suggested that the double-logarithmic model performed better. The other two functional forms also resulted in more variables having coefficient signs inconsistent with theory.

5 Dropping the Greek price variable from the United Kingdom import demand equation did not change the estimation results in any significant manner.

6 However, the associated t-statistic indicated that the estimated coefficient was not statistically significant.

7 Because of the logarithmic functional form, export promotion expenditures in the two countries were set to the equivalent of $1.00 in this scenario since the log of zero is undefined.

8 The average annual export quantity of 23,263 metric tons for 1985–1998 can be compared with annual average exports of 5,878 metric tons for the earlier period, 1965–1984.

9 Kinnucan (1999) discusses the case of a small open economy where the supply curve is horizontal so there is no price response to the promotion-induced demand increase. In that case, export promotion does not yield benefits to growers because the additional revenue is perfectly offset by the additional production cost arising from the increased quantity. However, this is not the situation for RAC, which basically operates a surplus disposal program to enhance growers' revenue. At the beginning of the crop year, RAC estimates total crop production and the so-called "trade demand," which includes domestic and foreign demand. Almost always, production is higher than trade demand given the relatively high, negotiated, administered grower price, and hence RAC must deal with a surplus. Without additional outlets (e.g., additional sales in export markets), the surplus would eventually be dumped as low-value cattle feed or even discarded. Production costs on this surplus have already been incurred regardless of whether there is additional export demand. However, with additional export demand due to promotion, there would clearly be additional revenue to be distributed back to growers. Granted, it is conceivable that the increased grower revenue would eventually lead to more planted acreage and perhaps greater yields per acre. Due to the perennial nature of raisin grape crops and the fact that most of the production is located in a limited geographical area in the San Joaquin Valley, the potential supply response is likely to be minimal. Attempts to model domestic supply and demand conditions as done by Nuckton, French, and King (1988) were beyond the scope of this analysis.

10 The computed benefit-cost ratio pertains to the promotion program as a whole (including both industry and FAS money). It would be interesting to assess the impact of each individual program. However, the estimated model precludes this type of simulation because the various promotion expenditures were aggregated into one variable in the regression models. Attempts at including disaggregated export promotion expenditures as separate variables in the regression did not yield satisfactory results.

References

Federal Agricultural Improvement and Reform Act. U.S. Code. Vol. 7, sec. 7401 (1996).

Halliburton, K., and S. Henneberry. "The Effectiveness of U.S. Nonprice Promotion of Almonds in the Pacific Rim." *Journal of Agricultural and Resource Economics* 20 (1995):108–121.

Kaiser, H.M., and D.J. Liu. *An Econometric Analysis of California Raisin Export Promotion.* Fresno CA: California Raisin Administrative Committee, 2001.

Kinnucan, H.W. "Demographic Versus Media Advertising Effects on Milk Demand: The Case of the New York City Market." Department of Agricultural Economics Staff Paper 82-5, Cornell University, 1982.

——. "Media Advertising Effects on Milk Demand: The Case of the Buffalo, New York Market. Department of Agricultural Economics Research Bulletin 83-13, Cornell University, 1983.

——. "Evaluating Advertising Effectiveness Using Time Series Data." *Research on Effectiveness of Agricultural Commodity*. W.J. Armbruster and L.H. Meyers, eds., pp. 105–122. Oak Brook IL: Farm Foundation, 1985.

——. "Advertising Traded Goods." *Journal of Agricultural & Resource Economics* 24 (1999): 38–56.

Lanclos, D., S. Devadoss, and J. Guenther. "Impacts of Foreign Direct Investment and Advertising on the Export Demand for U.S. Frozen Potatoes." *Agribusiness: An International Journal* 13 (1997):273–284

Le, C.T., H.M. Kaiser, and W.G. Tomek. "Export Promotion and Import Demand for U.S. Red Meat in Selected Pacific Rim Countries." *Agribusiness: An International Journal* 14 (1998): 95–106.

Nuckton, C.F., B.C. French, and G.A. King. *An Econometric Analysis of the California Raisin Industry*. Berkeley CA: University of California Division of Agriculture and Natural Resources, Giannini Foundation Research Report 339, 1988.

Onunkwo, I.M., and J.E. Epperson. "Export Demand for U.S. Pecans: Impacts of U.S. Export Promotion Programs." *Agribusiness: An International Journal* 16 (2000):253–266.

Pindyck, R.S., and D.L. Rubinfeld. *Econometric Models & Economic Forecasts*. New York NY: McGraw-Hill, Inc., 1991.

Pritchett, J.G., D.J. Liu, and H.M. Kaiser. "Optimal Choice of Generic Milk Advertising Expenditures by Media Type." *Journal of Agricultural & Resource Economics* 23 (1998):155–169.

Raisin Administrative Committee. *Raisin Administrative Committee 1999–2000 Analysis Report*. Fresno CA, 1999.

Schmit, T., and H.M. Kaiser. "Egg Advertising, Dietary Cholesterol Concerns, and U.S. Consumer Demand." *Agricultural & Resource Economics Review* 27 (1998):43–52.

11

The Red Edge: Demand-Enhancing Strategies for California Strawberries

Colin A. Carter, James A. Chalfant & Rachael E. Goodhue

The industrial organization of production and marketing of fresh produce in the United States is rapidly changing toward greater consolidation and vertical coordination. Fresh strawberries are no exception to this phenomenon. More than 80 percent of California strawberry acreage is now controlled by six major producers/shippers, which is a substantial increase from even five years ago. Wholesalers now account for an insignificant share of total fresh strawberry volume; retailers now obtain the vast majority of their strawberry volume directly from shippers. Shippers and retailers are increasing their coordination, and formal and informal contracts are replacing spot market transactions. In 2002, for instance, one major western retailer estimated that 80 to 100 percent of its strawberry purchases were obtained through contracts. One motivation for greater control by shippers over strawberry production is the ability to provide a high-quality product that fetches a premium price.

Recent changes in the organization of strawberry production and marketing have important implications for the role and effectiveness of generic advertising. A handful of shippers now control a substantial majority of California's strawberry production. Prior to 2003, an important generic promotion strategy used by the California Strawberry Commission (CSC) was to promote fresh California strawberries jointly with major retailers. Now, major shippers increasingly promote their specific brands as high-quality strawberries through packaging and joint promotions with retailers. Along with these changes in the industry, CSC has shifted the emphasis of its marketing to consumer-oriented advertising focused on the health benefits of strawberries. In this chapter, we review the commission's promotion strategies and econometrically estimate the benefits from increasing retailer promotions of strawberries.

Industry Background

California is the most important strawberry growing region in the nation in terms of acreage, volume, and value of production. Yields are relatively high in California compared with other states. California accounted for 60 percent of national strawberry acreage in 2003 and produced almost 90 percent of the nation's strawberries. California strawberry acreage increased from 8,500 acres in 1970 to more than 31,639 acres in 2004. The total value of annual U.S. strawberry sales reached $1.3 billion in 2003 (U.S. Department of Agriculture (USDA), Economic Research Service (ERS) 2004b) while the value of California's fresh and processed strawberry production exceeded $1.1 billion (USDA, National Agricultural Statistics Service 2003). Strawberries rank in the top ten agricultural commodities in California in terms of gross value of production.

About 70 percent of the California strawberry crop is sold as fresh strawberries and the remainder is sold in the processed market. Figures 11.1 and 11.2 plot monthly California production of fresh and processed strawberries, respectively, for 2002 and underscore the seasonal nature of the strawberry market. Although fresh and processed/frozen strawberries are distinct markets, changes in the price of processed strawberries affect the production of fresh strawberries. Most growers in California begin the season producing for the fresh market. Later, as warmer temperatures diminish the quality of the fruit (firmness and size), growers switch to processing sales. However, growers may delay the transition, despite poor-quality fruit, if fresh-market prices are sufficiently favorable (or, conversely, if prices in the processing market are unusually low). Alternatively, poor weather conditions may dictate an early shift of the harvest to the processing market.

California strawberries are grown primarily in several distinct regions along the coast, each with a unique season. Fresh production begins in the far south, moves north as the season progresses, and ends on the central California coast in the Watsonville area near Monterey Bay. As the harvest moves north, production from each region peaks successively later in the year, generating distinct markets for fresh strawberries that are closely related but with patterns and relationships between regions that shift as the season progresses.

Figure 11.1. California Fresh Strawberry Volume, 2002

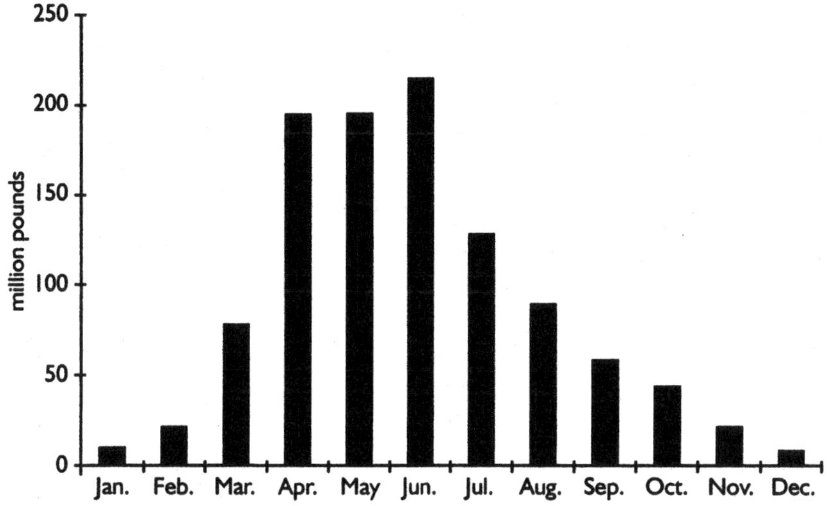

Source: USDA, Agricultural Marketing Service, *National Berry Report*, various issues.

Figure 11.2. California Processed Strawberry Volume, 2002

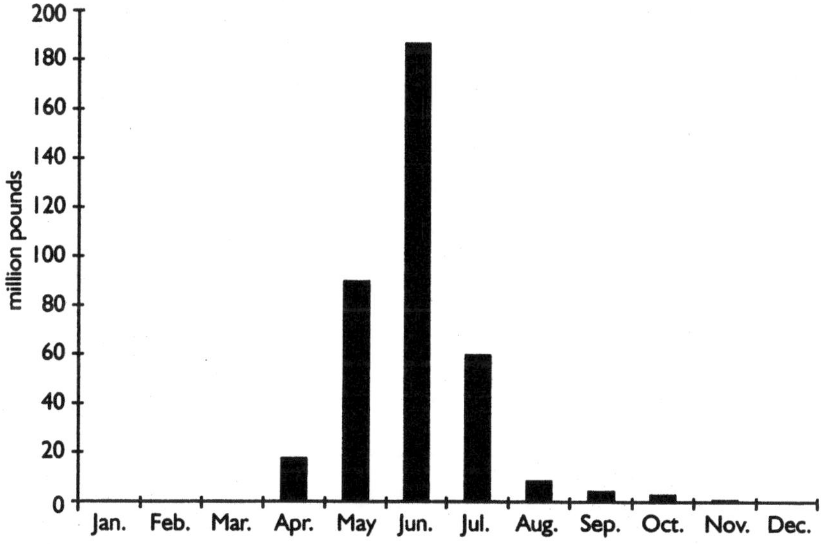

Source: USDA, Agricultural Marketing Service, *National Berry Report*, various issues.

For example, poor weather in Watsonville when production there normally dominates the fresh market, can reduce northern production and thus induce southern growers to continue to sell into the fresh market even as the quality of their fruit deteriorates.

Figure 11.3 shows the history of real prices for fresh and processed strawberries using annual averages for the past thirty years. Adjusted for inflation, prices trend downward. Considerable price variation exists within a season, however. Typically, prices have followed a U-shaped pattern each year–higher prices early in the season when only southern regions produce significant quantities for the fresh market and then again later in the year when the Watsonville area dominates. The middle of the season corresponds to high volumes of fresh strawberries along with peaks for some substitute fruits such as cherries and hence the lowest prices. These price patterns have led producers to increase off-season plantings in both southern and northern regions in recent years, which is one of the significant changes that have occurred in the industry's structure.

Harvesting occurs multiple times during a season because strawberry plants bear fruit continuously under proper conditions. Generally, berries for fresh use are harvested every three to five days and berries for processing

Figure 11.3. Average Grower Price for Strawberries in 1996 Dollars, 1970–2003

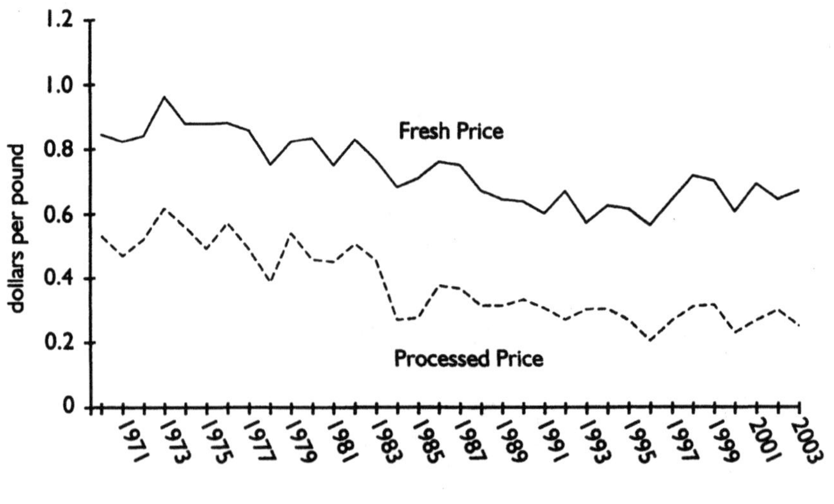

Source: USDA, ERS 2004b.

are harvested every five to seven days. Harvesting must be done by hand due to the fruit's fragility. While the harvest process is labor intensive and thus expensive, picking by hand protects both harvested and immature fruit from damage. Quality–reflected in size, shape, number of blemishes, firmness, and taste–is especially important for strawberries headed for the high-value fresh market. Damaged fruit left on the plants can attract pests and promote disease, harming subsequent harvests.

Once picked, berries for fresh use are placed directly into retail packages and then packed in the field in trays that hold twelve one-pint baskets, eight one-pound clamshells, or larger clamshell packages with fewer units per tray. After picking, the trays are immediately sent by truck from the field to a refrigerated warehouse for cooling and preshipping storage. Most fresh berries are shipped the next day, although they can be stored for a few days.

The one-pint basket was the standard shipping container for many years but is no longer very important. The one-pound clamshell was introduced in 1996, and its market share grew rapidly until about 2001, when it peaked at 81 percent of the market in dollar terms. The one-pound clamshell is popular with consumers. For retailers, it requires less labor but has higher packaging costs for shippers and growers than the one-pint basket. In 2003, the one-pound clamshell package accounted for 74 percent of the fresh-strawberry retail market (CSC 2004a). Market share that year for the one-pint basket was only 7 percent. Two- and four-pound clamshell packages have recently grown in popularity, and in 2003 they accounted for 8 percent and 7 percent of the market, respectively.

Unlike pint baskets, clamshell packages have space available for shipper advertising, so fresh strawberries are becoming more of a branded product. Now, some of the large shippers are competing on quality through proprietary varieties and more efficient handling.[1] Research is under way that will eventually allow shippers to attach a small radio chip on each clamshell package to facilitate improved quality control and thereby boost the quality of the strawberries on retail shelves.

Strawberries sold to processors are frozen, sliced, or pureed for juice stock directly after leaving the field. Demand for processed berries comes from the food-service industry, producers of food products such as jellies and smoothies, and government programs such as the federal school lunch program. Unlike fresh berries, processed berries are easily stored. Consequently, inventories of processed berries play a major role in the supply of processed fruit. They are the primary determinant of the processing market's impact on fresh-market prices. A large carryover from the previous season can diminish the willingness of the processed market to absorb residual fresh-market supplies. From 1980 to 2003, beginning inventories accounted for an average of 34 percent of total annual frozen supplies.

Marketing Channels

As previously mentioned, consolidation of market players and technological advances have led to changes in relationships among retailers, shippers, growers, and other industry members and to changes in the relative importance of these various agents and institutions. Actors at all levels of the production chain have consolidated–big players have become bigger. The retail sector in particular has become more concentrated during the last decade. Retail "super centers" such as Wal-Mart and Costco exemplify this latest shift. These retailers buy less of their produce through traditional terminal markets and brokers and instead deal directly with shippers to avoid intermediary margins and handling costs.

The importance of brokers in the strawberry marketing chain is thus declining rather quickly. Brokers typically purchase their supplies from shippers and resell to a variety of retail outlets, including smaller retail chains, independent grocers, institutional buyers, the food-service sector, and restaurants. Brokers often service the same clientele as shippers but volumes for each broker transaction are typically much smaller. Most brokers deal in a variety of fresh fruits and vegetables and are adept at procurement even when supplies are tight.

As the first handlers of strawberries, shippers provide most, if not all, of the marketing services for growers. Shippers locate buyers, determine grower prices, coordinate transportation, and guarantee product quality. In California, there are roughly sixty shippers that pack, cool, and market strawberries. The majority of total volume is shipped by six large firms, four of which are very large. The largest single shipper controls more than 20 percent of the state's acreage and the six largest typically market the strawberries from more than 80 percent of planted acres in the state. Large shippers provide strawberries to customers year round by supplementing their growers' supplies with berries from Mexico and elsewhere during California's off-season. Smaller shippers generally serve small, seasonal niche markets.

In 2001, there were approximately 400 strawberry growers in California and 25,000 acres of strawberries–a rough average of sixty-three acres of strawberries per grower. However, more than 70 percent of California strawberry producers plant fewer than fifty acres each year. Moreover, most strawberry production is controlled by a few large integrated operations that manage thousands of acres. Thus the share of state production originating from independent growers is shrinking rapidly. Growers who are part of a shipping firm (i.e., "in-house" growers) do not incur much financial risk; most of the risk falls on the shipper. However, one industry observer notes a countertrend: some shippers are increasingly using outside growers to reduce financial risk. Overall, however, the trend is toward greater shipper involvement in production, regardless of how financial risk is assigned. To meet retailer requirements, shippers want more control over production and the quality of the harvest. As mentioned earlier, shippers are now branding retail clamshell packages, so quality is also important for building and maintaining a reputation with final consumers.

Shipper-Retailer Relationships

Perhaps the most critical strategic relationship, as well as the one that is evolving most rapidly, is the shipper-retailer relationship. As a result of ongoing consolidation at the retail level, retailers increasingly prefer to deal with a smaller number of shippers who can provide them with strawberries year round. However, due to the seasonal and perishable nature of the fruit, the size of retail orders, and weather shocks, retailers can rarely rely exclusively on a single shipper. This is especially true during promotion periods. When shippers from one region are short of supply, a retailer with diverse shipper contacts can find enough supply to support its retail promotions.

Shippers have responded to these demands with one of two basic strategies, the choice of which depends on the size of the shipper. Smaller shippers are generally limited to specific seasonal niches since their production is more localized. These shippers rely on the tendency of some retailers to obtain strawberries from multiple shippers and are more likely to market a large share of their output through the spot market. The market share of small shippers has declined substantially over the past ten years. In contrast, large shippers have increased market share to become primary suppliers and have expanded production to different regions to cover all the seasonal niche markets. Some have invested in proprietary varieties and taken other steps to further differentiate their product in terms of quality.

Precommitment sales are an important feature of these long-term marketing relationships. Precommitments represent a type of forward contract. The shipper agrees to supply the retailer with an agreed-upon certain volume in the coming weeks at a price that is to be determined but typically falls within an agreed range. Precommitments are often linked to retailer promotions of strawberries. Improved coordination between shippers and retailers, whose demands are increasingly complex, is a common rationale (among shippers) for entering into these agreements. Shippers monitor their growers' fields and often communicate supply conditions to retailers. During peak demand for precommitments, buyers for large retail chains contact various shippers' sales staffs daily to check on supply conditions and arrange sales. As a result, precommitment sales may provide a means for consistent dialogue between shippers and retail buyers.

Shippers and retailers both have reason to seek precommitment agreements, though at different times during the season. Precommitment sales are often, but not always, sought by retailers to support a promotional advertisement that will appear in a forthcoming newspaper or flyer. Distinct periods of high prices in February, March, and April (when supplies are limited and uncertain due to weather and substitute fruits are less plentiful) and low prices at the season's peak (when supplies are generally larger and less uncertain and substitute fruits are more plentiful) characterize the industry. Retailers seem likely to value precommitments more during the early part of the season (Han 2003). Later, precommitments are more important to shippers, providing insight into why long-term relationships are so important to most of the larger shippers. Larger shippers that provide berries year round have to make sales during periods when strawberry prices are low. Periods with fewer promotions—typically the months of July, August, and September—are problematic for shippers because supplies are still substantial. Some shippers claim that their relationships with retailers are strong enough to allow them to occasionally call on "favors" owed to them by retailers to move some of the late-season volume. However, many of the same shippers also report that price reductions are typically required to move volumes during the summer. These price reductions may also provide an incentive for retailers to plan promotions. Even if price discounts are required, late season precommitment sales may still benefit shippers and growers. Since fresh strawberries are highly perishable, shippers must sell them quickly. Precommitments provide shippers with some assurance that they will find buyers for their perishable product.

Two recent developments are likely to have an important influence on the shipper-retailer relationship. First, there is the influence of Wal-Mart on the market. Wal-Mart's approach to produce procurement differs from that of most other retailers. Wal-Mart signs vendor-managed contracts, which means that shippers are responsible for keeping strawberries supplied to the retailer's distribution centers. Other retailers have not yet adopted the vendor-managed system for produce, but the trend is in that direction. Second, some retailers are now engaging in brand-specific promotions. Rather than feature California strawberries in their advertising circulars, they sometimes feature specific shipper brands. Both of these developments point toward closer shipper-retailer coordination and greater product differentiation.

California Strawberry Industry Organizations

CSC is a state-chartered marketing order originally founded as the California Strawberry Advisory Board in 1955. Its primary duties include "efficient creation and management of a research program to develop improved varieties of strawberries, an integrated approach to control pests and diseases common to strawberries and more efficient cultural practices." CSC also oversees "efficient development of a national and international advertising and promotion program which, combined with the research program, will enhance the competitiveness of the California strawberry industry within the national and international marketplace." (Food and Agricultural Code, Division 22, Chapter 17.5, Section 77402.)

Funding for CSC activities is derived from a 4.5 cent assessment levied on each tray of strawberries shipped. Growers and shippers each contribute 2.25 cents per fresh tray. In 2002, CSC's budget from assessments and grants was $6.8 million. Of that total, 54 percent funded marketing, 22 percent funded research, and 8 percent funded industry and public affairs. The remainder covered overhead and administrative costs. On average, CSC devotes $3.5 million to $4.5 million annually to marketing programs. This relatively small marketing budget means that CSC's marketing efforts cannot match the scale of advertising in industries such as dairy or beef. A different approach is required, as discussed in the next section.

The Processing Strawberry Advisory Board (PSAB) is a state-chartered marketing order that was established in 1960. PSAB is funded by a per-tray assessment paid by processor members at a rate in 2004 of $0.012 per tray. PSAB is authorized to set grade standards for the raw product, inspect and certify product quality, prevent unfair trade practices through price posting, and assess growers for research and advertising and promotion. In practice, PSAB leaves much of the research and promotion activities to CSC (Wells 1996).

Unlike marketing orders for some other fruits (e.g., kiwifruit, oranges), neither CSC nor PSAB controls supply. As a result, neither organization has a direct influence on prices. However, both organizations are industrywide vehicles for producer communication and collective action (Wells 1996). As such, they play an important role in governing individual behavior and shaping industry trends that can ultimately affect prices.

Trends and Factors Associated with Strawberry Demand

Americans have increased their per capita annual consumption of fruits and vegetables over the past thirty years from 578.1 pounds in 1970 (farm-weight basis) to 705.4 pounds in 2001 (USDA, ERS 2004a). Total fresh and processed fruit consumption per capita grew 0.54 percent annually from 1970 to 2001, culminating in per capita fruit consumption of 275.7 pounds (farm-weight basis) in 2001. Annual per capita consumption of fresh fruit grew almost twice as quickly, growing at an average annual rate of 1.05 percent over the period. Table 11.1 reports consumption growth rates for some individual fresh fruits. Fresh strawberries and grapes enjoyed annual per capita consumption growth rates well above the growth rate for all fresh fruits—3.7 percent and 3.9 percent, respectively. Peach and nectarine consumption grew at a rate of 0.145 percent, which was below the category average but still positive. Consumption of both cherries and oranges declined, posting growth rates of −0.64 percent and −0.829 percent, respectively.

Per capita consumption of strawberries more than doubled between 1970 and 2001. Per capita consumption of frozen strawberries has remained relatively stable over the past thirty years at between 1.2 and 1.7 pounds, but annual per capita consumption of fresh strawberries increased from 1.73 pounds in 1970 to 4.42 pounds in 2001 (USDA, ERS 2004a). The marked increase for fresh strawberries is undoubtedly linked to the increase in total production discussed earlier and particularly to extension of the strawberry season. Today, fresh strawberries are available to U.S. consumers virtually throughout the year.

Increasingly health-conscious consumers have not only increased their consumption of fresh fruits and vegetables but also consume a wider variety. The strawberry industry has benefited from this increased demand. In addition, information on the health benefits of berries (they contain antioxidants that may retard the aging process) has likely shifted demand. Because strawberries account for a large share of total berry consumption, the industry tends to emphasize the overall health benefits of "red berries" and distinguishing strawberries from other berries does not seem to be important.

Table 11.1. Annual Growth Rate of Per Capita Fresh Fruit Consumption, 1970–2001

Dependent Variables	Strawberries	Grapes	Cherries	Peaches and Nectarines	Oranges and Temples
Constant	0.40	1.01	−0.49	1.69	2.70
Standard Error	(0.030)	(0.064)	(0.087)	(0.052)	(0.047)
Growth Rate (percent per year)	3.70	3.90	−0.64	0.15	−0.83
Standard Error	(0.0016)	(0.0034)	(0.0046)	(0.0028)	(0.0025)
R^2	0.95	0.88	0.060	0.0091	0.27
Adjusted R^2	0.95	0.81	0.029	−0.024	0.24

Source: The authors' regression estimations used data from USDA, ERS 2004a.

Demand Enhancement for California Strawberries

The challenge facing CSC is to develop effective promotion strategies that enhance demand for a growing volume of strawberries. Production and yields are rising, and strawberries are now available in increasing amounts in months outside the historical peak season. One promotion tool available to CSC is generic advertising. However, the commission's annual budget for marketing and promotion activities ($3.5–$4.5 million) is not nearly sufficient for a significant, sustained generic advertising campaign at the national level. Recognizing this limitation, CSC has sought to leverage its budget by funding research on the health benefits of strawberries, which may generate results that will increase demand; by linking its direct promotions to other campaigns; by conducting promotion efforts targeted at retailers and food-service-industry members; and by researching handling methods that can improve product quality.

CSC leverages its expenditures by targeting retailers and food-service companies. For example, in 2003 the commission conducted a large survey of consumers regarding their fruit buying and consumption habits and used the information generated in web pages targeted at retailers and food-service companies to encourage them to buy strawberries. For example, survey participants described their response to a lack of strawberries where they typically shop: 17 percent reported that they would go to another store to find them, 20 percent would not choose to buy another fruit as a substitute, and 30 percent would always or often buy a substitute (CSC 2004a). Two additional statistics from the survey may prompt retailers to increase off-peak stocking and promotion of strawberries: 84 percent of respondents said they would be willing to purchase fresh strawberries year round and 85 percent indicated that they could eat strawberries every day.

In 2003, CSC introduced a new promotion campaign. The *Be Well–Get the Red Edge* (*Red Edge*) campaign focuses on the health benefits of strawberry consumption and links those benefits to the appearance and increased availability of fresh strawberries. The *Red Edge* campaign is linked to *Five-A-Day, The Color Way* by the Produce for Better Health Foundation, which is sponsored by CSC and other produce-industry members. The campaigns share a focus on encouraging healthy eating and are connected though website links and the Color Way logo.

The *Red Edge* campaign targets health and nutrition professionals and the consumer and trade media through trade events and media materials that communicate findings regarding the health benefits of strawberries from CSC-sponsored research. CSC is engaging in a partnership with the International Health Racquet and Sports Association called *Get Active America!* that promotes exercise and healthy eating. As part of this activity, CSC will distribute recipe books to participating clubs. By linking its promotion campaign to other ongoing campaigns and focusing on key individuals and groups with influence over consumer decisions, CSC has tried to leverage its direct expenditures and increase the impact of its spending on demand for strawberries.

While it is still too early to evaluate fully the effects of the Red Edge campaign on strawberry demand, the campaign is not the only advertising strategy CSC has used. Most strawberry advertising appears in grocery store circulars rather than as independent generic advertising. Retailer promotion of strawberries in these circulars and through in-store specials, end-of-aisle displays, and other placement choices may also influence demand. Prior to the Red Edge campaign, CSC sought to leverage its promotion dollars by providing retailers with incentives for promoting strawberries. CSC's contribution was identified by a California Strawberries logo somewhere in the circular ad. We used time-series data extending back to 1990 to evaluate the effectiveness of this strategy. Because we lacked a variable reporting total advertising expenditures in the industry, we captured the effect of advertising by using the share of retailers in the national market that promoted fresh strawberries in a given week.

Retailers' promotion decisions follow a seasonal pattern that is influenced by the harvest volumes of strawberries and other fruits and by holidays (Han 2003). One reason CSC sought joint promotion was to encourage retailers to promote strawberries at times when they otherwise would not have done so. Retailers prefer to promote fruits during their harvest peaks because the volume available is relatively high and the price is relatively low. For strawberries, the peak season is late spring and early summer, April through June. Retailers also prefer to promote "new" fruits (Jiang and Goodhue 2003). Apparently, they believe that advertising strawberries at the beginning of the peak season is more likely to bring customers into their stores than advertising done at the end of the season. As other fruits come into peak production (e.g., cherries and peaches), retailers are more likely to feature those fruits instead of strawberries.

Based on underlying weekly data, Figure 11.4 plots the average monthly percentage of the national market that advertised strawberries between 1990 and 2002.[2] The average percentage of the market "on ad" rises during March and April, peaks in May, and then slowly declines through the remainder of the season. Figure 11.5 plots the share of the national market that advertised strawberries each week in 1999, a representative year for our sample period.

Figure 11.4. Average Percent of National Market Promoting Fresh Strawberries by Month, 1990–2002

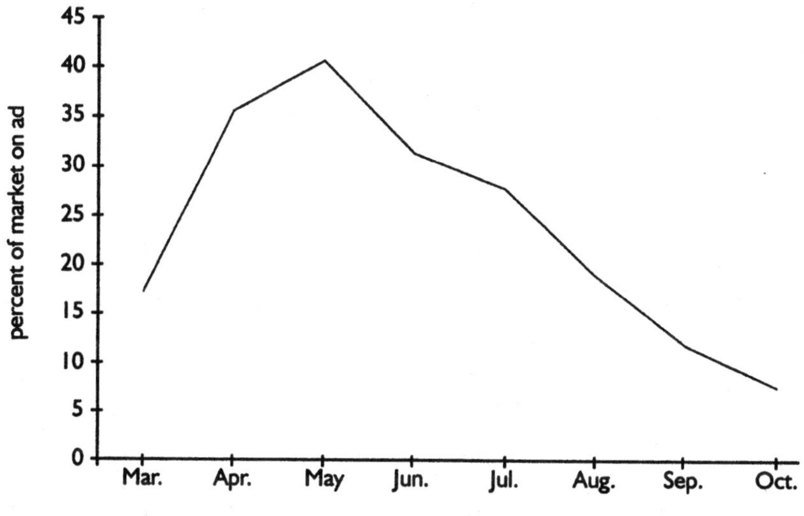

Source: Leemis Marketing database provided by CSC.

Figure 11.5. Percentage of National Market Promoting Fresh Strawberries by Week, 1999

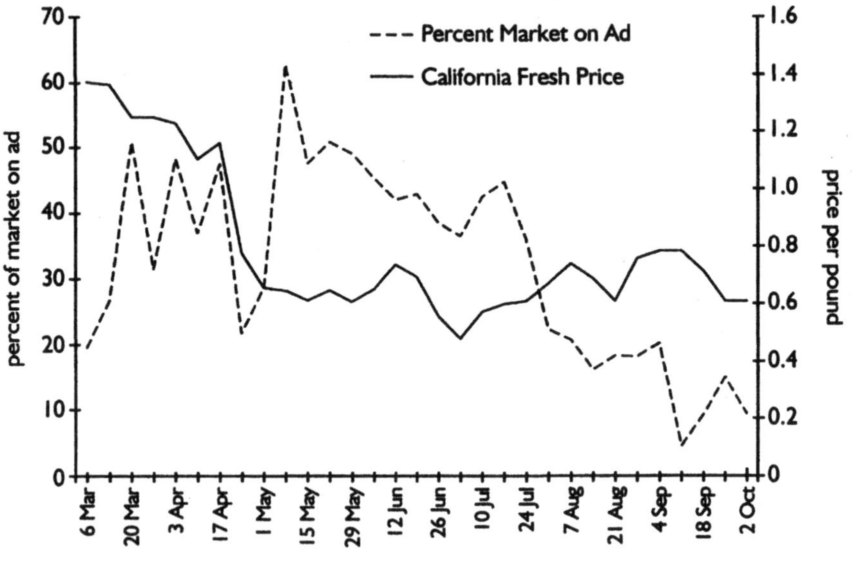

Source: Leemis Marketing database provided by CSC.

The percent of the market that was on ad rose during March, varied weekly during the peak May–June period but was high on average, and slowly declined through the latter part of the season. Figure 11.5 also plots the average F.O.B. price for fresh California strawberries. The price remained low after the peak of the season occurred even though volumes would have declined. Promotion effort was quite limited during the August–September period.

Retailers also like to promote strawberries just before holidays, especially spring holidays—Easter, Mothers' Day, and Memorial Day. Strawberry supplies are normally abundant for these holidays and there is a relatively limited volume of substitute fruits, especially for Easter and Mothers' Day. Figure 11.6 compares the average percentage of the national market that was on ad for holiday and nonholiday weeks. Easter weeks showed a noticeably higher percentage of markets on ad than did other weeks in March; in April, however, weeks other than Easter had the higher percentage of markets on ad. This is likely due to the seasonal increase in volume over the six calendar weeks in which Easter may occur. In years when Easter was in March, 16 percent of markets were on ad during the Easter week. When Easter occurred in April, 32 percent of markets were on ad. Strawberry promotion peaked during Mothers'

Day weeks, when an average of more than 50 percent of the market was on ad, a figure that is substantially higher than for other weeks in May and other times of the season. There were no more ads for the week of the Fourth of July than for subsequent weeks in July.[3] Labor Day showed a noticeably higher percentage of the market on ad than other weeks in September, although it was lower than other weeks in August. It was important for us to take into account the potential for holiday effects on demand in our empirical analysis. If demand for strawberries is different during holiday weeks than for other weeks, excluding the holiday effect from our analysis could cause us to mistakenly attribute their effect to the percentage of the national market on ad.

Empirical Analysis

Model

A price-dependent model for weekly demand seemed appropriate for a perishable crop such as strawberries, for which the total quantity available is determined by decisions made well before the current week. Economic theory predicts that a number of factors affect the price of fresh California

Figure 11.6. Average Weekly Percentage of National Market Promoting Fresh Strawberries during Holiday and Nonholiday Weeks, 1990–2002

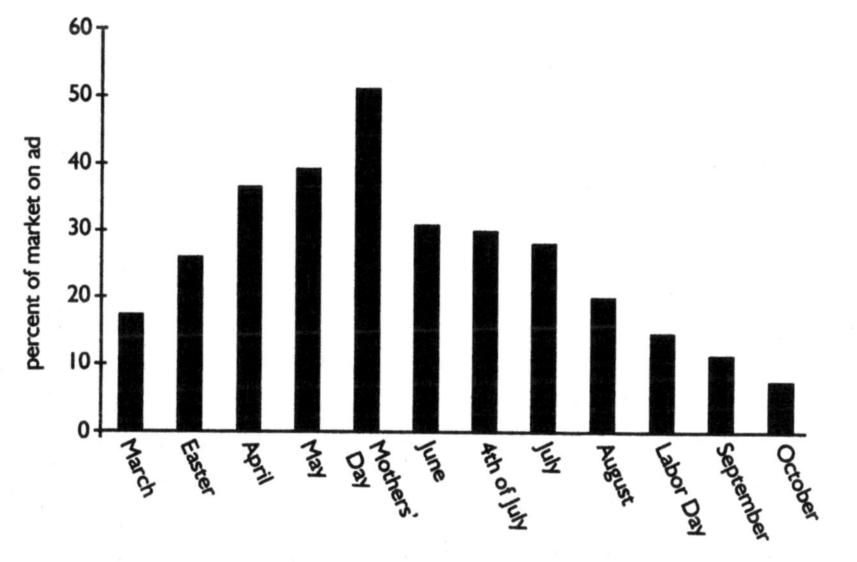

Source: Leemis Marketing database provided by CSC.

strawberries. To assess the effect of promotion on the price of strawberries, we had to control for the effects of these other factors. Prices should decline with increases in volume. When more strawberries are available on the fresh market, the price must fall to attract sufficient buyers. Accordingly, we included the volume of fresh strawberries harvested nationally each week as an explanatory variable for the price of fresh strawberries.

Increased volumes of substitute fruits should decrease the price of fresh strawberries as these substitutes become a relatively more attractive alternative to strawberries for consumers. Beyond the decisions of individual consumers, retailer strategies discussed earlier also suggest that an increase in the volume of substitute fruits will reduce the price of strawberries. Because retailers prefer to promote fruits at their seasonal peaks, high volumes of substitutes correspond to times when retailers choose to promote substitute fruits instead of strawberries. Other things being equal, this should correspond to a lower price for strawberries. The number of potential substitutes was large even if we restricted our attention to fresh fruit, complicating the analysis. Based on information from industry members, we included a few primary substitute fruits in our model: Florida citrus, California table grapes, Washington and California cherries, and California stone fruits (peaches, plums, and nectarines).

By the same token, if advertising of fresh strawberries is to be effective, it must increase the price, so we anticipated that our promotion variable would have a positive coefficient in the fresh-price equation.

We included other variables in our model to reflect additional relationships that seem to reflect the behavior of retailers and other industry members. To the extent that demand increases during holidays only because of retail promotions, there should be no separate holiday effect on strawberry prices—only an effect that is captured by the advertising variable. However, given the possibility that demand may shift due to the holidays themselves, we also used dummy variables for holiday weeks. To the extent that our advertising variable is not the correct one, these dummies could also capture some of the effect of strawberries being promoted around holidays. Similarly, if consumers are attracted to fresh fruits when it is relatively early in the season for those fruits and hence the fruits are "new" and appealing, as retailers tend to believe, then the time of the season could also influence price. A given volume early in the season should be associated with a higher price than that same volume after the peak harvest period. Such seasonal influences may not be captured fully by volumes of substitute fruits. Finally, we included a time trend to capture consumers' hypothesized increasing interest in eating a variety of fresh fruits and vegetables. Our initial model thus took the following form:

> California Fresh Price = *f*(*Fresh Volume, Processing Volume, Substitute Fruit Volumes, Time-of-Season Variables, Year, Percent Market On Ad*).

We examined the role of different variables representing the time of season, including holidays. In one model, we limited the variables to the week of the season and the week of the season squared. In a second model, we included a single holiday dummy variable. In a third model, we included dummy variables for individual holidays—Easter, Mothers' Day, the Fourth of July, and Labor Day.

Initial regression estimates demonstrated autocorrelation, so we included lagged values of the dependent variable and fresh volume in the model. Autocorrelation can, of course, be "pure" autocorrelation in which the lagged error terms have persistent effects on current prices. Or it can be "impure" autocorrelation, which is a symptom of some specification error. We included all observable variables we thought to be economically relevant, but it is possible that there is an unobservable influence on price. Alternatively, one of our data variables may not accurately represent the effect on price that we credited to it. We also may have induced autocorrelation by improperly aggregating daily data to weekly figures, particularly if the days do not match up perfectly in what is reported for prices and quantities. To account for such dynamic effects, we chose to include lagged values of fresh price and fresh volume, resulting in a single-equation model of the following form:

> California Fresh Price = *f*(*Lagged Fresh Price, Fresh Volume, Lagged Fresh Volume, Processing Volume, Substitute Fruit Volumes, Time-of-Season Variables, Year, Percent Market On Ad*).

These single-equation specifications failed Hausman tests for endogeneity. Consequently, we used two-stage least squares to model a system of three equations: fresh price, fresh volume, and processed volume. These tests results were supported by anecdotal evidence from industry members suggesting that, at least near the end of the fresh-market season for each region, growers' decisions to switch from supplying the fresh market to selling to the processed market depend on the relative prices in the two outlets. Even if the price-dependent model with exogenous quantities makes sense earlier in the season, a simultaneity problem might exist for the middle of the season, and the relationship between prices, volumes, and quality is more complicated. Quality is influenced by regional weather, and prices are influenced by supplies available from a grower's region and other regions. For example, as noted earlier, an unusually cool spring in the Watsonville area can prompt growers in a southern region such as Oxnard to remain in the fresh market longer than they would ordinarily.

We restricted our attention to relatively simple models of fresh and processed volumes. Growers were assumed to consider fresh and processed prices in choosing when to transition from the fresh to the processed market, so these prices were expected to influence both fresh and processed volumes. Volumes are also a function of the time of the season, as discussed previously. Finally, planted acres, yields, and the inventory of processing strawberries were included as determinants of processing volume.

Our system of three equations took the following general form.

(1) California Fresh Price = *f*(*Lagged Fresh Price, Fresh Volume, Lagged Fresh Volume, Processing Volume, Substitute Fruit Volumes, Time-of-Season Variables, Year, Percent Market On Ad*).

(2) National Fresh Volume = *f*(*Fresh Price, Processing Price, Time-of-Season Variables, Year*).

(3) California Processing Volume = *f*(*Fresh Price, Processing Price, Time-of-Season Variables, Acres Planted, Yields, Processing Inventory*).

Data

We estimated our models using data for the years 1990 through 2002 and for the weeks between March 1 and September 30. During the years studied, roughly 90 percent of fresh California strawberries and virtually all processing strawberries were produced during those weeks each year. The data for the analysis were compiled from a number of sources. All prices were real prices. The all-urban Consumer Price Index without seasonal adjustment from the Bureau of Labor Statistics was used to deflate nominal prices.

Prices and volumes for fresh strawberries were obtained from USDA's *National Berry Report*, which includes daily fresh strawberry volumes for each production region. For this analysis, the only regions we considered were the four California production regions plus Central Florida. In total, these volumes account for more than 90 percent of U.S. production. Daily volumes were aggregated into weekly volumes.

Processing prices were obtained from PSAB. Processors are required to post their initial prices with PSAB at the beginning of the season and report any price changes during the season. Regional prices were used to calculate weekly prices, which were aggregated into a single statewide processing price. These prices were for Grade Number 1 processing strawberries and did not include prices for strawberries used for juice, which accounted for 12 to 20 percent of processing-strawberry volume during the study period according to PSAB annual reports. Processing volumes were obtained from PSAB's annual reports and supplemented with information obtained directly from PSAB. Volumes were reported by region in terms of processed crates, which weigh approximately eighteen pounds. However, the average number of pounds per crate varied by region within a year and by year within a region. Region-year-specific pounds-per-crate conversion factors from PSAB were used to convert reported weekly volumes into pounds. Processing inventory, acreage, and fresh and processing yields were obtained from PSAB's annual reports.

The percent of the national market that advertised fresh California strawberries was obtained from the Leemis Marketing database and was provided by CSC. This percentage is constructed from data from forty-six city-regions collected by Leemis Marketing that capture strawberry advertisements by selected retailers for the period between March 1 and September 30 each year. Retailers are included in the data if their sales place them within the largest firms accounting for 80 to 90 percent of sales, depending on the market. The monitored markets accounted for 41 to 48 percent of national food sales annually. Advertisements reported here are only those for fresh California strawberries. Advertisements for frozen strawberries and strawberries from other than California were not included. The percentage of the market on ad each week was calculated by multiplying the market share for each retailer promoting strawberries in a given city-region by the city-region's national market share and summing those across all observed retailers.

Volumes of substitute fruits, including cherries, table grapes, stone fruits (peaches, plums, and nectarines), and Florida citrus, were obtained from commodity groups and USDA reports. Cherry volumes were obtained from the California Cherry Advisory Board, the report *Marketing of California Cherries* by USDA's Federal-State Market News Service, the Washington State

Fruit Commission, and Northwest Cherry Growers. Weekly California cherry volumes were constructed from the number of daily shipments of eighteen-pound cartons. Shipments from three production districts were aggregated to obtain the California total: Stockton-Linden-Lodi, Central San Joaquin Valley, and Gilroy-Hollister-San Jose. All four varieties of fresh sweet cherries reported by the Federal-State Market News Service–Bing, Rainier, Lambert, and Vans–were included in calculating California volumes. Washington cherry shipments were reported in tons per week and Rainier cherries were excluded from the Washington totals. Table-grape volumes were obtained from annual reports by the California Table Grape Commission. Volumes were reported by week in lug units, which vary in weight by year for major varieties. Stone-fruit volumes (peaches, plums, and nectarines) were obtained from annual reports by the California Treefruit Agreement. For some years, weekly pack-outs were available. For other years, the weekly pack-out was calculated from daily pack-out figures. All varieties were included. Data for two of the years were problematic. In 1990, plum shippers did not participate in the marketing order, so plums are not included in the data. In 1991, shippers' reporting of volumes was voluntary and it is estimated that only about 75 percent of the total volume of stone fruit was reported. Florida citrus volumes were obtained from annual statistical reports by the Citrus Administrative Committee, which reports weekly shipments of cars containing one thousand four-fifth-bushel (forty-pound) cartons.

Results

In Tables 11.2, 11.3, and 11.4, we report results from our general three-equation model only. Table 11.2 reports coefficients for the three alternative specifications of the fresh-price equation. Coefficient values and approximate significance levels were consistent across models. The lagged fresh-price variable had a positive and statistically significant effect on fresh price. Fresh volume had a negative and statistically significant effect on price. Lagged volume had an insignificant effect on price. Processing volume had a positive effect on price that was weakly significant (at the 10 percent level) in two specifications. Citrus volume had a significant and positive effect on price, which is the opposite of the anticipated effect. This may be because large citrus volumes are seasonally associated with small strawberry volumes during the first few months of the year. By the time strawberries reach their seasonal peak, the volume of citrus available is small. Cherry volume had a negative effect on price, as predicted. It was statistically significant in only two specifications, although its effect was

Table 11.2. Fresh-Price Equation Results (Standard Errors in Parentheses)

Variable	Model 1	Model 2	Model 3
Constant	−33.97[b]	−34.51[b]	−34.19[b]
	(8.82)	(8.84)	(8.42)
Lagged Fresh Price	0.76[b]	0.76[b]	0.76[b]
	(0.043)	(0.043)	(0.043)
National Volume	−0.016[b]	−0.016[b]	−0.016[b]
	(0.0054)	(0.0054)	(0.0051)
Lagged National Volume	0.0018	0.0017	0.0018
	(0.0041)	(0.0041)	(0.0038)
Processing Volume	0.0050[a]	0.0050[a]	0.0047
	(0.0030)	(0.0030)	(0.0030)
Citrus Volume	0.0042[b]	0.0043[b]	0.0041[b]
	(0.0015)	(0.0015)	(0.0015)
Stone Fruit Volume	−0.00047	−0.00044	−0.00043
	(0.000383)	(0.00038)	(0.00037)
Cherry Volume	−0.0039[b]	−0.0038[b]	−0.0030
	(0.0017)	(0.0017)	(0.0019)
Table Grape Volume	−0.00099	−0.0012	−0.0012
	(0.00096)	(0.00097)	(0.00095)
Percent On Ad	0.0039[b]	0.0040[b]	0.0040[b]
	(0.00075)	(0.00075)	(0.00074)
Year	0.017[b]	0.017[b]	0.017[b]
	(0.0044)	(0.0045)	(0.0042)
Week	0.014	0.013	0.014
	(0.013)	(0.013)	(0.012)
Week Squared	−0.00049	−0.00047	−0.00048
	(0.00036)	(0.00036)	(0.00034)
Holiday	–	0.053	–
	–	(0.033)	–
Easter	–	–	0.035
	–	–	(0.034)
Mothers' Day	–	–	0.074[b]
	–	–	(0.036)
Fourth of July	–	–	−0.035
	–	–	(0.038)
Labor Day	–	–	0.053
	–	–	(0.033)

[a] Significant at the 10 percent level.
[b] Significant at the 5 percent level.

consistent across all three models. Stone-fruit and table-grape volumes did not have statistically significant effects on the price of strawberries. The time-trend variable (*Year*) had a positive and statistically significant effect on price, which suggests that the combined effects of consumers' increased interest in consuming fresh produce and other trends have positively affected the price of strawberries. Week and week-squared variables were insignificant, which suggests that effects of the time of year, if any, were fully represented by volumes of strawberries and other fruits and perhaps by the percent of national markets on ad and holiday variables. The single holiday dummy variable was insignificant in Model 2. In Model 3, only the Mothers' Day dummy variable was positive and statistically significant. Easter and Labor Day had insignificant positive coefficients, and the Fourth of July had an insignificant negative coefficient. The percent of the national market on ad had a positive, significant effect on price. The magnitude of this effect was not affected by inclusion of either set of holiday variables.

Tables 11.3 and 11.4 report the results for the fresh-volume and processing-volume equations, respectively. For both equations, only one specification was tried. We report three separate sets of estimates only for consistency with the

Table 11.3. Fresh-Volume Equation Results (Standard Errors in Parentheses)

Variable	Model 1	Model 2	Model 3
Constant	-3,703.62[b]	-366.77[b]	-3,696.1[b]
	(234.9)	(233.9)	(233.8)
Fresh Price	-46.78[b]	-45.97[b]	-45.89[b]
	(6.98)	(6.91)	(6.89)
Lagged Fresh Price	20.00[b]	19.37[b]	19.31[b]
	(5.80)	(5.74)	(5.73)
Processing Price	9.07	8.96	8.95
	(5.97)	(5.94)	(5.94)
Year	1.88[b]	1.88[b]	1.88[b]
	(0.12)	(0.12)	(0.12)
Week	0.043	0.046	0.046
	(0.33)	(0.33)	(0.33)
Week Squared	-0.037[b]	-0.037[b]	-0.037[b]
	(0.0091)	(0.0090)	(0.0090)

[a] Significant at the 10 percent level.
[b] Significant at the 5 percent level.

Table 11.4. Processing-Volume Equation Results (Standard Errors in Parentheses)

Variable	Model 1	Model 2	Model 3
Constant	−1,209.19 (746.9)	−1,200.81 (745.9)	−1,216.92 (746.1)
Fresh Price	−14.41[a] (7.76)	−14.19[a] (7.70)	−14.62[a] (7.69)
Lagged Fresh Price	−5.79 (6.13)	−5.95 (6.087)	−5.64 (6.081)
Processing Price	41.76 (26.12)	41.58 (26.10)	41.92 (26.11)
Processing Price Squared	29.92 (89.99)	30.47 (89.94)	29.41 (89.99)
Acres Planted	0.000025 (0.00055)	0.000031 (0.00055)	0.00002 (0.00055)
Yield	−0.027 (0.0023)	−0.0027 (00.23)	−0.0028 (0.0023)
Yield Squared	−2.7[a] 10^{-8} −2.2[a] 10^{-8}	−2.7[a] 10^{-8} −2.2[a] 10^{-8}	−2.7[a] 10^{-8} −2.2[a] 10^{-8}
January 1 Inventory	0.015 (0.018)	0.016 (0.018)	0.015 (0.018)
Year	0.064[a] (0.39)	0.64 (0.39)	0.65[a] (0.39)
Week	1.77[b] (0.39)	1.77[b] (0.39)	1.77[b] (0.39)
Week Squared	−0.070[b] (0.010)	−0.070[b] (1.010)	−0.070[b] (0.010)

[a] Significant at the 10 percent level.
[b] Significant at the 5 percent level.

three price equations as changes in the sets of exogenous variables included in the price equation imply changes in the instrumental variables available for the two volume equations. Given the consistency of results across the price equations, it would have been surprising for results to have varied much between different volume equations. The statistical significance and magnitudes of the coefficients were indeed consistent across the models. As reported in Table 11.3, the fresh price had a negative and statistically significant effect on fresh volume while the lagged fresh price had a positive and significant effect.

The time-trend variable had a positive and significant effect. The week-squared variable had a negative and statistically significant effect, although the week variable was insignificant. The processing price had an insignificant effect.

In the processing-volume equations, the only statistically significant variables were the week, which had a positive effect, and the week squared, which had a negative effect. Fresh price had a negative effect that was weakly significant (at the 10 percent level) in all three specifications.

All of the three-equation models implied price flexibility with respect to volume of −0.65, corresponding to a demand elasticity of −1.54. This price flexibility is within the range of truncated-season flexibilities estimated by Han (2003) for the 1990–1998 period. These figures must be interpreted with caution because data from only a subset of the season, March to early October, were used for Han's analysis and for our own. However, most of the annual volume is harvested during these months, as shown in Figure 11.1.

We estimated a price flexibility of advertising of 0.16. While this number is small in absolute value, it is important to recognize that it represents an effect that is highly unlikely to be large in absolute terms. Intuitively, a 1 percent increase in the share of the national market on ad is unlikely to increase the national price of fresh California strawberries substantially on a percentage basis. Because the market for fresh strawberries is a regional rather than a true national market, the effect of the 1 percent increase in the share of the national market on ad could only influence consumers in the relevant market or markets, not consumers nationwide. Han (2003) obtained a somewhat smaller price flexibility of the percent market on ad—an estimated value of 0.101. He also evaluated the effect of another advertising variable, the number of ads nationally. This variable most likely represents a poorer proxy for promotion effort since it does not account for the number of consumers potentially exposed to the ad. For the same truncated season, Han's estimated price flexibility with respect to the number of ads was 0.235. Hence, our estimate fell roughly halfway between his two estimated flexibilities.

These estimates represent short-run values. Longer-run values could be calculated with an interpretation of the lagged dependent variable that implies a persistence of effects. Typically, elasticities from a model estimated with an autocorrelation correction would not be adjusted in that manner. Since we had no expectations about any partial adjustment of strawberry prices and included a lagged dependent variable due to apparent autocorrelation, we did not place sufficient confidence in dynamic effects to draw such a distinction between short-run and long-run effects of promoting strawberries. To the extent that there is persistence of promotion's effectiveness, therefore, our results understate the effects of promotion. The frequency with which produce is advertised in typical retailer newspaper circulars suggests that persistence of the effects from weekly advertisements is not important.

Overall, our empirical results confirmed some of our expectations but not all. Fresh price and fresh volume had the expected negative relationship. Fresh price had a negative relationship with processed volume, although the relationship was only weakly significant. Advertising had a positive relationship with fresh price. Of the holidays, only Mothers' Day had the predicted positive effect. The findings for substitute fruits were inconclusive. Finding a consistent pattern of substitution between fresh fruits is difficult, however. Their peak seasons do not coincide, so a good substitute for strawberries in some weeks may not be available in others. The relationship may also depend on the supply conditions for other fruits. Thus a stable relationship between fruits is unlikely to be present throughout the season, and intuition from commodities such as beef (for which pork and chicken serve as substitutes year round) may be misleading.

Marginal Benefit of Advertising

The effect of advertising expenditures on sales or growers' returns is the standard for evaluating the payoff from generic promotion. Statistical significance of advertising effects, of course, only implies that promotion *might* pay. It was difficult to construct a comparable rate-of-return measure in this study, however, because we did not know the cost to CSC of increasing the share of the national market on ad by a percentage point. While we had information about the amount CSC expended on promotion, we did not know the corresponding amount spent by retailers. If we were to use CSC expenditures to calculate a benefit-cost ratio, the ratio would be grossly overstated *as a measure of the benefit-cost ratio of advertising strawberries*. It would only capture the benefit-cost ratio for CSC if retailer spending was guaranteed to increase along with CSC spending. We did not have any way to assign causality to CSC's efforts for any particular promotion by retailers. Hence, our estimates of the marginal benefits from promotion require a slightly different interpretation from those in other chapters in this book.

The elasticity of fresh strawberry demand with respect to the percent of the national market on ad, which was derived from the price flexibilities we estimated, was 0.24 in Model 1 and 0.25 in Models 2 and 3. Thus, a 1 percent increase in the percent of the national market on ad increases the price of strawberries by an estimated 0.16 percent when holding the quantity of strawberries sold constant. Alternatively, a 1 percent increase in the percent on ad boosts the quantity of strawberries sold by 0.25 percent when holding price constant and other things being equal. Calculations of benefit-cost ratios for promotion require an assumption about the supply elasticity to capture the incidence of a levy on producers correctly. When supply is upward-sloping, some of the cost of promotion is passed along the marketing chain. We considered the case where there is a vertical supply curve. For fixed supplies, all of the incidence of a tax to fund promotion is borne by producers and then it is reasonable to compare the increase in industry profits to the cost of the promotion effort. The effect of promotion in our model seemed likely to consist of two separate effects. As noted earlier, we expected the direct effect of promotion to be an increase in the willingness-to-pay for a given quantity of strawberries. A second effect, also related to promotion, is that retailers promoting strawberries are likely to price them very competitively–perhaps even as loss leaders–so there is an additional increase in demand facing shippers. With fixed supplies, our estimate of the effect of promotion on the F.O.B. price can be interpreted as a combined effect from promotion's direct effect on willingness-to-pay for a given volume in markets where promotion occurs and a further increase in demand for strawberries faced by shippers because retailers on ad likely decrease prices as well. For fixed national wholesale volumes, this can only occur with greater competition for existing shipments and, hence, with an increase in the F.O.B. price.

For an increase in the percent of the national market on ad to be cost-effective, it must cost CSC no more than the increase in industry revenue associated with the increase in advertising. Our estimated elasticities implied a marginal revenue benefit associated with a 1 percent increase in the share of the national market on ad of somewhat more than $25,000, assuming that the weekly quantity of strawberries produced was perfectly inelastic.

These figures can be used to evaluate whether CSC's actual expenditures on joint promotions with retailers were cost-effective for the strawberry industry exclusive of retailer costs. For example, if CSC expenditures induced a retailer with 5 percent of the national market to promote strawberries in a week in which it would not otherwise do so, industry revenues would increase by about $129,000, assuming that the weekly supply of strawberries is perfectly inelastic and does not respond to price. If CSC's expenditures were less than this amount, then it was cost-effective for the industry to fund the promotion.

Another way of using our estimates to evaluate the cost-effectiveness of CSC expenditures on joint promotions was to estimate the cumulative change in industry revenue and compare that change to CSC's annual promotion budget. This approach assumes that CSC's entire promotion budget was spent on joint promotion and that all revenue effects were realized as increased prices for fixed weekly quantities. According to our estimates, CSC's total annual effect on promotion would have to be equivalent to between 68 and 87 percent of the national market going on ad for two weeks more than would have occurred in the absence of CSC efforts. This requirement could be met by inducing retailers that account for 20 percent of the national market to promote strawberries for seven to eight weeks more than they would have done in the absence of CSC expenditures.

Conclusions and Implications

Production and marketing of fresh strawberries has changed substantially over the past few years, and the next several years are likely to see even more dramatic changes. Retailers and shippers are increasingly using contracts with set prices or price ranges rather than relying on spot markets for price determination. Growers are increasingly integrated with shippers. The expanding use of one-pound clamshell packaging in place of traditional pint containers is providing shippers with the ability to brand their product at the consumer level, giving them a greater incentive to emphasize product quality. To protect quality, shippers are increasingly developing proprietary varieties and becoming more involved with production and harvesting. Shippers are beginning to team with retailers to engage in brand-specific promotion of strawberries.

Such developments are characteristic of an industry transitioning from producing a commodity to producing differentiated products. This transition will transform every aspect of the industry, including the nature of advertising. In addition to fresh strawberries competing with citrus, cherries, stone fruit, and other fruits, individual shippers will be competing with each other. While generic advertising is intended to increase total demand for fresh strawberries, individual shippers will also be interested in increasing their shares of the overall market. As in other industries, shippers who have invested in a brand may question the relative value of contributing to generic advertising.

Given that CSC leverages many of its promotion expenditures through coordination with other campaigns and with retailers, it is difficult to assess whether an increase in branded advertising associated with a decrease in generic advertising will benefit the industry as a whole. Our econometric analysis focused on retailer promotions that were done jointly with CSC. We found that an increase in the percent of the national market promoting fresh California strawberries increased the F.O.B. price of fresh strawberries. A shift from promoting California strawberries to promoting specific brands of strawberries through specific retailers could potentially have similar effects. Brand-specific advertising may increase the total size of the fresh strawberry market rather than just increasing the brand's market share. Such growth could occur simply because the advertising causes consumers to purchase strawberries at a time when they otherwise would not have done so, regardless of brand. Moreover, there may be some spillover effects. For example, promotion in store circulars may be seen by consumers who ultimately go to a different store and purchase strawberries other than those seen in the ad.

For generic advertising to complement branded advertising, it perhaps should focus on increasing the size of the market by promoting general strawberry attributes such as nutritional value rather than attributes such as firmness, color, or freshness that consumers likely link to a brand. The most effective generic message will presumably depend on the nature of any brand identities that are developed to the extent that shippers are successful in creating such identities. CSC's new promotion strategy focusing on health appears to be consistent with this broad recommendation. Monitoring and evaluating the effects of advertising in this rapidly changing industry present continuing challenges.

Notes

The authors thank Chris Christian, Nancy Ramm, Janice Sharp, Christopher Winterbottom, and the other industry members, including growers, shippers, and retailers, who graciously shared their knowledge of the strawberry industry. This chapter is based on research funded by the California Strawberry Commission, the Giannini Foundation of Agricultural Economics, and the Agricultural Marketing Center. Gregory McKee and Wensi Zhang provided valuable research assistance.

1 For the 2004 season, 30.8 percent of total California strawberry acreage was planted in proprietary varieties (CSC 2004b).

2 The data for Figure 11.4 do not span the entire season. We obtained these data from CSC and the data only cover March through September when California production is at its peak.

3 Because the data were weekly and labeled by the first day of the week, many July Fourth weeks were classified as part of June. Similarly, many Labor Day weeks were classified as part of August.

References

California Strawberry Commission. "Consumer News." Watsonville CA, 2004a. Available online in August 2004 at *www.calstrawberry.com/product/consumer.asp*.

—. *Strawberry Review: 2004 Acreage Survey Results*. Watsonville CA, 2004b. Available online August 2004 at *www.calstrawberry.com/filedata/docs/2004_acreage_survey.pdf*.

Han, F.M. *Seasonal and Weekly Price Determination in a Market for Perishables: An Econometric Model of the California Strawberry Industry*. PhD dissertation, University of California, Davis, 2003.

Jiang, J., and R.E. Goodhue. "Prices, Volumes and Promotions in the Fresh Strawberry Market." *ARE Update* 6(3) 2003:9–11.

U.S. Department of Agriculture, Agricultural Marketing Service. *National Berry Report*. Washington DC, annual issues.

U.S. Department of Agriculture, Economic Research Service. Food consumption (per capita) data. Washington DC, 2004a. Available online August 2004 at *www.ers.usda.gov/data/foodconsumption/datasystem.asp*.

—. *U.S. Strawberry Industry (95003)*. Washington DC, 2004b. Available online August 2004 at *www.ers.usda.gov/data/sdp/view.asp?f=specialty/95003/*.

U.S. Department of Agriculture, National Agricultural Statistics Service. *Agricultural Statistics: 2003*. Washington DC, 2003.

Wells, M. *Strawberry Fields: Politics, Class and Work in California Agriculture*. Ithaca NY: Cornell University Press, 1996.

12

National Benefit-Cost Estimates for the Dairy, Beef, Pork, and Cotton Promotion Programs

Henry W. Kinnucan & Yuqing Zheng

This chapter focuses on the costs and benefits of national promotion programs for dairy, beef, pork, and cotton. These programs are of interest because they account for the lion's share of the total funds raised for agricultural product promotion (Forker and Ward 1993, pp. 102–103; Vande Kamp and Kaiser 1999). In 2001, the programs in question generated expenditures in excess of $500 million and accounted for some 83 percent of the total funds available from national check-offs (Table 12.1). California is a significant contributor to all four programs and is the major contributor to the dairy program. The programs are of special interest from a research standpoint in that they have been the target of sophisticated econometric analysis and thus provide perhaps the best basis for determining whether promotion pays. This is especially true for dairy, since the 1983 federal legislation authorizing the program requires that it be evaluated annually. Hence, a rich history exists upon which we can draw inferences about program impacts.

Since 1990, all national promotion programs with mandatory assessments have been required by law to be independently evaluated every five years. This chapter focuses on evaluations submitted to the U.S. Department of Agriculture for the 1995–2000 evaluation period. Evaluations for the dairy, beef, and pork programs were conducted under contract by university researchers; the cotton evaluation was conducted by Research Triangle Institute in collaboration with university researchers. Each evaluation was subjected to some form of peer review. In the case of pork and cotton, peer review was built

Table 12.1. Funds Available under National Check-Off Programs and Implied Promotion Intensities, 2001

Program	Budget (million $)	Farm Value (million $)	Intensity
Dairy	244.0	21,950	0.0111
Fluid Milk	109.5	10,500	0.0104
Beef	87.9	25,050	0.0035
Soybeans	61.4	14,700	0.0042
Cotton	60.2	5,278	0.0114
Pork	54.6	10,267	0.0053
Eggs	18.8	4,380	0.0043
Peanuts	18.7	985	0.0190
Potatoes	8.6	2,660	0.0032
Honey	3.6	145	0.0248
Mushrooms	2.6	NA	NA
Watermelons	1.6	286	0.0056
Popcorn	0.4	NA	NA
All	671.9	-	-

Source: Budget data from Armbruster and Nichols (2002); farm value from government sources through the website *www.census.gov/prod/2004pubs/03statab/agricult.pdf.*

into the evaluation design. For dairy and beef, peer review was less formal and the dairy program's evaluation involved U.S. Department of Agriculture (USDA) economists. The three dairy evaluation studies submitted to USDA so far in fulfillment of the five-year evaluation requirement have been published in peer-reviewed scholarly journals, lending them added credibility.[1]

Our chief purpose is to provide a synopsis of and to critique mandated evaluations for the dairy, beef, pork, and cotton promotion programs. In doing so, we present pertinent results from the literature to provide context. The critique focuses on five issues in promotion evaluation that were addressed with varying degrees of success in the mandated evaluations: lag structure, functional form, omitted-variables bias, spillover effects, and robustness. Spillover, for example, refers to the effect of beef promotion on demand for pork and poultry and its consequent effect on relative prices in the meat group. Failing to take into account promotion-induced changes in relative prices can prejudice an analysis in favor of the program. Similarly, an incorrect functional form for the demand equation or an inappropriate lag structure for the promotion variable can cause estimated promotion effects to be biased, which in turn biases the estimated benefit-cost ratios. Our critique addresses these and related issues in some detail with a primary goal of suggesting areas for attention and potential improvement in future evaluations.

A striking aspect of the studies is the diversity of methods used to compute benefit-cost ratios. For example, only the cotton study takes into account the shifting of tax incidence (the tendency for a portion of a promotion assessment to be shifted to consumers in the form of higher prices). The pork study takes into account the time value of money, a variable that was ignored in the other studies. The dairy study defines cost as total check-off income while the remaining studies exclude administrative and other expenses. The upshot is that the benefit-cost ratios generated are not comparable across studies. To address this problem, and to provide a consistent basis for relating economic costs to benefits, we introduce two short-cut formulas for computing benefit-cost ratios that are consistent with economic theory. The formulas are applied to the information provided by the mandated studies to demonstrate usefulness and to indicate the relative profitability of the four programs.

We begin with a review of the mandated studies. We then present standardized benefit-cost ratios based on the short-cut formula, followed by our critique and a summary of key findings.

Review of the Studies

A critical aspect of any benefit-cost analysis is the quality of the underlying science. That, in turn, depends on the quality of the data and methods used to obtain estimates of the demand shifts associated with a promotion effort. Accordingly, prior to discussion of the benefit-cost ratios themselves, we review the methods used in obtaining the estimates. After summarizing their data, we summarize the elasticities and other estimates that served as key inputs in the benefit-cost calculations.

Data and Demand Specifications

Data and demand specifications for thirteen studies are summarized in Table 12.2. Five of the studies, listed in boldface, were submitted in fulfillment of the five-year reporting requirement; the remaining eight were published in professional journals or books.[2] For ease of discussion, the studies in boldface are hereafter referred to as "mandated studies" and the rest as "published studies."[3]

Focusing first on the promotion data, the major difference between the published and mandated studies is that the mandated studies have tended to use more recent data. Mandated studies of the beef and cotton programs included data through 2000, the dairy study data carried through 1999, and the pork evaluation data extended through 1998. The data used in the published studies for cotton and fluid milk terminated in 1993 and 1994, respectively. For beef and pork, where the published literature is more extensive, all but one of the studies used data that predated 1996.

In other areas, the mandated and published studies are similar. Specifically, both used a variety of data intervals that ranged from monthly to annual and considered both brand and generic advertising. A unique aspect of the beef evaluations is that one of the published studies used split-cable scanner data. These data permit a unique comparison of households exposed to advertising with households that were not exposed and thus mimic a controlled experiment.

Turning to consumption data used in the studies, we note that the mandated studies for dairy, beef, and pork were exemplary in that household data were used in combination with and independently from time-series data to estimate promotion responses.[4] The published studies, with the exception of one study of the beef program by Benson, Breidt, and Schroeter (2002), were based strictly on time-series analysis. Because time-series analyses are based on aggregate consumption data, they are inherently less informative than a household-level analysis. In this sense, the mandated studies (except for cotton) potentially provided a sounder basis for drawing inferences about promotion impact than the published studies.

In terms of demand specification, a basic difference between the mandated and published studies is that the mandated studies rely on single-equation models to estimate promotion response. In essence, these studies do not take into account spillover and cross-commodity substitution effects, an issue that we discuss in some detail later. With respect to functional form, the published studies used either flexible forms or forms that permitted diminishing returns to promotion. By contrast, the mandated studies for pork and cotton used a linear specification that implied constant returns, another issue that we discuss in more detail later. Finally, whereas the lag structures in the published studies permitted promotion to have carryover periods extending up to one year, the mandated cotton study restricted the promotion response to one month (i.e., no lag) and the pork study implied that promotion does not affect demand until eleven calendar quarters (nearly three years!) after the initial expenditure. The issue of lag structure, too, is discussed in more detail later.

Table 12.2. Characteristics of the Dairy, Beef, Pork, and Cotton Studies

Promotion Data			Consumption Data		Demand Model	
Period	**Interval**	**Type**	**Type**	**Source**	**Functional Form**	**Ad Lag Structure**
Kaiser 2000 – Fluid Milk and Cheese						
1984–1999	Quarterly	Brand and Generic	Aggregate	Government	Double Log	PDL with 2- to 5-Quarter Lag
Schmit et al. 2002 – Fluid Milk and Cheese						
1996–1999	Monthly	Brand and Generic	Household Panel	A.C. Nielsen	Linear Square Root	PDL with 6-Month Lag
Kinnucan et al. 2001 – Fluid Milk, Juices, Soft Drinks, Coffee, and Tea						
1970–1994	Annual	Brand and Generic	Aggregate	Government	Flexible	No Lag (annual data)
Ward 1996, 2001 – Beef						
Market Model						
1987–2000	Quarterly	Generic Only	Aggregate Panel	Government	Double Log	Flexible, 1-Quarter Lag
Household Models						
1984–2000	Bimonthly	Generic Only	Aggregate Panel	NPD Group	Linear Square Root	None
1993–2000	Monthly	Generic Only	Aggregate Panel	NPD Group	Linear Square Root	None
Benson et al. 2002 – Beef						
1985–1987	Monthly/ Split Cable	Generic Only	Household Panel	IRI	Linear	12-Month Distributed Lag with Fixed Weights
Brester and Schroeder 1995 – Beef, Pork, Poultry, and Other Goods						
1970–1993	Quarterly	Brand and Generic	Aggregate	Government	Flexible (Rotterdam)	QEL with 1- to 3-Quarter Lag

Continued on following page

Table 12.2 (continued)

Promotion Data			Consumption Data		Demand Model	
					Functional	Ad Lag
Period	Interval	Type	Type	Source	Form	Structure
Kinnucan et al. 1997 – Beef, Pork, Poultry, and Fish						
1976–1993	Quarterly	Generic Only	Aggregate	Government	Flexible (Rotterdam)	Unrestricted Distributed 1-Quarter Lag
Piggott 1997 – Beef, Pork, and Poultry						
1979–1995	Quarterly	Generic Only	Aggregate	Government	Semi-Log (AIDS)	Unrestricted, 5-Quarter Lag
Boetel and Liu 2003 – Beef, Pork, Poultry, and Fish						
1976–2000	Quarterly	Generic Only	Aggregate	Government	Semi-Log (AIDS)	Restricted, 1-Quarter Lag
Davis et al. 2000 – Pork						
Market Model						
1987–1998	Quarterly	Generic Only	Aggregate	Government	Linear	Simple Lag, 11 Quarters
Household Model						
1994–1996	Annual	Brand and Generic	Household Surveys	USDA	Linear-Log	Simple Lag, 1 to 4 Quarters
Hyde and Foster 2003 – Pork						
1970–1993	Quarterly	Brand and Generic	Aggregate	Government	Linear VAR	Geometric Lag, 1 and 4 Quarters
Murray et al. 2001 – Cotton						
1986–2000	Monthly	Brand and Generic	Aggregate	Government	Linear	PDL with 0- to 3-Month Lags
Ding and Kinnucan 1996 – Cotton						
1976–1993	Quarterly	Generic Only	Aggregate	Government	Double Log	Geometric Infinite Lag

Note: Studies in shown **boldface** were submitted in fulfillment of the five-year congressional reporting requirement. PDL: Polynomial Distributed Lag. QEL: Quadratic Exponential Lag. AIDS: Almost Ideal Demand System.

Elasticity Estimates

The key scientific question to address in promotion evaluation studies is whether one can reject the null hypothesis that promotion has no effect on market demand. If an affirmative result is obtained, attention then focuses on the magnitude of the demand shift, which is generally measured by the promotion elasticity combined with the observed increase in promotion expenditure. In assessing the first question, our study used the Rule of Two, which states that the promotion effect is not zero if the estimated coefficient (or sum of coefficients in the case of a finite lag structure) is at least twice as large as its estimated standard error. The reason for using the Rule of Two rather than reported significance levels is that the latter generally overstate actual significance due to pretest estimation.[5]

On this basis, the main difference between the mandated and published studies is that the published studies tended to show a higher incidence of insignificant effects (Table 12.3). In particular, of the ten coefficients estimated in the mandated studies, only one (for cheese in the Kaiser study (2000)) failed to satisfy the Rule of Two. By contrast, among the published studies, only two of fourteen estimates satisfied the rule. Too much should not be made of these results as all but two of the published studies related to beef and pork. Still, it is interesting to note that the one published study showing a significant effect (for cotton) used a single-equation demand model. The remaining published studies, with the exception of Hyde and Foster (2003), which used a vector autoregressive (VAR) model, were based on demand systems. Because demand systems have a stronger theoretical foundation for modeling promotion effects than single-equation models do (Baye, Jansen, and Lee 1992; Brown and Lee 1992; Duffy 2001, 2003; Piggott et al. 1996), it is troubling that the results from single-equation models tended to contradict those from demand systems. At a minimum, the conflicting results raise concerns about whether the findings from the single-equation models are real, an issue we address later.

Perhaps not surprising in light of the preceding discussion, the promotion elasticities estimated in the mandated studies tended to be much larger than corresponding elasticities in the published literature (Table 12.3). In particular, the mandated studies reported promotion elasticities ranging up to 0.11 in the time-series models and up to 0.28 in the household models. In the published studies, which related strictly to time-series data, only two elasticities were above 0.03 (one for pork at 0.034 and one for cotton at 0.066). The figures for the remaining published estimates were minute, less than 0.007. For fluid milk, the most heavily advertised farm commodity in the group, the

Table 12.3. Elasticity Estimates and Related Information from the Dairy, Beef, Pork, and Cotton Promotion Studies

		Elasticity Estimates			Simulated Farm-Price Effect	Benefit-Cost Ratios	
Author	Product	Own-Adv'g (α)	Own Price ($-\eta$)	Ratio (α / η)		Marginal	Average
Kaiser 2000	**Fluid Milk**	0.051*	-0.170*	0.300	6.5%	NA	4.29
	Cheese	0.015	-0.320	0.047			
Schmit et al. 2002	**Fluid Milk**	0.150*	-0.243*	0.617	NA	NA	NA
	Cheese	0.240*	-0.654*	0.367			
Kinnucan et al. 2001	Fluid Milk	0.0018	-0.192*	0.0094	NA	NA	NA
Ward 1996, 2001	**Beef - Market Model**	0.028*	-0.670*	0.042 -5.2%	2.25%	5.7-9.7	5.70
	Beef - Household Model	(-)*	(-)*	-	NA	NA	NA
Benson et al. 2002	Ground Beef	(-)	-0.685*	-	NA	NA	NA
	Roasts	(-)	-2.750*	-			
	Steaks	(-)	-1.362*	-			
Brester and Schroeder 1995	Beef	0.006	-0.56*	0.011	NA	NA	NA
	Pork	-0.0005	-0.69*	-			
Kinnucan et al. 1997	Beef	-0.0003	-0.44*	-	NA	NA	NA
	Pork	0.00006	-0.68*	0.00009			
Piggott 1997	Beef	0.004	-0.950	0.004	NA	15.5	NA
	Pork	0.034*	-0.694	0.049	NA	237.0	NA

Continued on following page

Table 12.3 (continued)

		Elasticity Estimates			Simulated Farm-Price Effect	Benefit-Cost Ratios	
Author	**Product**	**Own-Adv'g (α)**	**Own Price $(-\eta)$**	**Ratio (α / η)**		**Marginal**	**Average**
Boetel and Liu 2003	Beef	−0.00004	−0.44*	–	NA	NA	NA
	Pork	0.0067	−0.48*	0.014			
Davis et al. 2000	**Pork – Market Model**	0.11*	−0.76*	0.140 or 4.3%	$1.99/cwt	NA	16.0
	Pork – Household Model	0.11*– 0.28*	NA	–	NA	NA	NA
Hyde and Foster 2003	Pork	0.005	NA	–	NA	NA	NA
Murray et al. 2001	**Cotton – Promotion**	0.023*	−0.41/ −0.69	0.026	NA	4.6	NA
	Cotton – Nonag. Research	0.351	–	0.404	280.3	NA	NA
Ding and Kinnucan 1996	Cotton	0.066*	−0.29*	0.230	NA	1.4–5.1	NA

Notes: Studies in **boldface** were submitted to fulfill the five-year congressional reporting requirement. Promotion elasticities are for generic promotion only. An asterisk (*) indicates that the elasticity is "significant" in the sense that the estimated coefficient is at least twice its estimated standard error. Parentheses (–) mean that the variable was included in the model but no elasticity estimate was provided; if the parentheses include an asterisk, the estimated coefficient was significant in the preceding sense and had the right sign. Simulated farm-price effects are for the total program, not product-specific impacts. Piggott estimates are those summarized in Alston, Chalfant, and Piggott (2001). The α / η ratio for cotton is based on total quantity elasticities; see text for additional explanation and qualifications.

mandated studies reported a promotion elasticity of 0.051 from the time-series model and 0.15 from the household model; the corresponding estimate from the (published) demand-systems model was 0.002. The discrepancies in estimates between the mandated and published studies for beef and pork were even larger.[6]

Given that the published studies were based on time-series data, it is of some significance that the results based on household data in the mandated studies supported the hypothesis that promotion has shifted the demand curves for the respective commodities (Table 12.3). What is disconcerting, however, is the size of the promotion effects implied by the household-level analyses. In particular, the dairy study estimated promotion elasticities of 0.15 and 0.24 for fluid milk and for cheese, respectively, and the pork study estimated an elasticity for all pork of between 0.11 and 0.28. (No elasticities were reported in the beef study using household data, and the cotton study used only time-series data.)

To appreciate the economic implication of the elasticity estimates from the household models, consider the Dorfman-Steiner theorem. This theorem, which applies to a competitive industry when promotion is funded by a check-off, as well as to a monopoly (Alston, Carman, and Chalfant 1994), may be stated as follows:

(1) $\theta^* = \alpha / \eta,$

where $\theta^* = A / PQ$ is the ratio of promotion expenditures to industry revenue that maximizes industry profit or producer surplus, α is the promotion elasticity, and η is the absolute value of the demand elasticity.[7] In the dairy study, the estimated values for η obtained from the household model were 0.24 for fluid milk and 0.65 for cheese. Substituting these values into Equation (1) along with the aforementioned promotion elasticities (0.15 and 0.24) suggests that, to maximize profit, the fluid-milk industry should spend 62 percent of its revenue on promotion and the cheese industry should spend 37 percent![8] Although the pork study does not provide an estimate of η from its household model, using the estimate of η = 0.76 from the time-series model for pork yields similarly implausible optimal intensities. ("Intensity" is defined as the ratio of promotion expenditures to industry revenue, A / PQ.) For this reason, the household estimates must be interpreted with caution.[9]

Returning to the models estimated with times-series (as opposed to household) data in the mandated studies, the elasticity estimates tended to be more reasonable when viewed from the perspective of the Dorfman-Steiner theorem. For example, the α / η ratios for cheese, beef, and cotton were between 0.03 and 0.06, which suggests an optimal intensity of about 5 percent for these commodities. By way of comparison, observed intensities were 1.1 percent each for dairy and cotton and 0.35 percent for beef (Table 12.1). That observed intensities were less than optimal intensities is not unexpected since collective goods such as generic promotion tend to be underprovided relative to the economic optimum (Hardin 1982). Still, the estimates of α / η for fluid milk and pork of 0.30 and 0.14, respectively, in the mandated studies are troubling. Not only do they vastly exceed the corresponding observed intensities of 0.016 and 0.005,[10] but the implied price enhancement is implausible. For example, the estimate $\alpha / \eta = 0.30$ implies that a 100 percent increase in promotion expenditure would increase price by 30 percent when quantity is fixed. The notion that a doubling of promotion expenditures could cause fluid-milk prices to rise in the short run (when supply is fixed) by 30 percent seems far-fetched. (Either that or the fluid-milk industry is missing out on a golden opportunity to increase profits.) A similar inference applies to the pork estimate. Since the demand-system estimates of promotion and price responses reported in Table 12.3 imply optimal intensities more in line with observed intensities (the one exception is the Piggott (1997) estimate for pork), one might conjecture that single-equation demand models exaggerate promotion effects.

Benefit-Cost Estimates

With the foregoing caveat in mind, we now turn to the benefit-cost estimates themselves. In interpreting these estimates, it is important to bear in mind the distinction between *marginal* and *average* return. Marginal return is the profit generated by the *last* dollar spent on promotion. It is typically measured using elasticities estimated from the model or by simulating the model for some small increase in promotion expenditure. The marginal return does not indicate profitability per se but rather whether the level of promotion spending is too high or too low. In particular, industry profit is maximized (ignoring opportunity cost) when the marginal benefit-cost ratio (BCR) is 0. Thus, a small value for the marginal BCR does not mean low profits. In fact, if the demand function permits diminishing returns, as is the case in the mandated studies for dairy and beef, profits will *increase* as the marginal BCR approaches 0.

Average return, by contrast, measures the profit for *every* dollar spent on promotion. It is typically measured by simulating the model for the full range of promotion expenditures and then dividing the associated total profit by the total promotion outlay. Like the marginal BCR, the average BCR falls as promotion expenditures approach the economic optimum (given diminishing marginal returns). However, at the economic optimum, where the marginal BCR equals 0, the average BCR is generally positive and must exceed 1 for the program to be profitable.

With these distinctions in mind, we first focus on the estimates of average return that were provided in three of the four mandated studies–for dairy, beef, and pork. In all cases, the reported average BCRs suggest that promotion is highly profitable (Table 12.3). For the dairy study, which referred strictly to the dairy-farmer assessment (the fluid-milk-processor assessment was not included due to data limitations), the average BCR was 4.29. This estimate implies that dairy farmers receive $4.29 in profit per check-off dollar expended. The 4.29 figure excludes nonmedia components of the check-off program (e.g., nutrition education and new-product research) that may have contributed to profits. Moreover, it assesses the full cost of the promotion program (about $244 million; see Table 12.1) against media promotion. Since media promotion represented less than 60 percent of the $244 million outlay, the estimate of 4.29 understated the actual return to dairy farmers from media promotion and may have understated total program return as well. (The latter depends on the extent to which nonmedia activities strengthened consumer demand for dairy products or lowered production costs.)

For beef, the reported average BCR was 5.7. A caveat in interpreting this estimate is that it was a *net* average BCR–promotion costs were subtracted from profit before calculating the ratio. Hence, the proper benchmark for determining profitability in this case was that the average BCR exceeded 0 (not 1). As with dairy, estimates of the average BCR for beef pertained strictly to media-advertising and consumer-information components of the program; other activities funded by the beef check-off were not reflected. With these caveats in mind, it appears that beef promotion is highly profitable.

For pork, the reported average BCR was 16 with a range of between 4.8 and 26.2 (Davis et al. 2000, p. II-36). These estimates referred to all activities funded by the pork check-off, including farm-level production research and foreign-market development. They were *net* estimates–program costs were subtracted from profit (producer surplus) prior to computing the ratios. They differed from the ratios reported for dairy and beef in that a discounting formula was used to take the time value of money into account. The assumed annual interest rate was 5 percent. In addition to total program benefits, the pork study provided project-specific benefit-cost ratios (Davis et al. 2000, p. II-36): 18 for pork-demand expenditures, 157 for postfarm-research expenditures, and −5 for farm-research and education expenditures.[11] From these results, it appears that, although the total program was highly profitable, returns would have been even larger with a budget allocation that favored postfarm research over farm-level research and education.

Since an average return was not reported in the mandated study for cotton, we focused on the estimate for the marginal return. Recalling that the demand function used in the cotton study was linear, the marginal estimate implicitly assumed that marginal returns were constant. With this caveat, the total program's marginal BCR was estimated at 44.5, which means that the *last* dollar expended generated $44.50 in producer surplus at the farm level (measured in constant 2000 dollars). Hence, based on this estimate, expenditures for promotion and nonagricultural research–the evaluated program elements–were too low in the sense that larger expenditures would have increased total producer profit. Project-specific marginal BCRs indicated that nonagricultural research (280.3) was more profitable at the margin than promotion (4.6) (Murray et al. 2001, pp. 6–7). This suggests that a reallocation of the budget in favor of nonagricultural research would have resulted in a larger total return.

In addition to being a marginal return, the cotton estimate differs from the estimates for dairy, beef, and pork in that the former takes into account the shifting of tax incidence. That is, the marginal BCRs for cotton include only the portion of the check-off expenditure actually borne by producers; the portion of the assessment shifted to consumers via the price mechanism was excluded from the calculation. This explains, in part, the very high marginal return for the total cotton-promotion program.

Returning to beef, which was the only study that provided estimates of *both* marginal and average returns, the marginal BCR ranged from 5.7 to 9. This estimate was based on the household "servings" model (Ward 2001) and thus is not strictly comparable to the average BCR, which was based on the time-series model. In addition, the beef estimate did not take tax shifting into account and thus is not comparable to the cotton estimate. Given these caveats, the marginal return for beef appears to exceed the average return, which was estimated at 5.7. Recalling that the demand model for beef imposed diminishing returns, this suggests that the beef program is inefficient in the sense that a larger budget would increase average as well as total profits.

Standardized Benefit-Cost Estimates

Given the variety of methods used to compute the benefit-cost ratios shown in Table 12.3, it was not possible to rank the programs in terms of profitability. To address this issue, and to provide a more consistent basis for comparing economic costs with benefits, we computed benefit-cost ratios for the four programs using two alternative formulas. The formulas, which we derive in an appendix that is available upon request, were based on the principle that promotion must increase farm price by more than the per-unit check-off to be profitable (Alston, Carman, and Chalfant 1994). In other words,

$$\Delta P > T,$$

where ΔP is the absolute rise in farm price associated with the promotion effort and T is the per-unit assessment. Dividing the preceding equation by P and multiplying the right-hand side by Q / Q yields:

$$\Delta P / P > TQ / (PQ),$$

where $\Delta P / P$ is the *relative* increase in farm price and TQ is the total dollar amount raised by the check-off program. By setting $A = TQ$, the preceding equation can be written more simply as:

$$(2) \quad \Delta P / P > \theta,$$

where $\theta = A / PQ$ is promotion intensity.

According to Equation (2), promotion is profitable whenever the relative increase in farm price exceeds promotion intensity. This condition serves as the basis for the first formula:

$$(3) \quad BCR^* = P^* / \theta,$$

where $P^* = \Delta P / P$ and BCR^* is the associated benefit-cost ratio. Since the denominator of Equation (3) indicates the *total* cost of the promotion effort, the numerator is measured as the price increase associated with the *total* promotion outlay. As such, BCR^* is properly interpreted as an *average* (not marginal) benefit-cost ratio. Moreover, since the denominator assigns the full cost of promotion to the producer, P^* should be measured with supply fixed to avoid the need to account for tax shifting. (Farmers bear the full incidence of the promotion assessment only when supply is fixed.)

To apply Equation (3), one must know the percentage increase in farm price associated with the promotion program. In instances where this information is not available, the benefit-cost ratio can be approximated using elasticities as follows:

(4) $BCR' = (\alpha / \eta) / \theta,$

where α / η measures the relative vertical shift in the demand curve due to a 1 percent increase in promotion expenditure (see note 7). Equation (4) indicates that it is always more profitable to shift a less elastic demand curve, which is consistent with the Dorfman-Steiner theorem (see Equation (1)).[12]

Equations (3) and (4) were applied to the information given in the four mandated studies as indicated in Table 12.4. Because not all check-off income is spent on promotion, the ratios were computed for two measures of intensity: one based on promotion expenditures ($\theta = A / PQ$) and one based on check-off income ($\psi = TQ / PQ$). Although the latter measure overstates the actual cost of promotion, it is consistent with the measure used in the dairy and pork studies and thus provided a basis for comparison.

First, for the estimates based on Equation (3), three of the studies provided estimates of the farm-price effect of promotion: dairy, 5.5 percent to 9.8 percent; beef, 2.5 percent to 5.2 percent; and pork, 4.3 percent. The 5.5 percent estimate for dairy corresponds to that portion of promotion funded strictly by the dairy-farmer check-off while the 9.8 percent estimate corresponds to promotion funded by a combination of the dairy-farmer and fluid-milk-processor check-offs. The 2.5 percent estimate for beef was based on a portion of the sales response function from a midlevel of promotion expenditure to a higher level as observed over the evaluation period; the 5.2 percent estimate was based on the entire sales response function from zero to the observed level. With these caveats in mind, and using the ψ-based measure of costs, we calculated that *BCR** ranges from between 5.3 and 6.4 for dairy to between 7.1 and 14.9 for beef; pork was in the middle at 8.3. The 5.3 estimate for dairy corresponds to the "farmer only" program, which can be compared to the 4.3 estimate for the program provided in the mandated study. Similarly, the 7.1 estimate for beef corresponds to the 2.3 percent price enhancement, which served as the basis of the 5.7 estimate provided in the mandated study. Recalling that the estimate in the mandated study is a *net* benefit-cost ratio (promotion costs subtracted from profit before calculating the ratio), the implied *gross* BCR is 6.7,[13] which compares closely to the 7.1 estimate that was based on the standardized formula.

The *BCR** estimate for pork of 8.3, by contrast, is considerably below the estimate of 16 provided in the mandated study, especially since the latter is a net benefit-cost ratio. One explanation may be differences in the treatment

Table 12.4. Standardized Benefit-Cost Ratios for Dairy, Beef, Pork, and Cotton Based on Elasticity and Price Estimates Provided in the Mandated Studies and Promotion Intensities for 2000/01

Commodity	Estimated Price Effect (P^*)	Promotion Intensity ($\theta = A/PQ$)	Check-Off Intensity ($\psi = TQ/PQ$)	Ratio α/η	BCR* P^*/θ	BCR* P^*/ψ	BCR′ $(\alpha/\eta)/\theta$	BCR′ $(\alpha/\eta)/\psi$
Fluid Milk	NA	0.0160[1]	-	0.300	-	-	18.8	-
Cheese	NA	0.0065[1]	-	0.047	-	-	7.2	-
Dairy – Total Program	0.098	0.0092	0.0153	-	10.6	6.4	-	-
Dairy – Farmer Only	0.055	0.0050	0.0103	-	11.0	5.3	-	-
Beef	0.025	0.0013	0.0035	0.042	19.2	7.1	32.3	12.0
	0.052	-	-	-	40.1	14.9	-	-
Pork	0.043	0.0027	0.0053	0.140	15.9	8.3	51.8	26.4
Cotton – Promotion	NA	0.0076	0.0114	0.026	-	-	3.4	2.3
Cotton – Nonag. Research	NA	0.0018	0.0114	0.404	-	-	224.4	35.4
Cotton – Combined[2]	NA	-	-	-	-	-	45.4	8.6

[1] Assumes the farm value of milk used for fluid and cheese in each case is $9 billion.
[2] Weighted average of returns to promotion and nonagricultural research; see text for details.

of costs (recall that the pork study used a discounting formula). Indeed, the *BCR** estimate for pork based on the θ measure of cost is 15.9, which is nearly identical to the mandated study's estimate of 16.0. Since $\theta < \psi$, this suggests that the mandated study may have understated costs. (The pork study's estimate of 16 is a composite for all check-off-funded activities. Hence, the proper basis for cost is check-off income, not promotion expenditure per se.) In any event, it appears from benefit-cost estimates based on Equation (3) that, among the three programs, beef was the most profitable and dairy was the least, although all showed a substantial return. Replacing ψ with θ in the *BCR** formula magnified the estimated BCRs but did not alter the ranking.

Price estimates were not available for cotton so here we had to rely on Equation (4). Since the *BCR′* estimate was based on elasticities, it was important that they be defined so that the BCR could be interpreted properly. Two issues were salient: the market level at which the elasticities were measured and whether the elasticities reflected total market response. For cotton, α and η were both estimated at the mill level and, thus, *BCR′* applied to that level of the market. The estimate $\eta = 0.41$ pertained to the domestic market, as did the two estimates for α—one for promotion expenditures (0.023) and one for nonagricultural research expenditures (0.351). Although the cotton study included an export-demand equation, it did not attempt to determine whether international promotion activities affected market demand. Hence, the *BCR′* estimates for cotton refer strictly to demand enhancements in the domestic market at the mill level.

We have to reinterpret Equation (4) to accommodate the fact that, while the assessment is collected on all cotton, some is exported and the demand enhancements from promotion and nonagricultural research apply only to the domestic market. In this setting, the elasticity of overall demand response to promotion is equal to the elasticity of domestic demand response to promotion multiplied by the fraction of the U.S. cotton crop used by domestic mills (k), such that $\acute{\alpha} = k\alpha$ replaces α in Equation (4). The demand elasticity in the denominator is overall demand for U.S. cotton, which is equal to the share-weighted average of the domestic elasticity of demand (η) and the export elasticity of demand (η_x). Assuming that the Law of One Price holds approximately, $\acute{\eta} = k\eta + (1 - k)\eta_x$ replaces η. Using the estimate of η_x from the mandated study, $\eta_x = 0.69$ and $\eta = 0.41$. Setting $k = 0.60$, the approximate domestic quantity share over the evaluation period, yields $\acute{\eta} = 0.52$, the value used to compute the $\acute{\alpha} / \acute{\eta}$ ratio reported in Table 12.4. Finally, the intensity remains equal to expenditures on the program in question as a share of the total value of production.[14]

In 2000, expenditures for cotton promotion were $40.3 million, yielding an intensity of 0.0076. Substituting this value into Equation (4) along with the associated $\acute{\alpha} / \acute{\eta}$ ratio yields *BCR′* = 3.4 for the promotion portion of the campaign. The corresponding estimate for marginal return from the mandated study is larger, 4.6, which suggests that the promotion program was inefficient; a higher level of spending would increase average as well as total profit. The average profit function is concave downward and attains a maximum when *Average BCR* = *Marginal BCR* (see, e.g., Mansfield 1985, pp. 159–162). Hence, if *Marginal BCR* > *Average BCR*, as appears to be true for cotton, the industry is on the rising portion of the average profit function. To maximize total profit, the industry must be on the falling portion of the average profit function where *Average BCR* > *Marginal BCR*. A similar result was obtained for nonagricultural research, which in 2000 represented an expenditure of $9.4 million. Substituting the implied intensity of 0.0018 into Equation (4) along with the associated $\acute{\alpha} / \acute{\eta}$ ratio yields an average BCR of 224.4, which is smaller than the marginal BCR of 280.3 reported in the mandated study.

Since promotion accounts for 81 percent of the combined promotion and nonagricultural-research expenditures for cotton, the implied combined average BCR is 45.4, which is comparable to the mandated study's estimated combined marginal BCR of 44.5 (Murray et al. 2001, pp. 6–7). Overall, it appears from these estimates that the cotton promotion program, like the dairy, beef, and pork programs, is highly profitable. Comparing the combined return for cotton to estimates for dairy, beef, and pork based on Equation (4), it appears that pork is the most lucrative program (51.8), followed by cotton (45.4), beef (32.3), and fluid milk (18.8). Replacing θ with ψ causes beef to supplant cotton in the ranking but does not alter the basic conclusion that benefits exceed costs for all programs.

Critique

Because no single study is definitive, the benefit-cost ratios shown in Tables 12.3 and 12.4 should not be accepted uncritically. Here we suggest areas for attention and possible improvement in future evaluations. Our critique focuses on five issues: lag structure, functional form, omitted-variables bias, spillover effects, and robustness.

Lag Structure

Clarke (1976), in his survey of the econometric literature on promotion response, introduced the concept of a "duration interval": "A *'p%' duration interval* is defined as that number of periods during which *p* percent of the expected cumulative advertising effect has taken place" (Clarke 1976, p. 346). Based on a review of seventy published studies, Clarke (1976, p. 355) reached the following conclusion:

> The published econometric literature indicates that 90% of the cumulative effect of advertising on sales of mature, frequently purchased, low-price products occurs within 3 to 9 months of the advertisement. The conclusion that advertising's effect on sales lasts for months rather than years is strongly supported.

While the results of the mandated dairy and beef studies were consistent with this finding, the results of the pork and cotton studies were not. In particular, the cotton study, which used monthly data to estimate the mill-demand equation, indicated no lag for promotion and a three-month lag for nonagricultural research (Murray et al. 2001, pp. 5–10). This implies no carryover for promotion expenditures and a three-month carryover of research expenditures, neither of which is realistic. For one thing, cotton products are semi-durable, which implies relatively long interludes between repeat purchases, a basic explanation for the duration interval (Clarke 1976, p. 347). Thus, one would expect the promotion lag for cotton to be *longer* than Clarke's three- to nine-month norm, not shorter. That research expenditures exhaust their effect on demand within four months of the initial expenditure is inconsistent with lags for agricultural research and development (Alston et al. 2000) and appears to be too short even in sectors where innovation cycles are short, such as for computer chips.

An opposite problem exists with respect to the pork case. There, the duration interval for promotion is estimated to encompass eleven calendar quarters with no response in the first ten quarters. The estimated pork-demand equation, with quarterly dummy variables suppressed, was (Davis et al. 2000, p. II-45):

$$(5)\quad Q_t = 10.58 - 1.32^{**}\, PP_t + 0.62^{**}\, PB_t - 0.10\, PC_t - 0.024^{**}\, GBA_t + 0.014^{**}\, GPA_{t-11} + \mu_t,$$

where $t = 1, 2, \ldots, 40$ (1989:4–1998:4); Q_t is per capita U.S. consumption of pork in quarter *t*; PP_t, PB_t, and PC_t are U.S. price indices for pork, beef, and poultry, respectively, in quarter *t*; GBA_t is generic promotion expenditures for beef in quarter *t*; GPA_t is generic promotion expenditures for pork in quarter $t - 11$; and μ_t is the regression residual in quarter *t*. The coefficients in Equation (5) marked with a double asterisk are significant at the 5 percent level or lower.

As can be seen from the equation, the promotion variables are significant and have correct signs. That is, pork promotion increases pork demand and beef promotion decreases pork demand. What is odd is that the cross-effect occurs immediately (within the same quarter as the promotion expenditure) while the direct effect does not take hold until nearly three years (eleven quarters) after the expenditure. The latter contradicts Clarke's finding (1976) and fails the common-sense test.

An additional problem with Equation (5) is that it did not include income (or meat expenditures), a variable clearly specified by demand theory. This, coupled with its linear form (addressed next), suggests that Equation (5) was misspecified. Since Equation (5) served as the basis for the benefit-cost estimates for pork provided in Table 12.3, it is difficult to have much confidence in the results, especially since most of the published studies were consistent in showing generic promotion of pork to have no effect on pork demand.[15] The implausible lag structures for cotton and pork suggest that the benefit-cost estimate for these commodities must be regarded with some skepticism.

Functional Form

In their study of the sales-advertising relationship, Simon and Arndt (1980) state:

> Our survey of the literature shows that almost all writers on advertising subscribe to one or the other of the following two proposed shapes of the response function: (1) the concave-downward function and (2) the S-shaped logistic function. The concave function [implies] monotonically diminishing returns...the S-shaped advertising response function...first has increasing returns and then, after an inflection point, diminishing returns...The theoretical root of the concave-downward curve is in the microeconomic "law" of diminishing returns to productive inputs...On the aggregate level, the argument is that as the amount of advertising is increased, the unreached prospects will have progressively weaker buying predispositions. (1980, p. 12)

This statement is followed by a review of more than one hundred published studies, after which Simon and Arndt concluded that the advertising response function is characterized by diminishing returns. In particular, the hypothesis that advertising exhibits increasing returns was rejected by the empirical evidence.

The implication is that promotion variables should enter the demand model in an appropriate nonlinear form. This was done in the dairy and beef studies but not in the pork and cotton studies. In particular, the latter two studies specified the demand equation in linear form, which implies constant returns. Constant returns may be plausible in a situation where the response function is S-shaped and promotion expenditures over the estimation period are near the inflection point. (The inflection point refers to the region of the sales-response function where it turns from convex to concave. It indicates the transition from increasing to decreasing marginal returns.) But this is a special case and appears to be especially inappropriate for cotton and pork given the large changes in those industry's promotion expenditures over the estimation periods. The cotton study, for example, used data for 1986 through 2000 to estimate the model, a period in which nominal promotion expenditures increased from $12.3 million to $40.3 million. Moreover, as shown in Kinnucan (1983, p. 15), a linear form implies that the promotion elasticity increases with the level of promotion expenditure, which contradicts the notion of a satiation effect. That is, the linear form implies that consumers become more, rather than less, responsive to promotion as the expenditure level increases. The upshot is that the demand models in the pork and cotton studies likely were misspecified due to incorrect functional form (see, e.g., Theil 1971, pp. 540–555), which may explain, at least in part, the implausible lag structures discussed earlier.

Omitted-Variables Bias

In Equation (5), the coefficient of the beef promotion variable, GBA_t, is negative and significant. This means that, if beef promotion is statistically correlated with pork promotion as has been suggested (e.g., in analyses by Alston, Freebairn, and James (2001) and by Kinnucan and Myrland (2002)), then omitting beef promotion from the pork-demand equation would cause the estimated pork promotion effect to be biased.[16]

This observation is important because the pork study is the only mandated study that considered cross-commodity promotion effects. For example, the beef study excluded pork advertising and the dairy study excluded soft drink advertising in its milk-demand specification. This may explain the large promotion elasticities estimated in the mandated studies relative to estimates in the published studies, which were based on demand-system specifications, most of which included multiple commodity-promotion variables. For example, the promotion elasticity estimate from the mandated beef study was 0.028; corresponding estimates from the published studies were all less than 0.007 (Table 12.3). For fluid milk, the mandated study indicated a promotion elasticity of between 0.051 and 0.15; the corresponding estimate from the published study that included equations for soft drinks, juices, coffee, and tea was 0.002. Overstating the own-promotion effect will, in general, cause the benefit-cost ratio to be overstated, a fact that must be borne in mind when interpreting the estimates provided in Tables 12.3 and 12.4.[17]

Spillover Effects

Spillover refers to the cross-commodity impacts of promotion. For example, an increase in generic beef promotion has been shown to decrease demand for poultry (Brester and Schroeder 1995, p. 977; Kinnucan et al. 1997, p. 21). The decrease in poultry demand lowers the price of poultry, which erodes the demand for beef due to second-round or "feed-back" effects. The failure to take into account these feedback effects tends to cause the returns to promotion to be overstated. This is especially true when supplies of the promoted and substitute goods are fixed, as was assumed in the mandated study for beef.[18] A second problem with ignoring spillover effects is that one obtains only a partial measure of promotion impact. For example, the decrease in demand and price of poultry caused by beef promotion implies a welfare loss in the poultry sector. This loss is not reflected in an analysis that considers only beef demand. The upshot is that ignoring spillover effects can cause returns to promotion to be overstated, both for the promoted commodity and, in a broader context, when the welfare of competing producers is taken into account.

None of the mandated studies took into account spillover effects because they were all based on single-equation demand models. For example, in the pork study we know from Equation (5) that beef promotion affects pork demand. What we do not know is whether pork promotion affects beef demand since the study did not include a demand equation for beef. Consequently, the benefit-cost ratios shown in Tables 12.3 and 12.4 are likely biased estimates of the private returns to promotion (setting aside the issues raised earlier regarding lag structure, functional form, and omitted-variables bias) and say little about the social returns.[19]

Robustness

Tomek (1993) argues that replication (i.e., estimating the same model with new or updated data) is essential for sound policy, learning, and research innovation. The only mandated study to take advantage of this wisdom was the one for beef, where the same time-series model was re-estimated using three sample updates that encompassed a three-year period (Ward 1996, p. 63). In that particular case, the promotion effect remained significant in all three updated models. Other studies have found the opposite. For example, in a replication experiment involving fluid milk, Tomek and Kaiser (1999) found that sample updating caused the promotion effect to vanish. A similar finding was obtained by Kinnucan et al. (1997) with respect to beef promotion. Coulibaly and Brorsen (1999) found that estimated beef promotion effects are sensitive both to model specification and the promotion-data source (whether provided by the industry or obtained from a commercial tracking service). These findings are important because they corroborate findings based on U.S. and Australian meat consumption data that suggest that estimated advertising effects may be fragile (Piggott 1997; Piggott et al. 1996), a fact that decision-makers must consider when interpreting the results of econometric studies.[20]

A related issue is specification search–adding and deleting variables from the model until a satisfactory result is obtained. In the mandated cotton and pork studies, for example, brand promotion was found to be insignificant and was deleted from the models. One problem with such "data dredging" is that a nominal significance level of, say, 5 percent overstates the actual level of significance. (This conclusion is based on statistical decision theory; for a good discussion, see Wallace 1977.) That nominal significance levels overstate actual significance levels is particularly important in the cotton study because the promotion variable's t-ratio ranged from 1.58 to 2.00, depending on the estimation method (Murray et al. 2001, p. 5-1), which means that the effect could well be insignificant when specification search is taken into account. Brester and Schroeder (1995) found brand promotion of pork to be significant, which suggests that the model in the mandated study for pork may suffer from omitted-variable bias.

Specification search raises the specter that an estimated model reflects sample properties rather than actual market behavior. One way to guard against this possibility is to make replication an integral part of the research design. A strength of the mandated studies for dairy, beef, and pork is that household data were used in conjunction with time-series data to test for promotion effects. Household data typically are plentiful, which permits the use of separate samples for specification search and final-model estimation. Unfortunately, this was not done. Hence, we do not know the extent to which the estimated promotion effects were reflective of household behavior or the product of specification search.

Concluding Comments

Promotion responses associated with farm-commodity check-off programs are difficult to estimate. The basic reason is that the responses are tiny *a priori*. Specifically, a basic insight from Dorfman and Steiner's classic study (1954) is that the absolute value of the market demand elasticity sets the upper limit on promotion elasticity. Since the demand for most agricultural products is price-inelastic, especially at the farm level, it follows that promotion responses must be inelastic. In fact, a good rule of thumb is that the estimated promotion elasticity should be no larger than about 5 percent of the commodity's demand elasticity.[21] Thus, for example, if the demand elasticity for fluid milk is −0.2, an estimate in line with the studies reviewed in this chapter, the expected upper limit on the fluid-milk promotion elasticity would be 0.01. A parameter of this magnitude is not easy to estimate, even in the best of circumstances where promotion and related data are measured without error. It should not be surprising, therefore, that studies differ in their conclusions about the efficacy of farm-funded promotion schemes.

The good news is that, given the tiny expenditures in relation to farm value indicated in Table 12.1, it does not take much of a demand shift for programs to be profitable. This is especially true for beef and pork, where check-off income represents less than 0.40 percent and 0.60 percent, respectively, of farm revenue. In these instances, a price rise of 0.41 percent and 0.61 percent, respectively, at the farm level would be sufficient for the programs to be remunerative. The actual price increases estimated in the mandated studies reviewed in this chapter were 2.25 percent or higher, which provides some latitude for overstatement without reversing the conclusion that benefits exceed costs. A similar inference applies to the dairy program, where the estimated farm-price effect of between 5.5 percent and 9.8 percent was substantially higher than the 1.1 percent intensity level indicated in Table 12.1. The cotton study did not provide an estimate of the farm-price effect, so the margin for error there is less certain. However, given cotton's assessment rate of 1.1 percent (Table 12.1), the program would need to increase the farm price by only 1.2 percent to be profitable. Moreover, since about one-third of the check-off income comes from imported cotton products, from the perspective of the domestic producer, a price increase of 0.9 percent might suffice. The point here is that modest price effects are compatible with profitable programs, which should be of some comfort to producers and program managers.

In summary, the benefit-cost estimates reviewed in this study must be treated with caution owing to outstanding issues with respect to model specification and estimation. Still, there is reason to believe that the private benefits generated by the programs exceed costs. What is less certain is the extent to which the programs are socially beneficial from the perspective of consumers who help finance the programs through tax shifting and of the total farming community. To address the latter issue, a more encompassing approach to program evaluation is necessary. A useful first step in this direction would be to insist that the programs be evaluated using multiple-equation rather than single-equation demand models.

Notes

Appreciation is expressed to Julian Alston, George Davis, Paul Patterson, and Richard Sexton for incisive comments on earlier drafts of this chapter and a related paper presented at the NEC-63 meetings in Baltimore, Maryland, March 2004. Responsibility for final content rests strictly with the authors.

1 The studies in question are Kaiser (1997, 2000) and Schmit et al. (2002, 2003). The results of these studies are summarized in USDA's July 2001 *Report to Congress*, which also contains an analysis of fluid-milk promotion by a marketing research firm. Because the latter analysis is subjective, it was not included in this review.

2 The latter studies were selected for their recentness and relevance. By relevance we mean studies that relate to national programs as opposed to state programs and that focus on aggregate effects as opposed to product-, media-, or market-specific effects.

3 A reviewer objected to this nomenclature, arguing that it is ambiguous (some mandated studies have been published) and may convey the impression that published studies are better because they have undergone formal peer review. Acknowledging these limitations, suffice it to say that the latter connotation is unintended.

4 Household data provide a rich context for promotion evaluation (Quilkey 1986, p. 47; see also Leontief 1993, p. 4), especially when the data include information on the consumer's awareness of the advertising, known as "ad recall" (Jensen and Kesavan 1987; Kinnucan and Clary 1995; Kinnucan and Venkateswaran 1990). Such data, however, are costly to obtain and often proprietary, which explains why so few evaluation studies based on these data appear in the literature.

5 Pretest estimation refers to selecting the final model on the basis of a test outcome, a common occurrence in empirical research, as is evident from a reading of the mandated studies. Wallace (1977, p. 443) states that "standard errors, t-tests, etc., as reported in a final regression after pretesting...are misleading and are surely biased in the direction of indicating significance when in fact there may not be any." The Rule of Two is a modest attempt to guard against such bias.

6 Because the mandated beef study did not report an advertising elasticity, the 0.028 value in Table 12.3 was derived using an arc-elasticity formula and estimated price effects reported in the study. Details are in an appendix available upon request from the authors.

7 Equation (1) also indicates the *vertical* shift in the demand curve due to a 1 percent change in advertising expenditure (i.e., the shift in the price direction while holding quantity constant). To see this, let $Q = D(P, A)$ where Q is quantity demanded, P is price, and A is advertising expenditure. Taking the logarithmic total differential yields $d\ln Q = -\eta d\ln P + \alpha d\ln A$. Solving for price change yields $d\ln P = -(1 / \eta)d\ln Q + \theta d\ln A$ where $\theta = \alpha / \eta$ is the vertical shift, as claimed.

8 As pointed out by a reviewer, the household data exclude away-from-home consumption, which accounts for about 35 percent of commercial disappearance of fluid milk and perhaps 65 percent for cheese. Depending upon the away-from-home segments' responses to price and promotion, the optimal intensities implied by the cross-section elasticity estimates may understate or overstate the true optima.

9 An additional caveat in interpreting results from the household models is that the promotion effects were estimated using extraneous information. Specifically, because ad-recall data were not available in the panels, this information had to be proxied using aggregate promotion-expenditure data (see, e.g., Schmit et al. 2002, 2003), which reduces, and in some cases eliminates, cross-section variation. Thus, the information gains in the household models may be more apparent than real, at least with respect to the promotion variables.

10 The 1.6 percent intensity for fluid milk included some \$45 million invested in fluid-milk advertising from the dairy-farmer check-off (see Table 12.1).

11 The authors attributed the negative returns to farm research and education to a "very inelastic" demand at the farm level. Specifically, they stated (Davis et al. 2000, p. II-36) that "shifting out the supply curve on such inelastic demand can result in lower returns at the farm level." Strictly speaking this is true only if the shift is not parallel.

12 Equation (4) was derived by applying the Theorem of the Mean (Hadar 1971, p. 128) to Equation (3) (details are available in the aforementioned appendix). An identical formula was derived by different methods in Alston et al. (1998) and used in their evaluation of the California prune promotion program.

13 To see this, let gross $BCR = \Delta PS / \Delta A$, where ΔPS is the gain in producer surplus or "profit" associated with advertising increment ΔA. By definition, net $BCR = (\Delta PS - \Delta A) / \Delta A = \Delta PS / \Delta A - 1$. Hence, net BCR = gross BCR − 1, as claimed.

14 Because about one-third of the check-off income for cotton comes from an assessment on the foreign-cotton-fiber content of imported textiles, the question arises as to whether the intensity parameter in Equation (4) should be redefined to include the value of the assessed foreign cotton. Although we believe the correct answer is "yes," we chose not to make the adjustment since the resulting return estimate would apply to foreign as well as domestic producers and the focus here is on the domestic return. The adjustment would decrease the numerator as well as the denominator of Equation (4); hence, the practical effect of the adjustment may be negligible.

15 Brester and Schroeder (1995) found *brand* advertising of pork to be effective at shifting the demand curve for pork, a result that the mandated study could not reproduce, which explains why only the generic advertising variable appears in Equation (5). Piggott's study (1997) is the only published work to find a significant own-effect for generic pork advertising.

16 The size and direction of the bias depends on the correlation between the two advertising expenditures and the size and sign of the effect of beef promotion on demand for pork. For example, if beef and pork advertising are negatively correlated, as suggested by the analysis by Alston, Freebairn, and James (2001), omitting beef advertising in the pork demand equation would cause the pork advertising effect to be overstated. A similar inference applies to beef. That is, if pork advertising decreases beef demand, omitting pork advertising in the beef demand equation, as was the case in the mandated study, would cause the beef advertising effect to be overstated.

17 Caution is required here because the price effect of advertising depends on the *ratio* of the advertising elasticity to the price elasticity of demand, α / η, not on the advertising elasticity α per se. Hence, if the single-equation model overstates η as well as α, the actual effect on the benefit-cost ratio may be less than supposed.

18 The implications of advertising spillover are discussed in some detail in Kinnucan (1996). A basic result is that, when supplies are fixed, advertising's own-price effect is indeterminate, depending on the extent to which the advertising decreases the demand and price of substitute goods.

19 As pointed out by a reviewer, there is a third issue with respect to cross-commodity effects–induced changes in the promotion of substitutes. For example, cola advertising might increase in response to an increase in milk advertising and vice versa. Competitor reactions are beyond the scope of the present analysis; the interested reader is referred to Alston, Freebairn, and James (2001) and Kinnucan and Myrland (2002).

20 This is not to say that econometric studies should be avoided, as other methods typically do not control for confounding influences. For example, the marketing research firm's assessment of fluid-milk advertising contained in the 2001 *Report to Congress* (USDA 2001, pp. 43–51) failed to take into account the effect of changes in relative prices and consumer income. As a consequence, the analysis almost surely suffered from the *post hoc ergo proctor hoc* fallacy. In particular, just because sales are rising during the promotion period, this does not mean that promotion caused the increase, as other forces affecting demand may be at work.

21 This rule is based on the assumption that the level of promotion that would maximize producer surplus at the farm level is no higher than 5 percent of farm revenue, an assumption that is consistent with the intensity ratios given in Table 12.1. However, it is also based on a further assumption that the program is operating close to the optimum. The fact that benefit-cost ratios associated with the observed intensities are generally substantially greater than 1 means that any "rules" based on a premise of optimality are at best indicative.

References

Alston, J.M., H.F. Carman, and J.A. Chalfant. "Evaluating Primary Product Promotion: The Returns to Generic Advertising by a Producer Cooperative in a Small, Open Economy." *Promotion in the Marketing Mix: What Works, Where and Why*. Proceedings from the NEC-63 Spring 1994 Conference. E.W. Goddard and D.S. Taylor, eds., pp. 145–167. Guelph Ontario: University of Guelph, 1994.

Alston, J.M., H.F. Carman, J.A. Chalfant, J.M. Crespi, R.J. Sexton, and R.J. Venner. *The California Prune Board's Promotion Program: An Evaluation*. Berkeley CA: University of California Division of Agriculture and Natural Resources, Giannini Foundation Research Report 344, March 1998.

Alston, J.M., J.A. Chalfant, and N.E. Piggott. "Advertising and Consumer Welfare." Mimeo. Department of Agricultural Economics, University of California, Davis, February 2001.

——. "The Incidence of the Costs and Benefits of Generic Advertising." *American Journal of Agricultural Economics* 82 (2000):665–671.

Alston, J.M., J.W. Freebairn, and J.S. James. "Beggar-Thy-Neighbor Advertising: Theory and Application to Generic Commodity Promotion Programs." *American Journal of Agricultural Economics* 83 (2001):888–902.

Alston, J.M., M.C. Marra, P.G. Pardey, and T.J. Wyatt. "Research Returns Redux: A Meta-Analysis of Agricultural R&D Evaluations." *Australian Journal of Agricultural and Resource Economics* 44 (2000):185–216.

Armbruster, W.J., and J.P. Nichols. *Commodity Promotion Policy*. College Station TX: Farm Foundation and Texas A&M University, 2002. Available online at *www.farmfoundation.org/2002_farm_bill/armbruster.pdf*.

Baye, M.R., D.W. Jansen, and J. Lee. "Advertising Effects in Complete Demand Systems." *Applied Economics* 24 (1992):1087–1096.

Benson, J.T., F.J. Breidt, and J.R. Schroeter. "Television Advertising and Beef Demand: Bayesian Inference in a Random Effects Tobit Model." *Canadian Journal of Agricultural Economics* 50 (2002):201–219.

Boetel, B.L., and D.J. Liu. "Evaluating the Effect of Generic Advertising and Food Health Information within a Meat Demand System." *Agribusiness: An International Journal* 19 (2003): 345–354.

Brester, G.W., and T.C. Schroeder. "The Impacts of Brand and Generic Advertising on Meat Demand." *American Journal of Agricultural Economics* 77 (1995):969–979.

Brown, M.G., and J. Lee. "Theoretical Overview of Demand Systems Incorporating Advertising Effects." *Commodity Advertising and Promotion*. H.W. Kinnucan, S.R. Thompson, and H. Chang, eds., pp. 79–100. Ames IA: Iowa State University Press, 1992.

Clarke, D. "Econometric Measurement of the Duration of Advertising Effect on Sales." *Journal of Marketing Research* 13 (1976):345–357.

Coulibaly, N., and B.W. Brorsen. "Explaining Differences between Two Previous Meat Generic Advertising Studies." *Agribusiness: An International Journal* 15 (1999):501–516.

Davis, G.C., O. Capps, Jr., D.A. Bessler, J.H. Leigh, J.P. Nichols, and E. Goddard. *An Economic Evaluation of the Pork Checkoff Program*. College Station TX: Texas A&M University, 2001.

Ding, L., and H.W. Kinnucan. "Market Allocation Rules for Nonprice Promotion with Farm Programs: U.S. Cotton." *Journal of Agricultural and Resource Economics* 21 (1996):351–367.

Dorfman, R., and P.O. Steiner. "Optimal Advertising and Optimal Quality." *American Economic Review* 44 (1954):826–836.

Duffy. M. "Advertising in Consumer Allocation Models: Choice of Functional Form." *Applied Economics* 33 (2001):437–456.

——. "On the Estimation of Advertising-Augmented, Cointegrating Demand Systems." *Economic Modeling* 20 (2003):181–206.

Forker, O.D., and R.W. Ward. *Commodity Advertising: The Economics and Measurement of Generic Programs*. New York NY: Lexington Books, 1993.

Hadar, J. *Mathematical Theory of Economic Behavior*. Reading MA: Addison-Wesley Publishing Company, 1971.

Hardin, R. *Collective Action*. Baltimore MD: John Hopkins University Press and Resources for the Future, Inc., 1982.

Hyde, J., and K. Foster. "Estimating Relationships between Pork Advertising and Revenues." *Review of Agricultural Economics* 25 (2003):279–293.

Jensen, H., and T. Kesavan. "Generic Advertising of Food Product Characteristics." *Proceedings of the 33rd Annual Conference of American Council of Consumer Interests.* V.L. Hampton, ed. Austin TX: American Council of Consumer Interests, 1987.

Kaiser, H.M. *Impact of Generic Fluid Milk and Cheese Advertising on Dairy Markets 1984–99.* Ithaca NY: Cornell University National Institute for Commodity Promotion and Research Bulletin 2000-02, July 2000.

——. "Impact of National Generic Dairy Advertising on Dairy Markets, 1984–95." *Journal of Agricultural and Applied Economics* 29 (1997):303–314.

Kinnucan, H.W. "A Note on Measuring Returns to Generic Advertising in Interrelated Markets." *Journal of Agricultural Economics* 47 (1996):261–267.

——. "Media Advertising Effects on Milk Demand: The Case of the Buffalo, New York Market." Department of Agricultural Economics Research Bulletin 83-13, Cornell University, 1983.

Kinnucan, H.W., and C.R. Clary. "Brand versus Generic Advertising: A Conceptual Framework with Application to Cheese." *Agribusiness: An International Journal* 11 (1995):355–369.

Kinnucan, H.W., Y. Miao, H. Xiao, and H.M. Kaiser. "Effects of Advertising on U.S. Non-Alcoholic Beverage Demand: Evidence from a Two-Stage Rotterdam Model." *Advances in Applied Microeconomics: Advertising and Differentiated Products.* M.R. Baye and J.P. Nelson, eds., Vol. 10, pp. 1–30. Oxford England: Elsevier Science Ltd., 2001.

Kinnucan, H.W., and Ø. Myrland. *Relationship between Partial and Total Responses to Advertising with Application to U.S. Meats.* Ithaca NY: Cornell University National Institute for Commodity Promotion and Research Bulletin 2002-05, 2002.

Kinnucan, H.W., and M. Venkateswaran. "Effects of Generic Advertising on Perceptions and Behavior: The Case of Catfish." *Southern Journal of Agricultural Economics* 22 (1990):137–151.

Kinnucan, H.W., H. Xiao, C. Hsia, and J.D. Jackson. "Effects of Health Information and Generic Advertising on U.S. Meat Demand." *American Journal of Agricultural Economics* 79 (1997): 13–23.

Leontief, W. "Can Economics be Reconstructed as an Empirical Science?" *American Journal of Agricultural Economics* 75 (1993):2–5.

Mansfield, E. *Microeconomics: Theory and Applications.* Fifth Edition. New York NY: W.W. Norton & Company, 1985.

Murray, B.C., R.H. Beach, W.J. White, C. Viator, N. Piggott, and M. Wohlgenant. *An Economic Analysis of the Cotton Research and Promotion Programs.* Research Triangle Park NC: Research Triangle Institute Project 8024, September 2001.

Piggott, N.E. "The Benefits and Costs of Generic Advertising of Agricultural Commodities." PhD dissertation, University of California, Davis, 1997.

Piggott, N.E., J.A. Chalfant, J.M. Alston, and G.R. Griffith. "Demand Response to Advertising in the Australian Meat Industry." *American Journal of Agricultural Economics* 78 (1996): 268–279.

Quilkey, J.J. "Promotion of Primary Products–A View from the Cloister." *Australian Journal of Agricultural Economics* 30 (1986):38–52.

Schmit, T.M., D. Dong, C. Chung, H.M. Kaiser, and B.W. Gould. "Identifying the Effects of Generic Advertising on Household Demand for Fluid Milk and Cheese: A Two-Step Panel Data Approach." *Journal of Agricultural and Resource Economics* 27 (2002):165–186.

Schmit, T.M., B.W. Gould, D. Dong, H.M. Kaiser, and C. Chung. "The Impact of Generic Advertising on U.S. Household Cheese Purchases: A Censored Autocorrelated Regression Approach." *Canadian Journal of Agricultural Economics* 27 (2003):15–37.

Simon, J.L., and J. Arndt. "The Shape of the Advertising Response Function." *Journal of Advertising Research* 20 (1980):11–28.

Theil, H. *Principles of Econometrics*. Santa Barbara CA: John Wiley & Sons, Inc., 1971.

Tomek, W.G. "Confirmation and Replication in Empirical Econometrics: A Step toward Improved Scholarship." *American Journal of Agricultural Economics* 75 (1993):6–14.

Tomek, W.G., and H.M. Kaiser. "On Improving Econometric Analyses of Generic Advertising Impacts." *Agribusiness: An International Journal* 15 (1999):485–500.

U.S. Department of Agriculture. *Report to Congress on the National Dairy Promotion and Research Program and the National Fluid Milk Processor Promotion Program.* Washington DC, July 2001.

Vande Kamp, P.R., and H.M. Kaiser. *Commodity Promotion Programs in the United States*. Ithaca NY: Cornell University National Institute for Commodity Promotion and Research Bulletin 99-03, 1999.

Wallace, T.D. "Pretest Estimation in Regression: A Survey." *American Journal of Agricultural Economics* 59 (1977):431–444.

Ward, R.W. "Beef Demand and Its Response to the Beef Checkoff." UF#NCA2001#1, University of Florida, Gainesville, 2001.

——. "The Beef Checkoff Impact on U.S. Beef Demand." UF#NCA96.1, University of Florida, Gainesville, 1996.

Part III

Other Case Studies

13

Evaluation of Nutrition Education by the Dairy Council of California

Julian M. Alston, James A. Chalfant & Jennifer S. James

In 1988, the Surgeon General reported that more than $200 billion was being spent annually for treatment of diet-related illnesses (U.S. Department of Health and Human Services 1988). Frazao (1995) put the cost at $250 billion in 1995. One way of reducing these costs is through nutrition education that successfully improves eating habits. Various agencies sponsor nutrition education programs, often targeted at specific ages or demographic groups. One important example is the Dairy Council of California (DCC), which provides California schools with nutrition education materials and lesson plans.

If effective in altering eating habits, nutrition education programs provide many benefits. Perhaps the least tangible of these is the improved quality of life that results from improved nutrition (e.g., increased energy and mental capacity, fewer illnesses, reduced absences from work or school). Second is reduced health care costs as a result of improved nutrition that makes diet-related illnesses such as osteoporosis less likely.[1] In addition, benefits accrue to producers of nutritious foods that are otherwise typically underconsumed, such as dairy products, fruits, and vegetables, when education programs increase their consumption. Similarly, consumption of less healthy foods such as sodas and candy bars should decrease, with resulting costs to their producers. Finally, because nutrition education programs provided to schools include materials and lesson plans, they effectively subsidize schools by saving them the costs of materials and teacher preparation time, and this is a further benefit to society from the programs.

DCC provides a number of nutrition education programs to about 80 percent of California's schools. Programs include *Healthy Choices, Healthy Me!* for lower elementary grades, *Shaping Up My Choices* for upper elementary grades, *Exercise Your Options* for junior high school students, and *Now We're Cooking!* for children in kindergarten through sixth grade. This study focused on the *Exercise Your Options* (*EYO*) program, which provides education materials to teachers to assist them in teaching about nutrition topics such as the basic food groups and the United States Department of Agriculture's (USDA's) food pyramid. *EYO* is aimed at sixth, seventh, and eighth grade students, a particularly important age group for dairy consumption; at this age, children begin building peak bone mass, which critically affects their likelihood of developing osteoporosis later in life. For this reason and because *EYO* was DCC's newest program, it was selected for evaluation.

DCC's activities are funded through a per-unit assessment on California milk, and one issue for analysis is the costs and benefits that accrue to the dairy industry from the *EYO* program. This study compared the benefits to the dairy industry with the costs of the program borne by the industry. In a benefit-cost analysis from the industry's perspective, only private benefits to milk producers (i.e., increases in the net value of milk sales resulting from increased milk consumption generated by *EYO* instruction) are relevant. These private producer benefits understate the total benefits of the program because they exclude many of the other benefits previously noted, some of which can be considered positive externalities or public benefits. On the other hand, the analysis also excludes private and public costs and benefits resulting from any induced changes in consumption of other goods–for instance, reduced sales by candy manufacturers.

Data Collection

To evaluate the *EYO* program, we estimated its effects on students' food consumption patterns. Food records were completed in autumn 1996 by sixth, seventh, and eighth graders in more than one hundred California classrooms. The classrooms were selected by DCC regional managers to cover a representative range of income levels, ethnic groups, and regions of the state and urban, suburban, and rural areas. Participating classes were of three types: intervention, nonintervention, and control. In intervention classes, DCC regional managers collected food records, while in nonintervention classes, teachers collected food records. These two class types comprised the treatment group, as all students in them received *EYO* instruction. Nonintervention classes were included to test whether the results in the intervention classes were affected by the presence of DCC staff, rather than measuring just the effects of the *EYO* information itself on the behavior of students. Finally, students in the control classes did not receive *EYO* instruction during the period of study but completed food records during roughly the same time period, thus providing a control for the effects of season, price changes, and any other influences on consumption besides the *EYO* program. Differences between students in treatment and control classes were used to measure the effects of the *EYO* program.

Three sets of food records were collected in each classroom. The first record (Record 1) was completed before *EYO* instruction began. The second (Record 2) was completed immediately after the program was finished, and the third (Record 3) was completed one month later (on average). For each food record, students were asked to record their first name, gender, birth date, and the ethnic group(s) with which they identify. Students were then asked to record all foods and beverages they recalled consuming in the twenty-four-hour period ending with the class session in which the records were completed. Directions for completing food records, whether conducted by DCC staff or by class teachers, were given following a detailed script and with the aid of specific visual props to help students estimate serving sizes and the same sample food records as a reference for all classes. The script and specific visual props were used to ensure that the settings in which food records were completed were kept as similar as possible across participating classrooms. In an effort to avoid biasing responses, the purpose of the study and the occupation and employer of the DCC regional managers were not revealed. Once the food records were collected, a team of dietitians coded all of the food and beverage entries into numbers of servings of each of the following food groups: Dairy, Meats, Fruits, Vegetables, Grains, Extras (which included sweets and fat but not sodas), and Sodas.[2] Records were coded according to agreed-upon standards designed to ensure consistency across coders and over time. Serving numbers were then summed within each food record so that the data from each food record were converted to total servings consumed of each of the food groups during the twenty-four-hour period of interest.

Each student's class, name, and birth date were used to match his/her three food records. However, owing to absences from class or incomplete or inaccurate food records, one or two food records were missing for several

Table 13.1. Numbers of Observations with Usable Records

Class Type	Usable Records 1 and 2	2 and 3	1, 2, and 3
Intervention	1,557	1,473	1,318
Nonintervention	538	521	480
Treatment	2,095	1,994	1,798
Control	941	882	827
Total	3,036	2,876	2,625

Note: Treatment = Intervention + Nonintervention

students. Table 13.1 summarizes how many food records in each class type had usable Records 1 and 2, how many had usable Records 2 and 3, and how many had usable Records 1, 2, and 3. In the discussion that follows, a change in total servings for each food group from Record 1 to Record 2 is attributed to *EYO*, and a change from Record 2 to Record 3 measures "depreciation" of the effects of the program.

The Effects of *EYO* on Consumption Patterns

Table 13.2 lists the mean number of servings of each food group consumed by students in each class type (intervention, nonintervention, all treatment, control, and all observations). The means in the upper panel of Table 13.2 are based on all of the observations with usable Records 1 and 2. The similarity of Record-1 means across class types for all food groups suggests that there were no significant differences in eating behavior among the different class types before *EYO* instruction began. For comparison, the number of servings recommended for each food group in USDA's food pyramid is also listed in Table 13.2.

The means in the upper panel of Table 13.2 were used to calculate the percentage changes in serving numbers from Record 1 to Record 2. These changes can be interpreted as "short run" or immediate effects of the *EYO* program. The means in the lower panel of Table 13.2 were based on all observations with usable Records 2 and 3 and were used to calculate percentage changes from Record 2 to Record 3. These changes represent depreciation of the effects of the program over the month or so afterward.

Finally, changes from Record 1 to Record 3 are simply sums of the mean changes from Record 1 to Record 2 and from Record 2 to Record 3. This net change from Record 1 to Record 3 can be thought of as the "longer run" effect

Table 13.2. Mean Daily Consumption of Food Groups by Class Type

Food Group	USDA Food Pyramid Recommended	Number of Servings per Student per Day: Treatment: Inter-vention	Noninter-vention	All	Control	All Obser-vations
For Observations with Usable Records 1 and 2						
Record 1						
Dairy	2–3	1.82	1.64	1.77	1.83	1.79
Meats	2–3	2.15	1.87	2.08	2.07	2.08
Fruits	2–4	0.87	0.81	0.86	0.77	0.83
Vegetables	3–5	1.32	1.08	1.26	1.31	1.27
Grains	6–11	5.39	4.66	5.20	5.33	5.24
Extras	–	1.54	1.53	1.54	1.58	1.55
Sodas	–	1.48	1.53	1.49	1.57	1.52
Record 2						
Dairy	2–3	2.12	2.03	2.10	1.84	2.02
Meats	2–3	2.30	2.21	2.28	2.14	2.24
Fruits	2–4	1.13	1.12	1.13	0.90	1.06
Vegetables	3–5	1.62	1.48	1.59	1.40	1.53
Grains	6–11	5.55	5.12	5.44	5.29	5.39
Extras	–	1.45	1.42	1.44	1.61	1.50
Sodas	–	1.27	1.48	1.32	1.38	1.34
No. of Observations		**1,557**	**538**	**2,095**	**941**	**3,036**
For Observations with Usable Records 2 and 3						
Record 2						
Dairy	2–3	2.16	2.04	2.13	1.82	2.03
Meats	2–3	2.29	2.24	2.27	2.09	2.22
Fruits	2–4	1.16	1.12	1.15	0.86	1.06
Vegetables	3–5	1.60	1.46	1.56	1.39	1.51
Grains	6–11	5.52	5.17	5.43	5.23	5.37
Extras	–	1.43	1.40	1.42	1.61	1.48
Sodas	–	1.23	1.48	1.30	1.36	1.32
Record 3						
Dairy	2–3	2.04	1.84	1.98	1.80	1.93
Meats	2–3	2.24	2.15	2.22	2.24	2.22
Fruits	2–4	0.90	0.84	0.88	0.75	0.84
Vegetables	3–5	1.60	1.48	1.57	1.37	1.50
Grains	6–11	5.69	4.77	5.45	5.38	5.43
Extras	–	1.47	1.40	1.45	1.51	1.47
Sodas	–	1.22	1.39	1.26	1.23	1.25
No. of Observations		**1,473**	**521**	**1,994**	**882**	**2,876**

of the program. To the extent that the nutrition message persists, depreciation is expected to be smaller and the longer-run effect to be larger. Assuming an immediate response to the program and some depreciation, the longer-run effect is expected to be smaller than the short-run effect. On the other hand, if students adjusted their behavior in response to the program only gradually over time, the longer-run effects of the *EYO* program would be greater than the short-run effects, and the change from Record 1 to Record 3 would be larger than the change from Record 1 to Record 2.

To analyze the results, 95 percent confidence intervals were formed for each food group and for each of the changes previously described. If a confidence interval included zero, then the change in consumption was not statistically significant. If a 95 percent confidence interval did not include zero, the corresponding change (increase or decrease) in consumption was statistically significantly different from zero. If the program is effective, we would expect the treatment confidence intervals not to include zero and confidence intervals for the control and treatment groups not to overlap. Formal tests of the effectiveness of the *EYO* program are provided by t-tests of the null hypothesis that, for each food group, the mean change in serving numbers did not differ between control and treatment students.

The changes in consumption in absolute and relative terms, along with the confidence intervals previously described, are summarized in Tables 13.3 through 13.6. Before comparing the treatment and control groups, we compared results between intervention classes (Table 13.3) and nonintervention classes (Table 13.4) to see whether the results were affected by the presence of DCC staff in the classrooms. Table 13.5 combines the results from Tables 13.3 and 13.4, representing all treatment students, to compare them to the results from the control classes, which are shown in Table 13.6.

Comparison of Intervention and Nonintervention Groups

Table 13.3 presents daily changes in consumption for intervention students. Between Record 1 and Record 2, intervention students increased their average consumption of Dairy by 0.3035 servings per day, or 16.67 percent, a statistically significant increase. The best estimate of the depreciation effect for Dairy is −0.1204 servings per day, a statistically significant decrease of 5.58 percent from the Record 2 mean. The net longer-term effect, comparing Records 1 and 3, is thus an increase in dairy food consumption of 0.1831 (i.e., 0.3035 − 0.1204) servings per day (a 10.06 percent increase), which is statistically significantly different from zero. Consumption changes for other food groups can be interpreted similarly.

Table 13.4 presents the corresponding information for nonintervention classrooms. The results for the two treatment groups are qualitatively similar, and the confidence intervals in Tables 13.3 and 13.4 overlap for every food group except Grains between Record 2 and Record 3. In particular, for each group of students, the change in the number of servings of dairy products was statistically significant and positive between Records 1 and 2 and statistically significant and negative between Records 2 and 3.

The t-tests for the equality of the means between intervention and nonintervention students were calculated first under the assumption that the variance of changes in serving numbers was equal for the two types of students and then after relaxing this assumption. In every instance except Grains between Records 2 and 3, the differences between the two groups were not statistically significant. None of the longer-run changes from Record 1 to Record 3 were statistically significantly different between intervention and nonintervention students.[3] We concluded that DCC intervention in the data collection process did not significantly influence results and thereafter we regarded the two class types as one combined group called the treatment group.

Comparison of Treatment and Control Groups

Table 13.5 shows average changes in daily consumption for the combined treatment group. Between Records 1 and Record 2, consumption of every food group reflected a statistically significant change toward a healthier diet (i.e., statistically significant increases for Dairy, Meats, Fruits, Vegetables, and Grains and statistically significant decreases for Extras and Sodas). Only two depreciation effects were statistically significant–those for Dairy and Fruits. The longer-run effects (changes from Record 1 to Record 3) included statistically significant increases for Dairy, Meats, Vegetables, and Grains and a statistically significant decrease for Sodas. The longer-run effects were statistically insignificant for Fruits and Extras.

Table 13.6 shows changes in daily consumption for the control classes. From Record 1 to Record 2, only the increase in Fruits and the decrease in Sodas were statistically significant. Between Record 2 and Record 3, consumption of Meats increased while Soda consumption declined further. The statistically significant longer-run effects (from Record 1 to Record 3) in the control group included an increase for Meats and a decrease for Sodas.

Table 13.3. Changes in Consumption of Food Groups for Intervention Classes

Number of Servings per Student per Day				
			95 Percent Confidence Limit	
Food Group	Percent Change	Mean Change	Lower	Upper
Changes from Record 1 to Record 2				
Dairy	16.67	0.30[a]	0.20	0.41
Meats	7.11	0.15[a]	0.04	0.26
Fruits	28.99	0.25[a]	0.15	0.36
Vegetables	23.23	0.31[a]	0.19	0.42
Grains	3.02	0.16	−0.08	0.40
Extras	−5.75	−0.09	−0.19	0.01
Sodas	−14.46	−0.21[a]	−0.30	−0.13
Note: Based on 1,557 observations with usable Records 1 and 2.				
Changes from Record 2 to Record 3				
Dairy	−5.58	−0.12[a]	−0.23	−0.01
Meats	−2.10	−0.05	−0.17	0.07
Fruits	−22.21	−0.26[a]	−0.36	−0.16
Vegetables	−0.03	0.00	−0.12	0.12
Grains	3.00	0.17	−0.06	0.39
Extras	2.66	0.04	−0.06	0.14
Sodas	−1.22	−0.02	−0.10	0.07
Note: Based on 1,473 observations with usable Records 2 and 3.				
Changes from Record 1 to Record 3				
Dairy	10.06	0.18[a]	0.07	0.29
Meats	4.88	0.10	−0.02	0.23
Fruits	−0.37	0.00	−0.10	0.10
Vegetables	23.19	0.31[a]	0.18	0.43
Grains	6.10	0.33[a]	0.07	0.58
Extras	−3.28	−0.05	−0.15	0.05
Sodas	−15.48	−0.23[a]	−0.32	−0.14
Note: Reported changes in servings in the first two panels may not sum exactly to those reported in this panel because of rounding error.				

[a] The reported change is statistically significantly different from zero.
Note: Serving numbers are rounded to two digits to reduce clutter, but percentage changes were calculated with the greater precision of the numbers provided in the text.

Table 13.4. Changes in Consumption of Food Groups for Nonintervention Classes

Number of Servings per Student per Day

Food Group	Percent Change	Mean Change	95 Percent Confidence Limit Lower	95 Percent Confidence Limit Upper
Changes from Record 1 to Record 2				
Dairy	23.80	0.39[a]	0.22	0.56
Meats	17.91	0.34[a]	0.14	0.54
Fruits	38.76	0.31[a]	0.14	0.49
Vegetables	37.31	0.40[a]	0.23	0.57
Grains	9.84	0.46[a]	0.13	0.79
Extras	−7.38	−0.11	−0.27	0.04
Sodas	−3.49	−0.05	−0.21	0.11
Note: Based on 538 observations with usable Records 1 and 2.				
Changes from Record 2 to Record 3				
Dairy	−9.89	−0.20[a]	−0.37	−0.04
Meats	−4.01	−0.09	−0.29	0.11
Fruits	−24.74	−0.28[a]	−0.45	−0.10
Vegetables	1.37	0.02	−0.15	0.19
Grains	−7.72	−0.40[a]	−0.73	−0.07
Extras	0.25	0.00	−0.15	0.16
Sodas	−6.14	−0.09	−0.25	0.07
Note: Based on 521 observations with usable Records 2 and 3.				
Changes from Record 1 to Record 3				
Dairy	11.45	0.19[a]	0.02	0.35
Meats	13.11	0.25[a]	0.05	0.44
Fruits	4.43	0.04	−0.13	0.20
Vegetables	39.17	0.42[a]	0.25	0.59
Grains	1.28	0.06	−0.29	0.41
Extras	−7.15	−0.11	−0.28	0.06
Sodas	−9.41	−0.14	−0.33	0.04
Note: Reported changes in servings in the first two panels may not sum exactly to those reported in this panel because of rounding error.				

[a] The reported change is statistically significantly different from zero.
Note: Serving numbers are rounded to two digits to reduce clutter, but percentage changes were calculated with the greater precision of the numbers provided in the text.

Table 13.5. Changes in Consumption of Food Groups for Treatment Classes

Number of Servings per Student per Day				
			95 Percent Confidence Limit	
Food Group	**Percent Change**	**Mean Change**	**Lower**	**Upper**
Changes from Record 1 to Record 2				
Dairy	18.36	0.33[a]	0.24	0.42
Meats	9.61	0.20[a]	0.10	0.30
Fruits	31.36	0.27[a]	0.18	0.36
Vegetables	26.33	0.33[a]	0.24	0.43
Grains	4.59	0.24[a]	0.04	0.44
Extras	−6.16	−0.09[a]	−0.18	−0.01
Sodas	−11.57	−0.17[a]	−0.25	−0.10
Note: Based on 2,095 observations with usable Records 1 and 2.				
Changes from Record 2 to Record 3				
Dairy	−6.67	−0.14[a]	−0.23	−0.05
Meats	−2.59	−0.06	−0.16	0.04
Fruits	−22.86	−0.26[a]	−0.35	−0.18
Vegetables	0.31	0.00	−0.10	0.11
Grains	0.34	0.02	−0.17	0.21
Extras	2.04	0.03	0.06	0.11
Sodas	−2.69	−0.03	−0.11	0.04
Note: Based on 1,994 observations with usable Records 2 and 3.				
Changes from Record 1 to Record 3				
Dairy	10.37	0.18[a]	0.09	0.28
Meats	6.77	0.14[a]	0.04	0.25
Fruits	0.78	0.01	−0.08	0.09
Vegetables	26.71	0.34[a]	0.23	0.44
Grains	4.94	0.26[a]	0.05	0.47
Extras	−4.28	−0.07	−0.15	0.02
Sodas	−13.90	−0.21[a]	−0.29	−0.12

Note: Reported changes in servings in the first two panels may not sum exactly to those reported in this panel because of rounding error.

[a] The reported change is statistically significantly different from zero.
Note: Serving numbers are rounded to two digits to reduce clutter, but percentage changes were calculated with the greater precision of the numbers provided in the text.

Table 13.6. Changes in Consumption of Food Groups for Control Classes

Number of Servings per Student per Day				
			95 Percent Confidence Limit	
Food Group	**Percent Change**	**Mean Change**	**Lower**	**Upper**
Changes from Record 1 to Record 2				
Dairy	0.68	0.01	-0.11	0.13
Meats	3.60	0.07	-0.07	0.22
Fruits	16.16	0.13[a]	0.00	0.25
Vegetables	6.87	0.09	-0.05	0.23
Grains	-0.72	-0.04	-0.34	0.26
Extras	1.92	0.03	-0.11	0.17
Sodas	-12.39	-0.19[a]	-0.32	-0.07
Note: Based on 941 observations with usable Records 1 and 2.				
Changes from Record 2 to Record 3				
Dairy	-1.15	-0.02	-0.14	0.10
Meats	7.14	0.15[a]	0.01	0.29
Fruits	-12.84	-0.11	-0.23	0.01
Vegetables	-1.55	-0.02	-0.17	0.13
Grains	2.78	0.15	-0.15	0.44
Extras	-6.61	-0.11	-0.25	0.03
Sodas	-10.01	-0.14[a]	-0.24	-0.03
Note: Based on 882 observations with usable Records 2 and 3.				
Changes from Record 1 to Record 3				
Dairy	-0.47	-0.01	-0.14	0.12
Meats	10.84	0.22[a]	0.06	0.38
Fruits	1.95	0.02	-0.11	0.14
Vegetables	5.22	0.07	-0.08	0.21
Grains	2.00	0.11	-0.24	0.45
Extras	-4.83	-0.08	-0.22	0.07
Sodas	-21.04	-0.33[a]	-0.46	-0.21
Note: Reported changes in servings in the first two panels may not sum exactly to those reported in this panel because of rounding error.				

[a] The reported change is statistically significantly different from zero.
Note: Serving numbers are rounded to two digits to reduce clutter, but percentage changes were calculated with the greater precision of the numbers provided in the text.

As mentioned previously, it was important to see whether estimated differences in consumption patterns between the control and treatment students for a given food group were statistically significant. Statistically significant differences between control and treatment groups are suggested by nonoverlapping confidence intervals, which can be seen most clearly graphically. Figure 13.1 shows 95 percent confidence intervals for changes in daily consumption of each food group from Record 1 to Record 2 for the treatment and control groups. While each of the intervals for the treatment group excludes zero, they all overlap either partially (in the case of Meats, Fruits, Grains, and Extras) or completely (in the case of Sodas) with the corresponding confidence intervals for the control group with only two exceptions, Dairy and Vegetables.

The hypothesis that μ_1, the control group's mean change in servings, equals μ_2, the treatment group's mean change in servings, can be tested

Figure 13.1. Ninety-Five Percent Confidence Intervals for Changes in Daily Consumption from Record 1 to Record 2 for Treatment and Control Groups

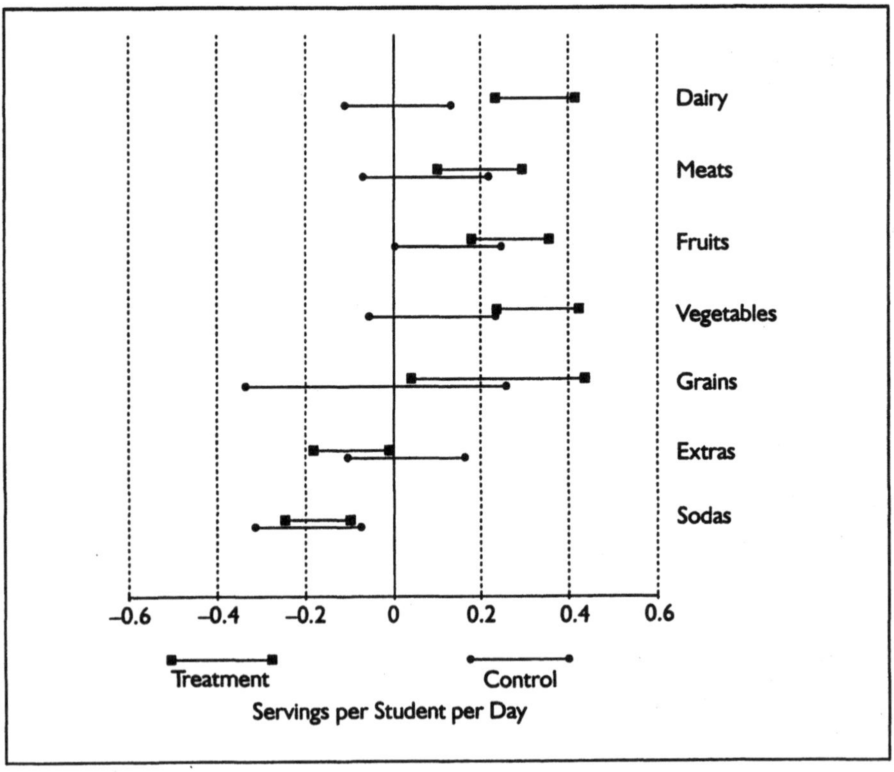

formally using the t-tests described in the notes. For the short-run effects, assuming equal variances of changes in serving numbers between treatment and control students, we found that the differences in mean changes in servings of Dairy and Vegetables were statistically significant even at the 99 percent level, with t-statistics of 3.91 and 2.76, respectively. We also found weaker evidence of a difference for Fruits, for which $t = 1.81$—statistically significant at the 90 percent level. No other differences were statistically significant. As was the case with our comparisons between intervention and nonintervention groups reported earlier, the conclusions never depended on the variance assumption. Upon relaxing the assumption that the variances in individual changes in serving numbers were the same for the treatment and control students, the corresponding t-statistics were 4.05 for Dairy, 2.74 for Vegetables, and 1.86 for Fruits.

Turning to changes from Record 2 to Record 3, the only statistically significant changes were for Dairy and Fruits, both negative, in the treatment group and for Meats (positive) and Sodas (negative) in the control group. The confidence intervals overlap between the two sets of students for all food groups, suggesting that the changes in consumption between Records 2 and 3 for the two class types did not differ much. Only for Meats is there little overlap, and this is reflected in the t-tests for differences between treatment- and control-group means: the t-statistics for Meats were significant at the 95 percent level ($t = 2.25$ if equal variances were assumed and $t = 2.31$ if not). We also found that the differences were statistically significant for Fruits with t-statistics of 1.99 for unequal variance and 1.95 for equal variance. For Extras, the evidence was weaker and depended on the variance assumption: the t-statistics were 1.63 (unequal variance) and 1.69 (equal variance), and only the latter was significant at the 90 percent level (using a critical value of 1.645, as is appropriate for such large samples).

The changes from Record 2 to Record 3 are more meaningful when combined with the changes from Record 1 to Record 2 to yield the longer-run effects shown in the third panels of Tables 13.5 and 13.6. Figure 13.2 shows the 95 percent confidence intervals for the longer-run changes. For the treatment group, the longer-run changes in servings of Dairy, Meats, Vegetables, and Grains remained statistically significant and positive, while the change for Fruits was positive but not statistically significant. The longer-run changes for Extras and Sodas were both negative, but the change for Extras was not statistically significant at the 95 percent level of confidence. For the control group, every longer-run change in servings was statistically insignificant except for

Figure 13.2. Ninety-Five Percent Confidence Intervals for Changes in Daily Consumption from Record 1 to Record 3 for Treatment and Control Groups

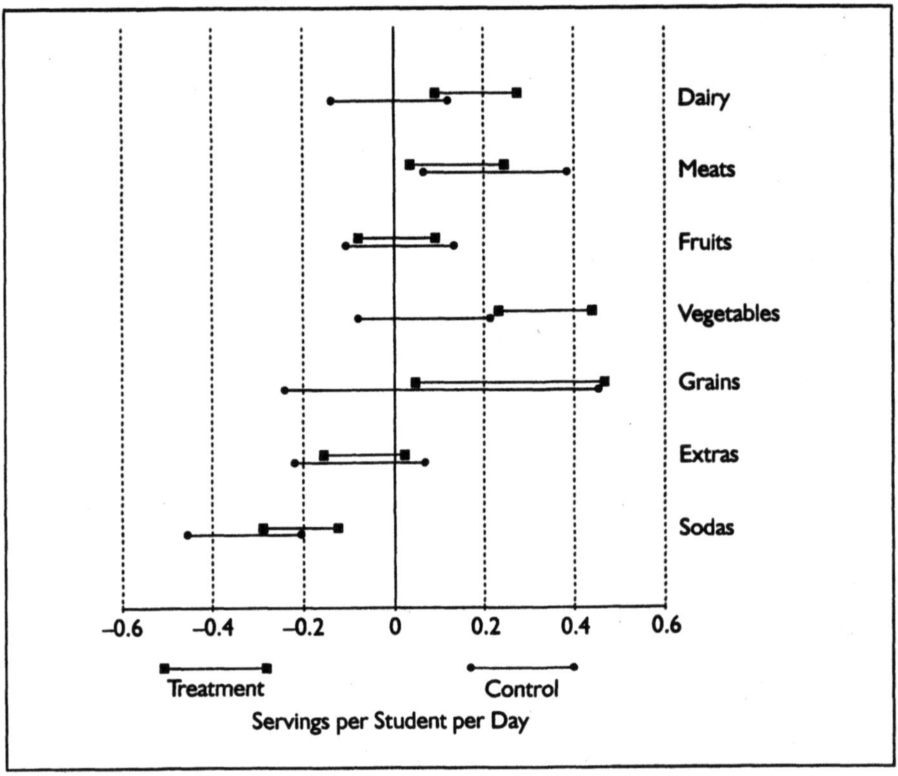

the positive change observed for Meats. Thus, the confidence intervals shown in Figure 13.2 suggested that the longer-run effects of *EYO* instruction were toward a healthier diet.

The formal tests of the hypothesis that the longer-run mean changes in servings differ between the treatment and control groups indicated that the observed differences were statistically significant for Dairy and Vegetables and insignificant for the other food groups. This was true whether we used data only for the students with all three food records or the results in the third panels of Tables 13.5 and 13.6, which were derived from the results in the first two panels of each table and include students who missed either Record 3 or Record 1. It was also true whether or not we assumed that the variances between the groups were equal. In summary, even after some depreciation was observed, *EYO* instruction had a statistically significant, positive effect on the numbers of servings consumed by students for Dairy and Vegetables.

The tests just described indicate whether a particular change in consumption is statistically significant. To estimate the effects of *EYO* on food consumption, the changes in serving numbers for treatment students had to be compared with those for control students. Behavior found in the treatment group but not in the control group could be attributed to the effects of the nutrition information in *EYO*. Thus, differences between the treatment and control groups in the changes in serving numbers from Record 1 to Record 2 measure the short-run effects of the *EYO* program, while differences in the changes in serving numbers from Record 1 to Record 3 measure the longer-run effects. For Dairy, the *EYO* effect is thus 0.3131 servings per day from Record 1 to Record 2 and 0.1924 servings per day from Record 1 to Record 3. These *EYO* effects are summarized in Table 13.7 and were used to determine the economic significance (i.e., the benefits to milk producers) of the *EYO* program. Any effects of *EYO* on other food groups were excluded from the analysis because they did not involve direct benefits to the dairy industry.

Theoretical Benefits and Costs of Promotion in California's Milk Market

As noted earlier, we measured only the benefits to milk producers from increased milk sales and compared our estimate of those benefits with the cost of the *EYO* program. To determine the benefits generated by a nutrition education program, we drew from literature evaluating the benefits of advertising but also had to account for dairy policies in California that control the pricing and allocation of milk in the state.

A great deal has been written about how to evaluate the producer benefits from a demand shift induced by promotion funded by commodity assessments, also known as check-offs. The same models are directly applicable to the present problem, where the demand shift is induced by nutrition education rather than advertising; for simplicity, we can refer to both types of investment in demand enhancement as "promotion."[4] A useful review and summary of issues in the literature and some extensions that are relevant to the present analysis are provided by Alston, Carman, and Chalfant (1994).

Promotion cannot be profitable for producers unless it leads to an increase in their average sales revenue (average price) per unit. Hence, it would not be profitable to promote many commodities in the absence of supply restrictions. This would be true of milk in most, if not all, of the state milk markets in the United States, where each state would be a price taker, unable to materially influence the price in the domestic and international markets for freely traded milk and dairy products. Profitable promotion is made possible by the

Table 13.7. Average Changes in Consumption for Treatment and Control Groups

Number of Servings per Student per Day			
	Mean Changes		
Food Group	**Treatment**	**Control**	***EYO* Effect**
Changes from Record 1 to Record 2			
Dairy	0.33	0.01	0.31
Meats	0.20	0.07	0.13
Fruits	0.27	0.13	0.14
Vegetables	0.33	0.09	0.24
Grains	0.24	−0.04	0.28
Extras	−0.09	0.03	−0.13
Sodas	−0.17	−0.19	0.02
Changes from Record 1 to Record 3			
Dairy	0.18	−0.01	0.19
Meats	0.14	0.22	−0.08
Fruits	0.01	0.02	−0.01
Vegetables	0.34	0.07	0.27
Grains	0.26	0.11	0.15
Extras	−0.07	−0.08	0.01
Sodas	−0.21	−0.33	0.12

Note: *EYO* Effect is the difference between the changes for the treatment and control groups. These differences may not appear exact because of rounding error.

creation of trade barriers that allow markets to be separated, both geographically and according to the end use of the commodity. For producers, fluid milk commands a premium over manufacturing milk, and it is within the power of milk-marketing authorities to determine whether any promotion-induced increase in demand is accommodated by an increase in price or an increase in volume.

The California milk market is represented in Figure 13.3. Subscripts 0 and 1 denote (1) supply, S, before and after a check-off of t per unit is collected from producers, respectively; (2) demand for fluid milk, D, before and after an increase in demand induced by successful promotion or nutrition education; and (3) quantity produced, Q, and quantity of fluid milk consumed, F, before and after the check-off and promotion are applied. The difference $Q - F$ represents the quantity of milk sold for processing.

Suppose California is a price taker in the markets for fresh milk and manufactured dairy products so that, in the absence of any policies, producers would receive and consumers would pay the "world" market price, P_W, for all milk, regardless of its final use. Then, in the absence of any commodity programs, Q_0 would be produced, F_0 would be sold for fluid milk, and $Q_0 - F_0$ would be sold for processing. In this setting, promotion of fluid milk *must* be unprofitable because the price remains at P_W regardless of any demand shift that might occur. However, suppose a quota is introduced, along with an embargo against imports of fluid milk, limiting fluid-milk sales to F_0'. Then the market-clearing price for fluid milk before promotion would be P_F, creating quota rents equal to $(P_F - P_W)F_0'$, or area P_FbeP_W. With this policy in place, producers might benefit when a check-off of t (at a cost to them of I_1fgI_0 and yielding check-off revenue of I_1fhI_0) is used to finance promotion that shifts demand from D_0 to D_1. Whether producers benefit depends on whether the increase in quota rents is greater than the cost of the check-off, which depends on how the increase in demand is accommodated in the market.

One way to accommodate the demand shift would be to keep the fluid-milk quota at F_0' and allow its price to rise to P_F', increasing quota rents by $(P_F' - P_F)F_0'$, or area $P_F'abP_F$. Alternatively, the quota could be expanded to F_1',

Figure 13.3. Effects of a Marketing Quota on Returns to Promotion

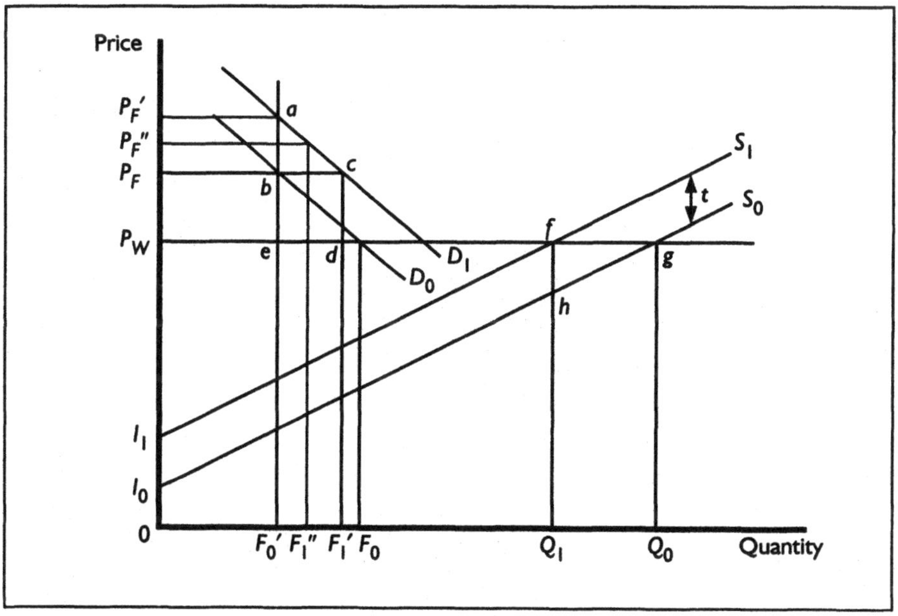

allowing an increase of quota rents equal to $(P_F - P_W)(F_1' - F_0')$, or area *bcde*. A third way would be to allow both price and quantity to rise, as indicated by the price P_F'' and quantity F_1''. Which of these the producers would prefer would depend on the elasticity of demand; which would happen in practice would depend on political constraints and the arrangements for fixing prices (and quantities) in the quota market. Notice that producers can gain even with the second option, which applies when the price of fluid milk does not rise. This is because their *average* revenue rises when a higher proportion of their production is sold on the higher priced market. Average revenue also rises when price rises in response to the increase in demand.

This last point is illustrated by Sumner and Wolf (1996), who presented a comprehensive description and analysis of California's milk marketing order system as it applied in 1994. Their analysis showed that California's marketing arrangements are quite complicated, involving a combination of (1) federal policy, including national import barriers and federal price supports for manufactured dairy products; and (2) a set of state marketing order regulations (other states participate in the federal milk marketing order rather than having state-specific orders). The California program includes minimum prices by end-use class (classified pricing), blend (or pooled) prices at the farm level, and a milk marketing quota. The details of operation of the program have changed a little since Sumner and Wolf published their study. In any event, the complete details are far too complicated to be accommodated in this analysis as we could not obtain estimates of changes in demand for individual dairy products. Instead, we used a simpler model as an approximation that was likely to be a good representation of the policy while retaining the virtue of simplicity.

Our analysis corresponded to the situation depicted in Figure 13.4, where we interpreted P_F as representing the price on the higher priced Class I milk market (essentially for fluid milk, which makes up about 30 percent of the total market for California milk) and P_W as representing the price for milk used for all other products. While prices of individual dairy products fluctuate, the premium for fluid milk is relatively stable and was assumed to be constant for this analysis. Given the pricing policy in the industry, the costs of the *EYO* program would be borne entirely by producers if the program was funded by an additional check-off on all milk (as shown in Figure 13.3) or if it was funded out of existing check-off funds at the expense of some other program. Hence, for simplicity, we omitted from Figure 13.4 any supply shift that might be associated with the program, depending on the details of how it was financed, which are irrelevant for the present analysis. When demand for fluid milk increases from D_0 to D_1, quantity consumed increases from F_0' to F_1' and producers gain an amount (B) equal to the price premium, $(P_F - P_W)$ times the increased quantity sold on the premium market (shown as area *abcd* in Figure 13.4):

$$B = (P_F - P_W)(F_1' - F_0').$$

Quantifying the Benefits

Under current program rules, the quota milk price premium, an estimate of the price premium for fluid milk, is $1.70 per hundredweight of milk. Thus, we set $(P_F - P_W)$ equal to $1.70 per hundredweight. To establish the gross producer benefits from the program, all that remained was to determine the size of the increase in consumption of fluid milk caused by the nutrition education program and then to multiply that by the estimate of the fluid-milk premium.

Figure 13.4. Benefits from Milk Promotion in California

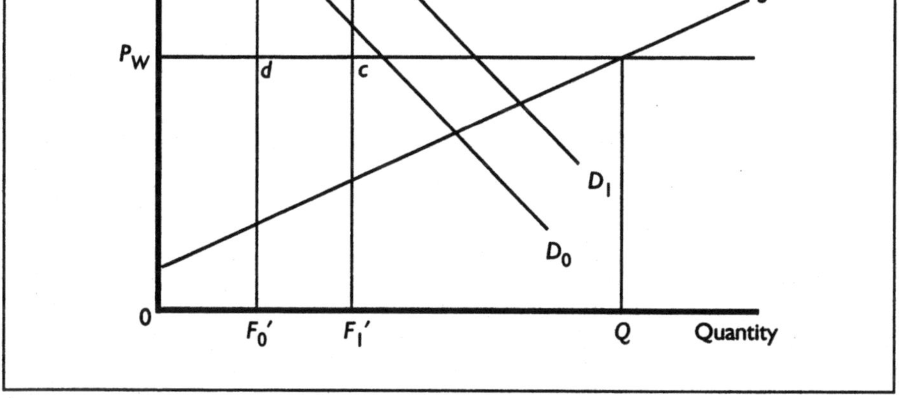

On any given day d, the increase in servings of fluid milk consumed as a result of the program is given by:

$$q_d = s_d nf,$$

where s_d is the increase in consumption of dairy products measured in servings (one serving is one cup or eight ounces) per child per day, n is the number of children affected, and f is the proportion of the consumption increase that is in the form of fluid milk. The sum of the daily fluid milk increases over a year, then, represents the total additional servings of fluid milk sold as a result of the nutrition education program that year. The servings can be converted to hundredweight by dividing by 195.6 (2 servings per pint, 8 pints per gallon, 8.18 pounds per gallon, and 100 pounds per hundredweight). So, the total gross producer benefit from *EYO* for a one-year period is equal to:

$$B = \$1.70 \frac{nf}{195.6} \sum_{d=1}^{365} s_d.$$

In 1996, there were approximately 1.4 million children in the sixth, seventh, and eighth grades in California. Since *EYO* is taught in roughly 80 percent of their classrooms, n equals 1.12 million. Because we aggregated the survey data into total dairy products and did not isolate fluid milk from other nonpremium products, f is unknown. In recent years, fluid milk has represented roughly 50 percent of total dairy consumption in the United States. In the absence of any other information, it was reasonable to assume that consumption patterns in California are similar to those of the nation and that the proportion of dairy consumption in the form of fluid milk did not change in response to the program. Thus, $f = 0.5$ seemed to be a suitable value, but we also varied f in order to explore the implications of different assumptions about the proportion of the total change in servings in the form of fluid milk.

We expected that dairy consumption would increase in response to the *EYO* program and that the magnitude of the increase would eventually depreciate over time. Thus, if we were to plot values of s_d against d, the number of days elapsed since the start of *EYO* instruction, we expected a path for s_d that depreciates toward zero as d increases. Our data included two point estimates along such a path, as reported in Table 13.7. Because the average number of days between Records 1 and 2 was twenty-one, the average increase in dairy consumption between those records, 0.3131, corresponded to s_{21}. Similarly, on average, thirty days elapsed between Records 2 and 3, so the average increase in dairy consumption between Records 1 and 3, 0.1924, corresponded to s_{51}. These point estimates provided no information regarding the shape of the path connecting the two points or the shape and height of the path before Day 21 and after Day 51. A measure of the longer-term effect of *EYO* beyond fifty-one days would be valuable information, but would require looking for changes in behavior in, for example, the year following *EYO* instruction, which was not possible with the available data since they were collected over a two-month period. Similarly, we do not know the time path of the effects of *EYO* on dairy servings between Record 1 and Record 2.

To model depreciation of daily increases in dairy consumption, we assumed that the path of the increases declined geometrically from the Day-21 estimate at a constant rate. We used the daily rate of depreciation observed between our two point estimates to estimate the daily increases in dairy consumption for Day 22 through the end of the year. The daily rate of geometric depreciation in the *EYO* effect was calculated as the value of δ that solves $s_{51} = s_{21} \times (1 - \delta)^{30}$ and is equal to $\delta = 0.0161$, 1.61 percent of the daily *EYO* effect per day. This rate of depreciation from Record 2 to Record 3 was used to interpolate the daily *EYO* effects between Records 2 and 3 and to project those beyond Record 3. For example, the estimated *EYO* effect on dairy consumption per child for the twenty-second day after instruction began would be 0.3081 ($s_{22} = 0.3131 \times (1 - 0.0161) = 0.3081$).

It was less clear how to treat the values of s_d prior to Day 21. We expected that increased dairy consumption began at some point between the beginning and the end of instruction. If we assumed that the program was immediately effective and that the *EYO* effect depreciated over the first twenty days at the same rate as it did after the twenty-first day, we could use the rate of depreciation described previously (1.61 percent) to calculate estimated increases in dairy consumption for Days 1 through 20. This method of calculating the changes in dairy consumption was likely to overstate the program's effects, though, since they probably started some number of days after initial instruction. We used this method to calculate an upper bound on *EYO* effects and the benefit from the program. A lower bound was calculated by assuming that consumption did not change until twenty-one days after instruction began. The true consumption change lies somewhere between these two alternatives. Almost certainly there was some change in consumption for some of the days between Day 1 and Day 20. Furthermore, children may have become more convinced of the benefits of improving their eating habits as the program progressed, so the daily increase in dairy consumption could have started at zero, increased over some number of days, and then decreased to the 0.3131 average that we observed on Day 21. As an intermediate estimate of the program's effects, we used s_{21} as the estimated increase in dairy consumption per child for each of the first twenty days, recognizing that this was likely to overstate the change in consumption for some days and understate it for others.

The changes in average dairy consumption per day for each of the first thirty days after instruction began were calculated according to each of the three methods previously described (the upper-bound, lower-bound, and intermediate estimates, respectively). Summing the intermediate estimates of the changes for the first thirty days yielded an increase in dairy consumption of 9.2 servings per child, or 10.3 million servings for all children in California affected by the *EYO* program. The corresponding lower and upper bounds were 2.9 to 10.4 servings per child or 3.3 to 11.6 million servings for all of California. If this entire increase was in the form of fluid milk, the intermediate estimate of gross producer benefits would amount to $89,317 for those first thirty days (with the lower-bound benefits equal to $28,362 and upper-bound benefits equal to $100,960). Regardless of the method used, a key parameter was the percentage of the increase that was in the form of fluid milk. As f, the proportion of the increase in dairy consumption in the form of fluid milk, decreases, producer benefits decrease proportionately. The lower-bound estimate of benefits was considerably lower than the intermediate or upper-bound estimates because there was assumed to be no effect on consumption for twenty of the first thirty days after instruction began. The benefits after the first month did not differ by method. Over the second interval of thirty days, dairy consumption under all three methods would increase by 6.4 servings per child relative to the preinstruction level, making a total increase of 7.1 million servings and a gross producer benefit of $62,040 (again for $f = 1$), and so on.

The upper-bound, lower-bound, and intermediate estimates of the benefits to producers from *EYO* were calculated for various assumptions about the proportion of the increase in consumption that was in the form of fluid milk and for various lengths of persistence of *EYO* effects.[5] These benefits were then compared to the costs of the program, which DCC records showed to be $150,126 per year. This cost figure included the cost of *EYO* materials, distribution, marketing, and personnel and excluded sunk costs such as rent, utilities, development, and research. The estimated benefit-cost ratios are summarized in Table 13.8 for the upper-bound (panel 1), intermediate (panel 2), and lower-bound (panel 3) estimates of the benefits in the first month.

Benefit-cost ratios greater than 1 indicate that the benefits to milk producers from the *EYO* program outweighed its costs. In each panel, scenarios above and to the right of the bold line represent combinations of assumptions under which the *EYO* program was profitable for milk producers. Those below and to the left of the line represent combinations of assumptions where the program was not profitable. Naturally, more of the combinations of assumptions were above the bold line (i.e., implying benefit-cost ratios greater than 1) when we used our upper-bound estimates of program benefits than when we used the lower-bound estimates.

Using our point estimates of changes in dairy consumption, benefits must have persisted into at least the second month for it to be possible for the program to have been profitable for producers. Then, whether it was profitable depended on the share of the increase in dairy consumption that was in the form of fluid milk and the number of months of persistence, as well as assumptions about the effects in the first twenty days. For instance, suppose we believe that 60 percent of the increase in dairy consumption was actually in the form of fluid milk. Using our upper-bound method of calculating the program benefits, if the effects of *EYO* lasted seven months, then the private benefits of the program just offset its costs. Using the intermediate estimate of program benefits, the *EYO* effect had to persist for at least eleven months for benefits to exceed costs. The benefit-cost ratio was always less than 1 in the lower-bound case, which required that at least 80 percent of the increase in dairy consumption be in the form of fluid milk for the program to pay for itself.

Conclusion

The message from our analysis is clear. The *Exercise Your Options* program did affect the eating habits of children who participated in it. Under a number of reasonable assumptions, the benefits to milk producers from increased milk consumption outweighed the costs of the program and the private benefit-cost ratio was greater than 1. But this outcome turns on three unknowns: the effects on consumption during the program, the fraction of the increase in dairy consumption that was fluid milk, and the persistence of the effect beyond the second month.

Table 13.8. Private Benefit-Cost Ratios for *EYO* Program under Various Assumptions

Percent Increase in Fluid Milk Form	Months of *EYO* Effect Persistence											
	1	2	3	4	5	6	7	8	9	10	11	12
Benefit-Cost Ratios Based on Upper-Bound Estimate of Program Benefits												
100	0.67	1.09	1.34	1.50	1.59	1.65	1.69	1.71	1.72	1.73	1.74	1.74
90	0.61	0.98	1.21	1.35	1.43	1.49	1.52	1.54	1.55	1.56	1.56	1.57
80	0.54	0.87	1.07	1.20	1.27	1.32	1.35	1.37	1.38	1.38	1.39	1.39
70	0.47	0.76	0.94	1.05	1.11	1.16	1.18	1.20	1.21	1.21	1.22	1.22
60	0.40	0.65	0.80	0.90	0.95	0.99	1.01	1.03	1.03	1.04	1.04	1.04
50	0.34	0.54	0.67	0.75	0.80	0.83	0.84	0.85	0.86	0.87	0.87	0.87
40	0.27	0.43	0.54	0.60	0.64	0.66	0.67	0.68	0.69	0.69	0.69	0.70
30	0.20	0.33	0.40	0.45	0.48	0.50	0.51	0.51	0.52	0.52	0.52	0.52
20	0.13	0.22	0.27	0.30	0.32	0.33	0.34	0.34	0.34	0.35	0.35	0.35
10	0.07	0.11	0.13	0.15	0.16	0.17	0.17	0.17	0.17	0.17	0.17	0.17
Benefit-Cost Ratios Based on Intermediate Estimate of Program Benefits												
100	0.59	1.01	1.26	1.42	1.51	1.57	1.61	1.63	1.65	1.65	1.66	1.66
90	0.54	0.91	1.14	1.28	1.36	1.42	1.45	1.47	1.48	1.49	1.49	1.50
80	0.48	0.81	1.01	1.13	1.21	1.26	1.29	1.31	1.32	1.32	1.33	1.33
70	0.42	0.71	0.88	0.99	1.06	1.10	1.13	1.14	1.15	1.16	1.16	1.16
60	0.36	0.60	0.76	0.85	0.91	0.94	0.97	0.98	0.99	0.99	1.00	1.00
50	0.30	0.50	0.63	0.71	0.76	0.79	0.80	0.82	0.82	0.83	0.83	0.83
40	0.24	0.40	0.50	0.57	0.61	0.63	0.64	0.65	0.66	0.66	0.66	0.66
30	0.18	0.30	0.38	0.43	0.45	0.47	0.48	0.49	0.49	0.50	0.50	0.50
20	0.12	0.20	0.25	0.28	0.30	0.31	0.32	0.33	0.33	0.33	0.33	0.33
10	0.06	0.10	0.13	0.14	0.15	0.16	0.16	0.16	0.16	0.17	0.17	0.17
Benefit-Cost Ratios Based on Lower-Bound Estimate of Program Benefits												
100	0.19	0.60	0.86	1.01	1.11	1.17	1.20	1.23	1.24	1.25	1.25	1.26
90	0.17	0.54	0.77	0.91	1.00	1.05	1.08	1.10	1.12	1.12	1.13	1.13
80	0.15	0.48	0.68	0.81	0.89	0.93	0.96	0.98	0.99	1.00	1.00	1.00
70	0.13	0.42	0.60	0.71	0.78	0.82	0.84	0.86	0.87	0.87	0.88	0.88
60	0.11	0.36	0.51	0.61	0.66	0.70	0.72	0.74	0.74	0.75	0.75	0.75
50	0.09	0.30	0.43	0.51	0.55	0.58	0.60	0.61	0.62	0.62	0.63	0.63
40	0.08	0.24	0.34	0.40	0.44	0.47	0.48	0.49	0.50	0.50	0.50	0.50
30	0.06	0.18	0.26	0.30	0.33	0.35	0.36	0.37	0.37	0.37	0.38	0.38
20	0.04	0.12	0.17	0.20	0.22	0.23	0.24	0.25	0.25	0.25	0.25	0.25
10	0.02	0.06	0.09	0.10	0.11	0.12	0.12	0.12	0.12	0.12	0.13	0.13

The social benefit-cost ratio is likely to be higher than that reported here. The target audience, teenagers, is at an age where peak bone mass is most affected, having implications for bone density and the likelihood of osteoporosis developing perhaps decades later. Likewise, increased consumption of fruits and vegetables is one of the goals of USDA's food pyramid (*Five A Day*), which is the basis of the nutrition education message presented in DCC's materials. Meeting the *Five A Day* goal is likely to reduce the incidence of cancer and other illnesses. Thus, improvements in the eating habits of today's teenagers could significantly reduce tomorrow's health care costs. These unmeasured cost reductions must be weighed against any economic losses in, for example, the candy industry.

An interesting issue for further work is whether, given the public-good nature of the benefits, there is an underprovision of nutrition education and whether government funds ought to be provided on some type of matching basis to encourage programs such as those by DCC. Complicating such an analysis, of course, is the fact that there are numerous other sources of health and nutrition information. Our study used a control group to hold those other effects constant. The continuing controversy over the use of mandatory checkoffs for commodity marketing programs and the new requirement in the 1996 Farm Bill for evaluation of such programs will make it necessary to broaden the analysis of typical commodity promotion to include programs such as DCC's nutrition education programs.

Notes

This chapter originally appeared as an article entitled "Doing Well by Doing a Body Good: An Evaluation of the Industry-Funded Nutrition Education Program Conducted by the Dairy Council of California" by Alston, Chalfant, and James in *Agribusiness: An International Journal* (© 1999 John Wiley and Sons). This material is used by permission of John Wiley and Sons, Inc. Several minor wording changes have been adopted in this chapter relative to the text of the original article for reasons of consistency with other chapters, correction of errors, and a preference for alternative wording. However, the bulk of the text and certainly all of the substance is identical to that of the original article.

1 Gray, Malla, and Stephen (1998) showed that the total social benefits from increased canola consumption in Canada were much greater than the direct consumer benefits because of the way the health care industry operates. Similar results would apply in the United States and for any product that has significant effects on private and public health care costs. Establishing the health benefits that come from changes in diet is difficult. Establishing the health benefits from increased dairy consumption is particularly difficult since positive effects (e.g., reducing the likelihood of osteoporosis) must be weighed against negative ones (e.g., increased risk of coronary heart disease). Gray and Malla (1998) focused on negative health effects associated with increased dairy consumption and incorporated the reduction

in health care costs associated with coronary heart disease into a welfare analysis of dairy production quotas.

2 These categories correspond to those on the USDA food pyramid except that we have separated what they call Extras into Extras (excluding sodas) and Sodas.

3 An exact t-test may be calculated under the assumption of equal variances for the two groups, using the usual pooled estimate of the variance (s_p^2):

$$t = (\bar{X}_2 - \bar{X}_1) \Big/ s_p\sqrt{\frac{1}{n_1} + \frac{1}{n_2}}\ .$$

Alternatively, an approximate t-test can be used when the assumption of equal variances is not made:

$$t = (\bar{X}_2 - \bar{X}_1) \Big/ \sqrt{\frac{s_1^2}{n_1} + \frac{s_2^2}{n_2}}\ .$$

As noted earlier, the estimated longer-run effects are simply the sums of the mean changes reported in the first two panels in each table. The standard errors of these estimated longer-run effects were obtained by taking into account an observed negative correlation between changes from Record 1 to Record 2 and those from Record 2 to Record 3 for the students for whom all three food records were available. Recall from Table 13.1 that these are the majority of the students in the sample. The same standard errors were used to construct the longer-run confidence intervals. We repeated the tests for statistically significant differences between groups using only the students with usable records for all three dates and found no change in our conclusions.

4 Indeed, an important component of many promotional messages in campaigns more commonly analyzed by agricultural economists is nutrition information—for instance, *Milk: It Does a Body Good.*

5 The same rate of depreciation was used regardless of the persistence of the *EYO* effect considered.

References

Alston, J.M., H.F. Carman, and J.A. Chalfant. "Evaluating Primary Product Promotion: The Returns to Generic Advertising by a Producer Cooperative in a Small, Open Economy." *Promotion in the Marketing Mix: What Works, Where and Why.* Proceedings from the NEC-63 Spring 1994 Conference. E.W. Goddard and D.S. Taylor, eds., pp. 145–167. Guelph Ontario: University of Guelph, 1994.

Alston, J.M., J.A. Chalfant, J.E. Christian, E. Meng, and N.E. Piggott. *The California Table Grape Commission's Promotion Program: An Evaluation.* Berkeley CA: University of California Division of Agriculture and Natural Resources, Giannini Foundation Monograph 43, November 1997.

Carman, H.F., and R.D. Green. "Commodity Supply Response to a Producer Financed Advertising Program: The California Avocado Industry." *Agribusiness: An International Journal* 9 (1993):605–621.

Frazao, E. *The American Diet: Health and Economic Consequences.* Washington DC: U.S. Department of Agriculture, Economic Research Service Agriculture Information Bulletin 711, 1995.

Freebairn, J. "Evaluating the Level and Distribution of Benefits from Dairy Industry Research." *Australian Journal of Agricultural Economics* 36 (1992):141–165.

Goddard, E.W., and A. Tielu. "Assessing the Effectiveness of Fluid Milk Advertising in Ontario." *Canadian Journal of Agricultural Economics* 36 (1988):261–278.

Goddard, E.W., and M.L. McCutcheon. "Optimal Producer Investment in Generic Advertising: The Case of Fluid Milk in Ontario and Quebec." *Canadian Journal of Agricultural Economics* 41 (1993):329–347.

Gray, R., and S. Malla. "A Note on Evaluating Agricultural Policy in the Presence of Health Care Cost Externalities: Dairy Production Quotas and Coronary Heart Disease Costs." *Canadian Journal of Agricultural Economics* 46 (1998):247–256.

Gray, R., S. Malla, and A. Stephen. "Canadian Dietary Fat Substitutions, 1955–93, and Coronary Heart Disease Costs." *Canadian Journal of Agricultural Economics* 46 (1998):233–246.

Kaiser, H.M., D.J. Liu, T.D. Mount, and O.D. Forker. "Impacts of Dairy Promotion from Consumer Demand to Farm Supply." *Commodity Advertising and Promotion*. H.W. Kinnucan, S.W. Thompson, and H.S. Chang, eds., pp. 40–57. Ames IA: Iowa State University Press, 1992.

Lee, H., J.M. Alston, H.F. Carman, and W. Sutton. *Mandated Marketing Programs for California Commodities*. Berkeley CA: University of California Division of Agriculture and Natural Resources, Giannini Foundation Information Series 96-1, August 1996.

Nerlove, M., and F.V. Waugh. "Advertising without Supply Control: Some Implications of a Study of the Advertising of Oranges." *Journal of Farm Economics* 43 (1961):813–837.

Piggott, N.E., J.A. Chalfant, J.M. Alston, and G.R. Griffith. "Demand Response to Advertising in the Australian Meat Industry." *American Journal of Agricultural Economics* 78 (1996): 268–279.

Sumner, D.A., and C.A. Wolf. "Quotas without Supply Control: Effects of Dairy Quota Policy in California." *American Journal of Agricultural Economics* 78 (1996):354–366.

Thompson, S.R., and D.A. Eiler. "Determinants of Milk Advertising Effectiveness." *American Journal of Agricultural Economics* 59 (1977):330–335.

U.S. Department of Health and Human Services. *The Surgeon General's Report on Nutrition and Health*. Washington DC, 1988.

Ward, R.W., and B.L. Dixon. "Effectiveness of Fluid Milk Advertising Since the Dairy and Tobacco Adjustment Act of 1983." *American Journal of Agricultural Economics* 71 (1989):730–740.

Wohlgenant, M.K., and Clary, C.R. "Development and Measurement of Farm-to-Retail Price Linkage for Evaluating Dairy Advertising Effectiveness." *Journal of Agricultural Economics Research* 44 (1992):18–27.

14

The Economic Value of Spin Control: Food Safety and the Strawberry Case

Timothy J. Richards & Paul M. Patterson

Growers and their commodity associations are fully aware that an outbreak of a food-borne illness can cause irreparable harm to the public's perception of the wholesomeness of their products. However, little is known about the exact cost to producers of such incidents, and less is known about the value of efforts used to counteract the negative press that invariably results. For example, California strawberry growers, who experienced incidents in 1996 and in 1997, calculated the cost of the 1997 incident at some $40 million, but this was admittedly only a rough estimate. Beyond the direct loss of not being able to sell contaminated product, any calculation of the true, long-run cost must include any damage done to a product's reputation as a safe and healthy food choice. Such changes in consumer perceptions may be long lasting, widespread, and very difficult to reverse. Complicating matters is the fact that the damage may emanate from one small firm in a fragmented, atomistic industry while all growers are made to suffer.

While the federal government considers imposing industrywide regulations, private growers are developing guidelines for self-regulation. Accurate measures of the economic losses they could potentially suffer are important to their efforts to evaluate the costs and benefits of adopting more intensive quality monitoring and process controls to prevent future outbreaks. And if self-regulation fails to prevent such outbreaks, the more detailed information can be used to quantify the benefits of any actions taken by commodity boards like the California Strawberry Commission (CSC) in the aftermath of an outbreak. To the extent that a product's "clean reputation" is a public good enjoyed by all producers at virtually no cost, an analysis of the costs and benefits of trying to dispel consumer concerns provides a measure of the amount that a group of growers, through their association, should be willing to spend on an industrywide system of contamination control irrespective of any new government regulation.

Growers already underwrite extensive advertising and promotion expenditures through commodity organizations such as CSC. Using traditional methods of marketing communication to counteract the effects of negative publicity, however, is not likely to be effective due to the different ways consumers process information from advertising and from the seemingly objective news media (Lord and Putrevu 1993). Although advertising offers the benefits of a high degree of message control and repetition (Kotler 1991), an increasing body of research questions the effectiveness of advertising because its inherent repetition tends to desensitize its audience to the underlying message (Tellis 1988; Lipman 1989). Sales promotion also has the potential to erode brand equity as consumers become conditioned to expect deals on their favorite products (Davis, Inman, and McAlister 1992). Publicity, on the other hand, is often believed to enjoy a credibility advantage, especially if the source is considered objective and unbiased (Levy 1989; Salmon et al. 1985). In fact, there are many possible explanations for the differential impact on behavior of positive and negative information.

Mizerski (1982) grouped the research on these explanations into three classifications: surprise and frequency of use, ambiguity and uncertainty, and differences in causal attribution. In a consumer-demand context, the first category reflects the notion that negative information reaches a market infrequently, so this "surprise value" tends to be regarded as more credible than a single positive message. However, this notion has little empirical support. The second explanation holds that negative information is somehow less ambiguous than positive information and consequently has a greater effect on behavior. This theory, again, has had only limited empirical support. Mizerski, however, provides more rigorous conceptual and experimental support for the third explanation–attribution theory.

Attribution theory (Kelley 1967) represents an explanation of the way decision-makers respond to behavior or signals provided by others who can be individuals, organizations, or any stimulus-provider. Such signals are regarded as credible representations of the true intentions of the provider if the cause of the signals can be attributed to interests of the provider that go beyond self-interest and reflect a wider concern–for, perhaps, other individuals or society as a whole (Bemmels 1991; Sparkman and Locander 1980). Determinations of causality are, in turn, made on the basis of three factors: consensus, distinctiveness, and consistency (Kelley 1967). If everyone in the provider's environment behaves the same way (a consensus) or gives the same signal, then the signals are likely to be viewed as credible. However, if the provider gives a signal that is unique (distinctive), then the cause of the signal is more likely to be attributed to self-interests and therefore viewed as less credible. Similarly, if a signal is inconsistent with prior behavior, then the cause is more likely to be viewed as self-interest and the signal as less credible. For example, if a firm consistently tries to cover up potentially harmful incidents, then claims to the contrary for all incidents in the future will be seen as lacking credibility. In the current application, therefore, attribution theory maintains that people are far more likely to believe and to act on information from a usually reliable source, particularly when the source could provide information that is counter to the firm's own interests. In contrast, mollifying information from a vested interest such as CSC would not be viewed as credible and would not be acted upon.

In early experimental work, Richey, McClelland, and Shimkunas (1967) found that positive initial impressions of individuals are easily reversed by receipt of some negative information, whereas negative initial impressions are both difficult to reverse and tend to persist despite new positive information. Further, Swinyard and Ray (1977) reported the results of an experiment wherein individuals who were labeled as "charitable" by Red Cross canvassers were more likely to donate to the cause in the future than those who were not so labeled. The analogy to negative media information about a potential food-borne disease outbreak is clear.

In this chapter, we quantify both grower losses due to food-borne disease outbreaks and the value of defensive media activities. For this purpose, we developed an econometric model that was based in attribution theory. Application of this approach to the disease outbreaks in strawberries that occurred in 1996 and 1997 demonstrates not only the value of avoiding future incidents but also the return to public relations expenditures intended to "control the spin" created by the media. In this sense, the source of the negative demand shock is not the outbreak itself but rather the public's exposure to information about strawberries that either is negative or that reduces the public's "perception of safety" in consuming strawberries. Similarly, defensive public-relations activities create potentially offsetting sources of information that can improve the perception of safety. Including measures of both of these variables in a dynamic model of demand provided estimates of both the long-run and short-run effects of each type of information.

First we provide a brief narrative of the events surrounding the strawberry disease outbreaks in 1996 and 1997 with particular attention paid to how these events were covered in the media. We then present a simple conceptual model of how consumers incorporate both positive and negative information provided by media reports into their decision-making. Next, we describe the empirical model used to estimate the effect of each information type on market demand and to calculate the resulting impacts on producer profits. In this model, negative and positive shocks to demand cause retail prices to change, which then shifts prices and quantities supplied at the grower level. Each of these variables was then incorporated in a simulation model of grower profits that was used to calculate the impact, both in the short run and in the long run, of the shocks to demand. Consequently, we were able to produce an estimate of the economic damage caused by the release of news that may have caused consumers to question the safety of eating strawberries. Because the model took into account the effect of both positive and negative news, it also provided some insight into the value of defensive public-relations activities taken by grower associations and, ultimately, the value of measures that could potentially be taken by growers to prevent future disease outbreaks. We conclude by drawing some implications from these results for the potential role of grower associations in overseeing and funding preventive measures.

Background on Disease Outbreaks Linked to Strawberries

Although outbreaks of food-borne diseases linked to produce have been reported by the Centers for Disease Control (CDC) periodically since the early 1980s, the first such incident concerning strawberries in the period covered by our study surfaced in May 1996. In late May, some 300 people in nine states complained of diarrhea, vomiting, weight loss, fatigue, and muscle aches–symptoms subsequently attributed to infection with the parasite *Cyclospora* (*Boston Globe* 1996). Initially, the Texas Department of Health proclaimed that California strawberries "were almost certainly the source" of the *Cyclospora*, but laboratory tests conducted by CDC could find no evidence of *Cyclospora* contamination in strawberries (*Portland Oregonian* 1996). It was not until some four weeks later, in July, that CDC and the Ontario Ministry of Health released a statement exonerating strawberries and placing the blame squarely on imported Guatemalan raspberries. However, much of the damage had already been done. Some stores reported reductions in strawberry sales of as much as 30 percent for at least three weeks following the Texas announcement (*Washington Post* 1996).

Almost exactly a year later, CDC reported another series of *Cyclospora* cases, but this time the link to imported raspberries was established much earlier. However, in March of 1997, 198 Michigan schoolchildren and teachers contracted hepatitis A from eating a frozen dessert made from contaminated strawberries. In this case, the source was clear and particularly easy to trace. Contrary to the laws governing the National School Lunch Program, the dessert's maker, Andrew and Williamson Sales Co., had acquired strawberries in Mexico for freezing and later sold them to the U.S. Department of Agriculture (USDA). As a result, the firm and its principal owner, Fred Williamson, were subsequently indicted on forty-seven charges of making false statements and one count of defrauding the United States (Kraul 1997). Initial claims by CSC put the cost of lost strawberry sales resulting from the negative publicity surrounding this case at $15 million—a figure that was later raised to $40 million but is still subject to much debate. Growers' chances of recovering economically from this incident and of preventing others in the future depend in large part on the ability of CSC to dispel consumers' perceptions of domestic strawberries as a food with a substantial risk of contamination. That ability, however, depends on consumer trust in CSC given its vested interest in selling strawberries. Attribution theory provided a logical basis from which to assess the ability of CSC to respond to these incidents in a credible way.

An Economic Model of Responses to Positive and Negative Publicity

Although the psychological and experimental bases for attribution theory are relatively well established (Mizerski 1982), the theory's implications for economic behavior are less clear. The usual approach is to assert that the demand for a commodity is a function of both positive and negative information received regarding its attributes and to leave the question of which type of information is dominant to empirical study (Swartz and Strand 1981). Chang and Kinnucan (1991) used attribution theory to motivate their expectation that negative information should exert a disproportionate influence on demand but did not develop an economic explanation for this asymmetry.

However, asymmetry is a natural outcome of an assumption economists usually make regarding consumer preferences—the law of diminishing marginal utility. Essentially, this "law" implies that the more units of a good a consumer has, the less another unit of that good will add to the consumer's well-being or utility. The same is true for a product reputation. If a certain good enjoys a generally positive reputation for safety, new information that reinforces this reputation is not likely to generate much of an incremental gain. However, information that diminishes this reputation is likely to have a relatively large negative effect on consumer utility. So, while Mizerski (1982) developed the psychological basis for attribution theory, its predictions naturally arise from well-behaved consumer preferences. Few of the empirical models of negative and positive publicity in the literature, however, take this preference structure into consideration.

Empirical Model of the Value of Disease Information

A proliferation of legislation designed to protect growers of fruits and vegetables from unsubstantiated claims of chemical and bacterial contamination suggests that there is a need for economic research on the effects of negative publicity on grower returns. Although the legislation is a relatively recent phenomenon, there have been significant cases in which contamination, a disease outbreak, or simply the release of new scientific research results have adversely affected consumer demand for a product.[1] Swartz and Strand (1981) were among the first investigators with their evaluation of the impact of purely negative information regarding a keptone contamination of the James River oyster beds in the 1970s. Their proxy variable for negative information consisted of a weighted sum of negative print-media articles appearing in the local Baltimore market regarding the contamination where the weights were estimates of the probability that a given story was read. This variable, however, did not allow for the potential differential effects of positive and negative media exposure.

Recognizing that such incidents are usually met with strategic responses from either government agencies interested in maintaining the safety of the food supply or private firms with a vested interest in preventing lost sales, Smith, van Ravenswaay, and Thompson (1988) estimated the effects of both negative and positive media articles on demand for milk following the 1982 heptachlor contamination in Oahu. Using a statistical procedure to determine the appropriate news-variable specification, they rejected models containing both types of information together in favor of including only negative publicity. This result, they argued, supported Weinberger and Dillon's (1980) claim that negative information dominates a similar amount of positive information–an asymmetry similar to the one we described previously. They also explained the insignificance of positive information as evidence that dairy processors lacked credibility due to their vested interest in maintaining product sales. Chang and Kinnucan (1991) took the separation of positive and negative information one step further in a model of Canadian butter demand. By considering separate variables measuring negative information about the effect of dietary cholesterol on the likelihood of heart disease (Brown and Schrader 1990) and positive information from generic butter advertising, they estimated the ability of promotion to overcome negative media reports. Although it is difficult to compare the relative impacts of two variables measured in different ways, the negative information parameter (elasticity) was an order of magnitude greater than the advertising parameter (elasticity) in absolute value, suggesting that industry efforts were only partially successful in counteracting reductions in demand from adverse media exposure. However, despite defining both negative and positive information in terms of stocks rather than flows, Chang and Kinnucan (1991) did not consider the relative dynamic effects of either. Nonetheless, the methods used to quantify information flows in these studies offered great potential for valuing investments in publicity and media relations.

In one of the first attempts to create a quantitative measure of information, Brown and Schrader (1990) constructed an information index regarding the connection between cholesterol and cardiovascular health by counting the number of medical journal articles on the topic over a multiyear period. The number of articles that argued against a relationship between cholesterol and health were subtracted from the number of those claiming otherwise to produce a cumulative "net" information variable. This index was used in a model of shell-egg demand to show the effects of cholesterol information on the demand for eggs. The information index used in our study differed from Brown and Schrader in several respects. First, we considered the number of articles in popular media outlets rather than in the academic literature. Information disseminated through medical journals is likely to take a long time to reach consumers and even then may be confusing and inconclusive. Media articles, however, convey information to consumers (even when it is incorrect) immediately, thereby changing their behavior shortly after dissemination. Second, Chang and Kinnucan (1991) recognized that information must be regarded as a stock variable because it accumulates over time and decays with memory loss and obsolescence. Negative information accumulates as a weighted sum of each period's number of net negative articles with the weights in proportion to the total number of articles in each period that are defined as negative. Given the performance of this variable in their research, we constructed similar indices for both positive and negative media articles. Although this procedure implicitly assumed that articles from different media outlets provided equivalent information content, the errors that were introduced through this assumption were likely to be random, thereby producing unbiased estimates.[2] Third, we did not impose the assumption that positive information exactly offsets negative information. Other studies (Chang and Kinnucan 1991; Brown and Schrader 1990; Smith, van Ravenswaay, and Thompson 1988) developed information indices in which positive and negative articles offset one another to produce a net positive or negative index value. Attribution theory, on the other hand, suggests that consumers are likely to place more weight on negative than on positive information. Smith, van Ravenswaay, and Thompson (1988) suggested adding the number of positive and negative articles together, reasoning that a negative impression about a product is reinforced by all types of publicity regardless of whether the message itself is positive or negative. However, we did not adopt that approach because the assumption that consumers ignore a valuable source of information is not tenable if the underlying assumption is that consumers are rational. Using an index of net negative information allowed previous studies to capture the effect of negative publicity in a parsimonious way, but it constrained market responses to positive and negative articles by making them equivalent, so we constructed a separate index for negative and positive articles. We used these indices to estimate the impact of both positive and negative media exposure on the demand for strawberries and, ultimately, on grower returns.

To measure the effect on grower returns, we used a simulation model to first calculate the net grower-price effects of negative or positive news regarding strawberries and then to estimate the net loss in producer profits (Kinnucan, Xiao, and Hsia 1996). The approach used a simple but complete model of a strawberry market equilibrium consisting of equations for demand, supply, retail-farm price linkage, and market clearance. The solution to this system expressed the change in grower price resulting from changes in each exogenous demand factor–including releases of both positive and negative information. These shocks cause changes in grower price, commodity supply, and, ultimately, producer profits. To accurately estimate the change in profits with this model, however, it was important to have precise estimates of the impact of news articles on demand.

Our econometric model of demand reflected a number of critical facts regarding strawberry marketing: (1) prices adjust to clear the market, (2) any shock to demand is likely to persist beyond one period, (3) strawberry demand is likely to be closely related to demand for other potentially substitutable fresh fruits, (4) strawberry demand is highly seasonal, and (5) news regarding food-borne-illness outbreaks constitutes a significant shock to demand. Consequently, our model consisted of a dynamic system representing monthly U.S. fresh fruit consumption. It was assumed that price adjusts to clear the market since in monthly data supply is likely to be predetermined. As such, the model is formally described as an inverse demand system. Although supply was fixed for each period, consumer buying behavior tends to reflect habits, learning, and experience that persist over time (Ray 1984; Blanciforti and Green 1983; Chen and Veeman 1991; Pollak 1970). Specifying a dynamic demand model to account for these effects was particularly important in this case because the principal effect of adverse publicity could be to break consumption habits and instill new, unfavorable impressions of the good in consumers' minds. Produce items are generally thought to be "experience goods" (Nelson 1974), so establishing a brand reputation for consistently high quality is difficult. Therefore, when this image is broken, it is difficult to re-establish. Although there is no *a priori* information to determine the specific form of the dynamic effects of information on prices, it is likely that shocks to demand have an immediate impact that then recedes slowly over time as consumers either forget the information received or believe that the original problem has been corrected. Therefore, the demand model assumed that current-period consumption of each type of fruit depended in part on its consumption in the previous period in a manner similar to that described in Ray (1984).[3] This specification allowed estimation of both long-run and short-run responses to demand-side shocks.

A system of fresh-fruit-demand equations that included strawberries, apples, bananas, grapes, and oranges was estimated to ensure that the demand-response parameters were consistent with established economic theory on rational consumer behavior. The demand system also accounted for the observed seasonality in strawberry demand. Similarly, we incorporated indices of good and bad news on strawberry consumption. Sources for monthly measures of these and the other model variables are discussed in the following section.

Data and Methods

Estimates of the fresh-fruit-demand-system parameters were obtained using monthly price and quantity data for each of the major fresh fruits evaluated over a period that included both of the recent disease outbreaks attributed to strawberries. Specifically, the data for strawberry prices and quantities came from USDA and the Bureau of Labor Statistics (BLS). Because annual per capita consumption aggregates published by USDA cannot capture the immediate effects of information shocks to the market, monthly arrivals reported in the USDA Market News Service's *Fresh Fruit and Vegetables: Arrivals in Western Cities* and *Fresh Fruit and Vegetables: Arrivals in Eastern Cities* data set (hereafter referred to as *Arrivals*) for 1995–1997 provided the quantity data. Arrivals to six Eastern markets (Atlanta, Baltimore, Boston, Detroit, Philadelphia, and Pittsburgh) and six Western markets (Chicago, Dallas, Los Angeles, San Francisco, Seattle, and St. Louis) were recorded, giving 432 panel observations. Although *Arrivals* included additional markets, we excluded markets for which strawberry arrivals in several months were less than one standard unit of measurement (100,000 pounds). With the declining relative importance of terminal markets represented by the arrivals data, there was some concern about how well arrivals data would correspond to amounts destined for retail consumption. However, monthly CSC shipment data included movements of all strawberries on a national level and therefore equaled consumption plus loss in transport. The average correlation between arrivals to the eleven markets in the sample and national CSC shipments was 91.6 percent. Therefore, we are confident that arrivals represented a close approximation to the amounts ultimately consumed at the retail level. Moreover, because the arrivals data represented supply that was not previously committed under contract, fluctuations in these marginal shipments would cause changes in the market price. Monthly arrivals data also permitted estimation of any seasonal effects on demand. Strawberry prices, reported on a retail dollar-per-pound basis, for nine months of the year were supplied by the BLS data set *Consumer Price Index: Average Price Data* (hereafter referred to as *Average Price*). Shipments during October, November, and December were insufficient to determine a representative price, so only the other nine months per year remained in the sample for estimation purposes.

Assuming fresh-fruit consumption is independent from consumption choices regarding other foods and goods, we included apples, bananas, oranges, and grapes in the set of alternative products. Consumption of these other fruits was again approximated using *Arrivals* data, and retail prices were also found in BLS's *Average Price* database. As in the strawberry case, each price series was expressed on a real per-pound basis and was measured at the retail level.

With data on both farm (or F.O.B.) and retail prices, it also would have been possible, and preferable, to estimate the extent to which changes in retail demand passed through to changes in farm prices—the retail-farm price transmission elasticity. However, in the absence of data on F.O.B. prices, the transmission elasticity was calculated using a widely accepted formula (Gardner 1975). Because there was some question regarding how responsive producers or shippers are to changes in strawberry prices (i.e., the elasticity of supply) in a monthly time period, the change in producer returns due to favorable and unfavorable news articles was calculated for elasticity-of-supply values ranging from 0.01 to 5.00. At one extreme, a 1 percent increase in price would only elicit a 0.01 percent increase in supply. More responsive behavior was captured by the case where a 1 percent increase in price induces a 5 percent increase in supply.

The number of favorable and unfavorable articles per month was determined by searching the top fifty newspapers in terms of circulation in the U.S. through the *Dow Jones News Service* for the words strawberry, disease, hepatitis, and *Cyclospora*. By searching the top newspapers by circulation, this procedure not only ensured far broader coverage of U.S. consumers than that used in Brown and Schrader's (1990) approach but also minimized the likelihood that the medium would add a source of uncertainty to the credibility of the message. After reviewing either the full article or its abstract, we determined whether the article created a favorable or unfavorable perception of the safety of consuming strawberries.[4] Although it is well known that the impact of a story varies by its location within the paper and on the page, the *Dow Jones News Service* does not provide sufficient information to rate articles on that basis. Consequently, the articles were assumed to be relatively homogeneous in exposure. As attribution theory predicts, an article's impact would depend not only on exposure but also on the credibility of the source. However, any adjustments we might attempt to make for credibility while evaluating the articles would be arbitrary and subjective. Instead, each article was recorded as creating an unambiguously positive or negative impression of strawberry safety. Although the process of arriving at a definitive assessment of each article may seem arbitrary, upon reviewing the articles it became immediately clear which conclusion to draw. For example, the initial articles written in May 1996 following the *Cyclospora* outbreak were unambiguously negative; each cited the Texas Department of Health's claim that California strawberries were "almost certainly" the source of the disease. Later articles, however, cast doubt on this conclusion and pointed instead to imported raspberries as the cause. These were coded as positive because they so clearly contradicted the initial claims. By coding each article one way or the other, the econometric model determined statistically whether consumers act as if the message is credible or not and whether, as a result, producer returns should rise or fall.

Finally, the entire econometric model was estimated using data for all time periods across all markets. Once the parameter estimates were obtained, producer profits were calculated under two simulation scenarios using data over the entire sample period: the first assumed only unfavorable media exposure while the second allowed for both good news and bad news regarding strawberries. In each case, the amount of information was defined as the cumulative total number of articles from January 1, 1995, to October 31, 1997. The grower loss or benefit of each type of information was calculated for each month, converted to a present value on January 1, 1995, as an annuity at a 5 percent rate of interest, and then expressed in November 1997 dollars through a future-value calculation. By comparing the producer profits that resulted, we were able to isolate the value of providing "defensive" information to the market to counter the negative effects of adverse media information. The results of each of these simulations are presented and interpreted following a brief discussion of the demand-model estimates.

Results and Discussion

Although our primary concern is with the effect of positive and negative media exposure on strawberry-producers' returns, estimates of the media's effect on the demand for both strawberries and for other fruit was of some interest. From the broader perspective of the entire U.S. fruit and vegetable sector, spillover effects from strawberry news on the demand for substitute products could mean that the total damage to the sector is far less than to the strawberry industry in particular.

To that end, Table 14.1 shows all of the estimated demand model parameters. Although the sample period is relatively short, the model appeared to fit the data quite well, as shown by the R^2 values, which exceeded 90 percent for each equation in the system. Furthermore, all of the model parameters were statistically significant. As an inverse demand system, the estimated demand parameters were termed flexibilities and showed how prices change in response to changes in market quantities. This is roughly the inverse of elasticities, which show how demand quantities respond to price changes, that are commonly discussed by market analysts. With elasticities, negative cross-price elasticities indicate complementarity. In an inverse demand model, negative cross-price flexibilities suggest that, as the supply of one increases, the price of the other is expected to fall, albeit less than proportionately. Table 14.1 clearly shows that all pairs of commodities were indeed substitutes.

Analysis of the dynamics of price movements for the fruits in this system showed that shocks that affect price persist for about two months before returning to a long-run equilibrium price. One explanation for this "overshooting" may be the tendency for markets to overcompensate for unexpected news. Given an uncertain situation, risk-averse buyers and sellers tend to err on the side of conservatism and expect the damage to be worse than it actually is. On the other hand, initial reports may later be proven wrong. In the case of the 1996 *Cyclospora* scare, blame initially placed on contaminated strawberries was later redirected to the true culprit–imported Guatemalan raspberries. This situation explains a large price drop that is followed by a gradual return to some long-run equilibrium.

The flexibility estimates also showed that both negative and positive media exposure have significant effects on commodity prices and, as attribution theory suggests, that their impacts are not symmetrical. For example, the estimated strawberry-price flexibility for bad news was −0.65, while the good-news flexibility was 0.34. Hence, a 1 percent increase in bad news decreases price by 0.65 percent while a 1 percent increase in good news increases price by only 0.34 percent. Thus, consumers apparently react strongly to negative information about a product and are less responsive when presented with favorable news. Because many of the positive articles included in this study were corrections of previous misinformation, this suggests that the damage is done once bad news is made public and little can be done to reverse it. It may also be the case that favorable information provided by CSC is viewed as "nonunique" or as reflecting the commission's own interests and so is largely discounted by consumers. Rather than release defensive communiques directly, therefore, it is clearly in a commodity board's best interest to filter the information through organizations such as universities or government research agencies that consumers perceive as objective.

It is also interesting to compare the relative magnitudes of the short-run negative- and positive-information flexibilities based on the number of news articles. Note that elasticities and flexibilities typically measure the response of one continuous value to changes in another. However, our article index was an integer–a discrete measure–because an author cannot produce half an article. Given this, one more negative article represents a 45 percent increase

Table 14.1. Fresh Fruit Demand Flexibilities: Short-Run and Long-Run

	Short-Run Flexibility with Respect to:								
Price	Q_S	Q_A	Q_B	Q_G	Q_O	Total Q	Bad News	Good News	R^2
P_S	−0.3527	−0.1804	−0.1462	−0.1599	−0.1236	−0.9504	−0.6482	0.3431	0.9136
	(−14.0523)	(−18.1807)	(−12.6448)	(−13.7142)	(−13.6225)	(−37.2245)	(−11.5284)	(7.7227)	
P_A	−0.4251	−0.0885	−0.1384	−0.1970	−0.1942	−1.0576	0.3947	−0.2641	0.9295
	(−19.9402)	(−2.8818)	(−9.4437)	(−7.7015)	(−9.2068)	(−50.0441)	(8.5569)	(−7.2134)	
P_B	−0.2831	−0.1092	−0.3137	−0.1520	−0.1482	−1.0081	0.3499	−0.1524	0.9394
	(−12.8734)	(−8.5395)	(−17.1112)	(−10.4442)	(−12.3651)	(−40.0461)	(6.5114)	(−3.5401)	
P_G	−0.3877	−0.2022	−0.1935	−0.1019	−0.1327	−1.0241	0.3144	−0.1799	0.9233
	(−14.4021)	(−7.5445)	(−10.8088)	(−2.6783)	(−5.4036)	(−38.5034)	(5.3680)	(−3.8625)	
P_O	−0.3814	−0.2593	−0.2019	−0.1679	−0.2055	−0.9963	0.4386	−0.1906	0.9156
	(−14.7932)	(−9.4288)	(−12.0386)	(−5.4328)	(−4.2946)	(−40.3527)	(8.1635)	(−4.4961)	
	Long-Run Flexibility with Respect to:								
Price	Q_S	Q_A	Q_B	Q_G	Q_O	Total Q	Bad News	Good News	R^2
P_S	−0.3944	−0.2017	−0.1635	−0.1788	−0.1382	−1.0626	−0.7247	0.3836	0.9136
	(−13.0542)	(−17.9645)	(−12.5186)	(−13.6498)	(−14.1695)	(−32.3398)	(−11.8713)	(7.8053)	

Continued on following page

Table 14.1 (continued)

Price	Long-Run Flexibility with Respect to: Q_S	Q_A	Q_B	Q_G	Q_O	Total Q	Bad News	Good News	R^2
P_A	−0.4687	−0.0976	−0.1526	−0.2173	−0.2142	−1.1662	0.4353	−0.2912	0.9295
	(−18.7035)	(−2.8259)	(−9.4522)	(−7.4468)	(−9.2664)	(−29.1502)	(8.7174)	(−7.2605)	
P_B	−0.3005	−0.1159	−0.3331	−0.1614	−0.1573	−1.0704	0.3716	−0.1618	0.9394
	(−12.4656)	(−8.4280)	(−14.5151)	(−10.4392)	(−12.9569)	(−29.4812)	(6.5557)	(−3.5467)	
P_G	−0.4434	−0.2312	−0.2212	−0.1165	−0.1517	−1.1710	0.3596	−0.2057	0.9233
	(−13.9179)	(−7.3302)	(−11.3101)	(−2.6372)	(−5.3876)	(−27.5059)	(5.5465)	(−3.8928)	
P_O	−0.4127	−0.2035	−0.2184	−0.1816	−0.2166	−1.0778	0.4745	−0.2063	0.9156
	(−14.4122)	(−9.1573)	(−12.3309)	(−5.3802)	(−4.2885)	(−25.3521)	(8.2803)	(−4.4918)	

Key: S = **strawberries; A = Apples; B = Bananas; G = Grapes; and O = Oranges**.
Notes: The values in parentheses are t-ratios. All elasticity estimates are significant at the 5 percent level.The "Bad News" flexibility estimate represents the percentage change in price of the row commodity with respect to a 1 percent rise in the index of negative articles regarding strawberry consumption while the "Good News" estimate represents the price impact of positive articles.

in the bad news index while one more positive article is a 76 percent increase in the good news index, on average. Weighting the price-response flexibilities by these proportionate changes in each index means that, if one more negative article is countered with a single positive article, we would expect to see a net 3 percent reduction in prices (−29 percent due to the negative article and a 26 percent rebound due to the positive article). Such a "tit for tat" strategy, therefore, means that CSC can almost entirely negate the impact of one additional negative article by producing a positive one even though the estimated price-response parameters suggest that good news has less of an impact on price than bad news.

Given that both positive and negative messages are likely to be remembered only for a limited amount of time, the difference between the short-run and long-run effects of news stories is also of considerable interest. Our econometric estimates showed that price responses to market shocks occur relatively rapidly and then prices return to a long-run equilibrium within two periods (months). Consequently, there is little difference between the short-run and long-run flexibilities. However, both the short-run and the long-run estimates were significantly higher than those implied by demand elasticities reported for similar commodities (Huang 1994; Lee, Brown, and Seale 1992; Richards 1997). Some of the difference between this study and prior results probably comes from this study using monthly data as opposed to annual data. Further, quantity-dependent demand systems will suffer from misspecification bias when prices actually adjust to clear the market following a shock to supply. With a short-run strawberry price flexibility of −0.35 and a long-run flexibility of −0.39, large changes in supply are required to cause prices to change significantly. This rigidity is consistent with the strong substitute relationships that the cross-flexibility estimates suggest. If consumers regard other fruits as close substitutes for strawberries, they will switch to other products rather than pay much higher strawberry prices. The long-run bad-news price flexibility of −0.72 indicates that one more negative article at the height of the 1997 scare (a 6.7 percent increase) would have caused a long-term price reduction of approximately $0.31 per carton (1.9 percent), suggesting that strawberry prices are relatively sensitive to even a small amount of bad news. Such estimates of the incremental cost of negative information are helpful in guiding growers' efforts to counter bad publicity through their own media programs.

Most of the impact of positive media exposure occurs in the period in which it is released. Similar to the negative-information case, prices are inflexible in response to positive news, having short-run and long-run flexibilities of 0.34 and 0.38, respectively. As explained previously, the fact that these flexibilities are lower than their negative counterparts suggests that equal percentage changes in each type of exposure will cause prices to fall. An examination of the cross-flexibilities in Table 14.1, however, shows that news of disease outbreaks related to strawberries did not necessarily reduce the demand for all fruits. In fact, the price of each alternative fruit rises on the news of a strawberry disease scare but falls in response to positive strawberry information. Produce growers as a whole, therefore, are not necessarily worse off as a result of negative media exposure. A more complete analysis of this issue, one that is beyond the scope of this study, would estimate a two-stage model that would include demand for the entire fruit category. If total fruit demand fell in response to negative information dominating positive, then perhaps industrywide programs would be required rather than commodity- and outbreak-specific measures.

With respect to strawberry growers alone, however, long-run cumulative impacts on grower returns are more clearly shown by calculating the change in producer profit rather than by simply looking at changes in price. To this end, the simulation results in Table 14.2 show both short-term and long-term effects on producer profits of both types of media information. To compare the change in profit that arises due to changes in the amount of each type of information, an increment was defined as a 10 percent rise in each information index while the other was held constant.

Prior to interpreting these results, however, we offer some caveats regarding their generality. In the short-run scenario, it was assumed that strawberry buyers do not completely adjust their buying behavior in response to the lower prices caused by a disease scare, nor do they remember negative media information for more than the current month. To the extent that these assumptions are unrealistic, this scenario represents an understatement of the true loss to growers. The cost estimates are also sensitive to the extent to which growers and shippers respond to lower prices by supplying fewer strawberries to the market. Specifically, the less elastic the supply is, the more price will fall in response to a shock to demand. Therefore, these simulations consider a range of supply elasticities from 0.01 (highly inelastic) to 5.0 (very elastic). Assuming a very low supply elasticity (0.01) and short-run price and information adjustments and focusing on the case where consumers are exposed to only bad news, growers would lose a total of $232.6 million over the sample period—19.6 percent of the total value of all shipments. Including the effects of positive media information reduces the total loss to $109.5 million, which implies that positive information is worth $123.2 million in this case. Under a scenario where growers and shippers are better able to adjust strawberry supplies in response to price changes (supply elasticity = 1.0), the loss due to negative information falls to $187.8 million and the net loss becomes $90.1 million, so the value of spin control in this case is $97.7 million. Public-relations officials at CSC should compare this value to the cost of mounting a concerted effort aimed at reassuring consumers of the safety and nutritional value of strawberries through mass media outlets. More realistic estimates of this value, however, would take into account consumers' long-run response to both positive and negative reports regarding the health implications of buying strawberries.

In fact, the results in Table 14.2 show that the long-run effects are somewhat greater than the short-run effects, as expected. Again, assuming that supply is virtually fixed (supply elasticity = 0.01, which is a less tenable proposition in the long run), the loss due to negative information rises to $273.4 million and the value of media efforts to ameliorate the damage rises to $144.8 million. Although these damage estimates are far higher than the rough estimates provided by CSC officials during the 1997 scare ($40 million), they become closer as the assumed elasticity of supply rises. Specifically, if the elasticity is 5.0 instead of 0.01, then the loss due to bad news falls to $83.5 million and a positive media response reduces this loss to $49.7 million. Therefore, the value of a campaign to either correct misinformation or to change consumer

Table 14.2. Loss of Producer Returns due to Unfavorable Media Exposure and the Value of Media Defense from a 10 percent Shock to the Number of Articles

	Short Run (million dollars)			Long Run (million dollars)		
Scenario Supply Elasticity	Bad News	With Defense	Value of Defense	Bad News	With Defense	Value of Defense
0.01	-232.656	-109.493	123.163	-273.424	-128.649	144.775
0.1	-227.585	-107.369	120.216	-266.978	-126.002	140.976
0.5	-207.424	-98.612	108.812	-241.975	-115.232	126.743
1.0	-187.800	-90.056	97.744	-218.559	-105.170	113.389
2.0	-161.250	-78.634	82.616	-189.173	-97.585	91.588
5.0	-140.828	-92.257	48.571	-83.518	-49.705	33.813

Notes: All cases represent the cumulative cost or benefit of all media exposure from January 1994 to October 1997. Producer returns are in millions of current dollars.

perceptions is $33.8 million. Given even this most conservative estimate, it appears that the amount spent by CSC during the 1997 outbreak, which officials estimated was roughly $250,000, was money very well invested. To be useful as a practical guide to investing in media programs, it is necessary to demonstrate how sensitive these results are to changing the definition of a publicity response or to how news is quantified.

Table 14.3 shows the change in producer profit that results from defining an increment to the stock of information due to a single negative or positive article in an average month. Given the relatively small number of articles reaching the media each month, these estimates are far higher than the scenario considered previously. Whereas both positive and negative media reports have very large independent effects on producer returns, the net effect is relatively small because a single-article change represents a far higher percentage change for good news than it does for bad news. Specifically, in the short run the net impact with inelastic supply is −$32.3 million and rises to −$14.1 million under an assumption of unitary supply elasticity (supply elasticity = 1). In the long run, inelastic supply produces an estimate very close to the "best guess" made by CSC officials ($38 million) and the assumption of unitary elasticity yields a net cost of $13.4 million to growers. Although these results also suggest that publicity is a good investment, the question remains whether growers' money could have been put to even better use.

Table 14.3. Loss of Producer Returns due to Unfavorable Media Exposure and the Value of Media Defense under a One-Article Shock Simulation

	Short Run (million dollars)			Long Run (million dollars)		
Scenario Supply Elasticity	Bad News	With Defense	Value of Defense	Bad News	With Defense	Value of Defense
0.01	-899.335	-32.347	866.988	−1,056.203	-37.980	1,018.223
0.1	-861.837	-30.527	831.310	−1,006.814	-35.530	971.284
0.5	-714.526	-22.532	691.994	-814.502	-24.761	789.741
1.0	-570.818	-14.143	556.675	-628.509	-13.389	615.120
2.0	-361.515	-0.486	361.029	-347.125	7.339	354.464
5.0	163.124	263.324	100.200	-126.225	-58.041	68.184

Notes: All cases represent the cumulative cost or benefit of all media exposure from January 1994 to October 1997. Producer returns are in millions of current dollars. Scenarios assume an average of one more positive and negative news item per month.

True to the cliche, in this case an ounce of prevention may indeed have been worth more than a pound of cure. Preventing such a large loss not just in revenue but in *profit* provides ample justification for investing in an industrywide safety-control program. The problem is that a reputation for being disease-free is a classic example of a public good. Because all growers benefit from this product image but no individual can be compelled to pay for the benefit, the private market fails to provide any protection at all. If it is rational for one grower to be a "free-rider"—to enjoy the benefits of the reputation without contributing anything to the cost of its protection—then it is rational for all to be free-riders. However, commodity organizations were initially established to provide another type of public good—generic commodity promotion. It seems natural, therefore, that commodity groups could treat contamination prevention as an equal and companion objective to promoting their products. Mandatory grower check-off fees could fund inspectors and industry-designed standards that would obviate the need for more government regulation of growers. In the strawberry industry, some of the largest vertically integrated growers have recognized the fundamental truth in this logic by investing in their own hazard analysis and critical control point (HACCP) programs. To the extent that their strawberries are seen as close substitutes for all others, however, they are still subject to the risk that one culprit will again damage the buying public's image of strawberries as a safe and healthy food.

In fact, as a result of the 1997 outbreak, CSC has taken steps in this direction by establishing a position for a food-safety director. While the director has no enforcement abilities, she has been successful in helping growers and shippers establish systems of tracing strawberries back to their source, has provided them with information on existing sanitation and hygiene standards, and is working toward industry-standard hazard analysis and critical control point programs. The creation of this position is but one part of an industrywide approach to maintaining food safety that is based on principles of grower-shipper self-interest and self-reliance, generating praise for CSC throughout the fruit and vegetable industry.

Conclusions and Implications

Whether it is beef growers arguing the amount of damage inflicted by Oprah Winfrey's comments or strawberry growers placing a value on one individual operation's indiscretion, estimates of profits lost due to a media-borne shock to an agricultural market have typically been little better than guesses. Accurate information on these costs is important to growers and their associations in establishing the value of a defensive response through promotion and/or news releases—spin control. This study developed a model by which to quantify the effects of media reports regarding a disease outbreak on the profits of commodity growers. An application of this model to the U.S. strawberry market from 1995 through 1997 demonstrates that the lost profits can be significant.

The approach taken in valuing negative media reports and efforts to counter those reports involved estimating a model of strawberry demand that allowed for the independent effect of negative and positive news. The psychology of "attribution" suggested that each type of news would have a different effect on consumers and would depend in part on the credibility of each source of information and the vested interests perceived as playing an important role in making news widely known. Quantitative measures of the amount of either type of news were developed by counting the number of positive and negative media articles appearing in the popular media on a monthly basis. These shocks were allowed to persist over time by specifying an econometric model that allowed for slow degeneration of "public memory" regarding adverse news stories. Estimates from this model showed that positive and negative media articles have the expected effects on price but that negative reports have a greater effect on price than positive ones—exactly the result predicted by attribution theory. Simulations of a producer-profit model that incorporated these estimated impacts found the loss due to bad news regarding strawberries to range between $273.4 million and $83.5 million in the long run depending upon the supply elasticity assumption made. Estimates of the value of defensive media efforts ranged from $144.8 million to $33.8 million, which were both many times the amount actually spent by CSC during the most recent outbreak.

This result has important implications for the role and function of commodity organizations in California and elsewhere. If a product's image among consumers as a safe and healthy choice is considered a public good, then grower associations have an incentive to broaden their missions to include not only generic commodity promotion but also industrywide efforts at preventing disease outbreaks. Individual firms' efforts to develop hazard analysis and critical control point programs or to adopt new production practices can be undermined by a single individual or operation taking one too many risks. Recognizing this fact, CSC has taken steps to remind growers and shippers of their responsibility for the integrity of the "California Strawberry" brand and to help the association develop food-safety regulations that cover all industry members.

Notes

This article is based on work previously presented in "The Economic Value of Public Relations Expenditures: Food Safety and the Strawberry Case" appearing in the *Journal of Agricultural and Resource Economics*, volume 24, issue 2, page 440–462. Readers seeking additional details are referred to that earlier article. An earlier version of this chapter benefitted from helpful comments by Henry Kinnucan.

1. This is commonly known as "veggie hate crime" legislation. Texas beef growers' unsuccessful lawsuit against Oprah Winfrey over her public comments regarding the safety of beef is a recent example, albeit in the context of meat rather than fruits and vegetables.
2. Other studies have attempted to define the information content specific to each article through a subjective rating system (Smith, van Ravenswaay, and Thompson 1988), but these calculations were arbitrary and likely created a systematic error because they attempted to mimic individuals' cognitive processes in receiving information. These processes are far too complex and diverse to reduce to a subjective rating variable. Although a complete analysis of the content of each article would be preferable to the approach adopted in this study, there was insufficient information for each article to conduct such an analysis.
3. Dynamic models of this type are referred to as "geometric lag" models because, by including the previous period's share of fruit expenditure in each demand equation, an initial shock to demand is assumed to decay slowly over time or "geometrically." These models are particularly easy to interpret because the estimated parameter on the lagged expenditure share gives the rate of adjustment toward a long-run equilibrium consumption level. Finding the long-run response parameter from the short-run estimate, therefore, required that the short-run estimate be divided by one–the estimated adjustment parameter. This specification was not only simple to estimate and interpret but was also consistent with the manner in which many dynamic phenomena appear and disappear–most shocks to demand have a large initial impact that fades after consumers forget, change their habits, or otherwise find ways to avoid the product in question.
4. This method approximates as closely as possible the impression a typical consumer will take from the article. Adding further specificity to this measure would have introduced bias to the extent that the researcher's interpretation differs from that of consumers. Any article that could not be classified as either positive or negative, such as a recipe or harvest report, was not included in the indices.

References

Bemmels, B. "Attribution Theory and Discipline Arbitration." *Industrial and Labor Relations Review* 44 (1991):548–561.

Blanciforti, L., and R. Green. "An Almost Ideal Demand System Incorporating Habits: An Analysis of Expenditures on Food and Aggregate Commodity Groups." *Review of Economics and Statistics* 65 (1983):511–515.

Boston Globe. "CDC Says Berries are Parasite Suspects." July 19, 1996, p. A3.

Brown, D.J., and L.F. Schrader. "Cholesterol Information and Shell Egg Consumption." *American Journal of Agricultural Economics* 72 (1990):548–555.

Chang, H.S., and H.W. Kinnucan. "Advertising, Information, and Product Quality: The Case of Butter." *American Journal of Agricultural Economics* 73 (1991):1195–1203.

Chen, P.Y., and M. Veeman. "An Almost Ideal Demand System Analysis for Meats with Habit Formation and Structural Change." *Canadian Journal of Agricultural Economics* 39 (1991): 223–235.

Darby, M.R., and E. Karni. "Free Competition and the Optimal Amount of Fraud." *Journal of Law and Economics* 16 (1973):67–88.

Davis, S., J. Inman, and L. McAlister. "Promotion Has a Negative Effect on Brand Evaluations–Or Does It? Additional Disconfirming Evidence." *Journal of Marketing Research* 29 (1992): 143–148.

Gao, X.M., and J.S. Shonkwiler. "Characterizing Taste Change in a Model of U.S. Meat Demand: Correcting for Spurious Regression and Measurement Errors." *Review of Agricultural Economics* 15 (1993):313–324.

Gardner, B.L. "The Farm-Retail Price Spread in a Competitive Food Industry." *American Journal of Agricultural Economics* 57 (1975):399–409.

Green, R.D., and J.M. Alston. "Elasticities in AIDS Models." *American Journal of Agricultural Economics* 72 (1990):442–445.

Hausman, J. "Specification and Estimation of Simultaneous Equations Models." *Handbook of Econometrics*. Z. Griliches and M. Intriligator, eds. Amsterdam: North Holland, 1983.

Huang, K.S. "A Further Look at Flexibilities and Elasticities." *American Journal of Agricultural Economics* 76 (1994):313–317.

Joreskog, K.G., and A.S. Goldberger. "Estimation of a Model with Multiple Indicators and Multiple Causes of a Single Latent Variable." *Journal of the American Statistical Association* 75 (1975):631–639.

Kelley, H.H. "Attribution Theory in Social Psychology." *Nebraska Symposium on Motivation*. D. Levine, ed., pp. 192–238. Vol. 15. Lincoln NE: University of Nebraska Press, 1967.

Kinnucan, H.W., H. Xiao, and C. Hsia. "Welfare Implications of Increased U.S. Beef Promotion." *Applied Economics* 28 (1996):1235–1243.

Kotler, P. *Marketing Management: Analysis, Planning, Implementation, and Control*. 7th ed. Englewood Cliffs NJ: Prentice-Hall, 1991.

Kraul, C. "San Diego Firm Accused of Fraud in Berry Scare." *Los Angeles Times*, June 11, 1997, p. D-1.

Lee, J., M.G. Brown, and J.L. Seale, Jr. "Demand Relationships among Fresh Fruit and Juices in Canada." *Review of Agricultural Economics* 14 (1992):255–262.

Levy, D. "What Public Relations Can Do Better than Advertising." *Public Relations Quarterly* 34 (1989):7–9.

Lipman, J. "TV Ads' Influence Found Wanting." *The Wall Street Journal*, February 15, 1989.

Lord, K.R., and S. Putrevu. "Advertising and Publicity: An Information Processing Perspective." *Journal of Economic Psychology* 14 (1993):57–84.

Mizerski, R. "An Attribution Explanation of the Disproportionate Influence of Unfavorable Information." *Journal of Consumer Research* 9 (1982):301–310.

Muth, R.E. "Derived Demand Curve for a Productive Factor and the Industry Supply Curve." *Oxford Economic Papers* 16 (1965):221–234.

Nelson, P. "Advertising as Information." *Journal of Political Economy* 82 (1974):729–754.

Pollak, R.A. "Habit Formation and Dynamic Demand Functions." *Journal of Political Economy* 78 (1970):745–763.

Portland Oregonian. "California Berries Blamed in Illness." June 21, 1996, p. A16.

Ray, R. "A Dynamic Generalisation of the Almost Ideal Demand System." *Economics Letters* 14 (1984):235–239.

Richey, M.H., L. McClelland, and A. Shimkunas. "Relative Influence of Positive and Negative Information in Impression Formation and Persistence." *Journal of Personality and Social Psychology* 6 (1967):322–327.

Richards, T.J. "A Dynamic Model of Fresh Fruit Promotion: A Household Production Approach." Paper presented at the American Agricultural Economics Association annual meeting, Toronto Ontario, July, 1997.

Rickertsen, K. "The Effects of Advertising in an Inverse Demand System." Paper presented at American Agricultural Economics Association meeting, San Antonio TX, 1996.

Salmon, C.T., L.N. Reid, J. Pokrywczynski, and R.W. Willett. "The Effectiveness of Advocacy Advertising Relative to News Coverage." *Communications Research* 12 (1985):546–567.

Smith, M.E., E.O. van Ravenswaay, and S.R. Thompson. "Sales Loss Determination in Food Contamination Incidents: An Application to Milk Bans in Hawaii." *American Journal of Agricultural Economics* 70 (1988):513–520.

Sparkman, R.M., Jr., and W.B. Locander. "Attribution Theory and Advertising Effectiveness." *Journal of Consumer Research* 7 (1980):219–224.

Swinyard, W.R., and M.L. Ray. "Advertising-Selling Interactions: An Attribution Theory Experiment." *Journal of Marketing Research* 14 (1977):509–516.

Swartz, D.G., and I.E. Strand, Jr. "Avoidance Costs Associated with Imperfect Information: The Case of Keptone." *Land Economics* 57 (1981):139–150.

Tellis, G.J. "Advertising Exposure, Loyalty, and Brand Purchase: A Two-Stage Model of Choice." *Journal of Marketing Research* 25 (1988):134–144.

U.S. Department of Agriculture, Agricultural Marketing Service, Fruit and Vegetable Division. "Fresh Fruit and Vegetable Arrivals in Western Cities" data. Washington DC, 1995–1997.

——. "Fresh Fruit and Vegetable Arrivals in Eastern Cities" data. Washington DC, 1995–1997.

U.S. Department of Agriculture, Economic Research Service. "Fresh Fruit Prices and Spreads" data. Washington DC, 1992.

U.S. Department of Labor, Bureau of Labor Statistics. "Consumer Price Index: Average Price" data. Washington DC, 1995–1997.

Washington Post. "Fear of Bacterial Illness from *Cyclospora* Parasite Stunts Strawberry Sales." July 4, 1996, p. A10.

Weinberger, M.G., and W.R. Dillon. "The Effects of Unfavorable Product Rating Information." *Advances in Consumer Research* 7 (1980):528–532.

15

Returns to the Dairy Check-Off Investment in Research for New Uses for Whey

Joseph V. Balagtas &
Daniel A. Sumner

In recent decades, the productivity of U.S. dairy farms has increased rapidly and milk prices have fallen. In this context, many see the prospects for enhanced returns to dairy producers depending largely on enhanced demand for dairy products. Although U.S. per capita consumption of cheese increased nearly fourfold in the last half of the twentieth century, per capita consumption of all dairy products fell 17 percent over the same period, in part because of reduced consumption of fluid-milk products (U.S. Department of Agriculture (USDA) 2000). This decline in consumption occurred despite a decline in real farm-level prices of 27 percent. With traditional consumption of dairy products flat or falling, finding new uses for milk is a way of increasing demand for dairy products.

Agricultural research and development (R&D) efforts have influenced food and agricultural markets in many ways, including lowering costs of production, reducing negative environmental consequences of agricultural activities, and developing new uses for farm products. Current ongoing research will affect tomorrow's markets. In that context, *ex ante* economic analysis can help anticipate market impacts of innovations and provide a basis for evaluating investments in research and setting research priorities (Alston, Norton, and Pardey 1995).

Scientists are developing new uses for agricultural products at research institutions around the world. New uses promise to add value to agricultural products, raise product prices, and, ultimately, increase returns to the grower.

For example, development in the latter half of the twentieth century of high fructose corn syrup and ethanol produced from corn increased demand for corn (House et al. 1993; Lee 1993; Alston and Beach 1996; Lee, Glauber, and Sumner 1994). Development of technology to make diesel fuel from oilseeds and biodegradable plastics from corn are two more examples of R&D creating potential for new markets for agricultural products (Duffield and Shapouri 1998; Coble et al. 1992). When the costs of R&D for new uses are funded by producer assessments (check-offs), the expense is distributed among various producer and consumer groups as opposed to the direct costs of publicly funded research, which are borne by general taxpayers. Analysis of the potential economic effects of new uses can provide a preview of the distribution of benefits and costs among these groups and allow evaluation of the payoff to each group from its investment in R&D.

Dairy check-off programs operate under federal and state legislation and regulation. These programs collect several hundred million dollars per year. Most is devoted to media promotion programs and relatively little to R&D. Economic analyses of the dairy check-off have focused on the effects of generic promotion (for example, see Kaiser and Chung 1999 and Kaiser and Schmit 2003); much less attention has been paid to the effects of dairy check-off expenditures on new-product R&D. Here, we project the potential market effects of new products that have yet to be marketed. Our results are useful for analyzing not only the absolute return on the investment in R&D but also the allocation of check-off revenue across alternative uses.

The California Dairy Research Foundation (CDRF) provides funds mainly to a handful of university researchers for R&D. CDRF receives approximately three-quarters of its budget from national check-off revenue generated in California and the remaining one-quarter of its budget from national check-off revenue generated in all states. CDRF is the main source of check-off funds for the research program considered here. We evaluated the economic returns to an ongoing research program at the University of California that has been funded primarily by the dairy check-off but also by general state and federal taxes that support the university. This research program is developing new uses for whey protein—a byproduct of cheese manufacturing—for films and coatings in food-related and other applications (for example, see Lee, Dangaran, and Krochta 2002; also Lee et al. 2002). For several of the technologies, progress in the laboratory indicates that commercial production is possible by 2005, and strong interest from potential industrial users suggests that commercial adoption is likely. Since current technology for films and coatings does not use whey or any other dairy product, successful marketing of the new technologies will create new demand for whey and for milk.

Balagtas et al. (2003) presented a simulation model to consider the likely market effects resulting from commercial adoption of the new whey-protein products and measure the expected return from investment in research accruing to U.S. milk producers. This chapter elaborates on the work in Balagtas et al. to analyze the distribution of costs and benefits across California and U.S. producers and to consider also the broader effects of the research program on the economy.

We used information from researchers and industry sources to estimate the potential for adoption of the new technologies and quantified the associated increase in demand for whey. We incorporated this information into a simulation model of the U.S. dairy economy, yielding expected changes in prices and quantities of milk and dairy products resulting from adoption of the technologies. We adopted a simulation approach similar to that used to measure the size and distribution of costs and benefits of R&D (studies of applications to producer-funded R&D include Mullen, Alston, and Wohlgenant (1989) and Zhao (2003). See Alston, Norton, and Pardey (1995) for an exposition of the general modeling approach). Based on our simulation results, we measured the expected benefit to dairy producers and others and evaluated returns to the investment in research. We then considered the distribution of costs and benefits and in this chapter we discuss the implications for the dairy check-off and for other U.S. dairy programs.

Supply and Demand for Whey and the Role of New Uses

When curds of protein and butterfat are extracted from milk to make cheese, whey—water, lactose, minerals, and whey protein—remains as a byproduct. In the recent past, commercial cheese manufacturers in the U.S. treated whey as waste or returned it to dairy farms, where it was fed to pigs and spread on fields. However, growth in demand for the components of whey and development of whey processing techniques, as well as rising environmental and financial costs associated with disposing of whey, have made it profitable to further process whey into its components, especially whey protein and lactose for use as food ingredients. For example, bakery products, infant formulas, and sport/nutrition products now use whey protein.

Though the market has developed rapidly in the U.S., some cheese processors still do not have the capacity to process whey. Smaller and/or older operations in particular have not been able to begin processing whey because of the cost of purchasing and installing the equipment. As a result, approximately 30 percent of the total amount of whey generated by cheese manufacture is not sold. The rest of the cheese industry, including most plants in major cheese-producing regions such as the upper Midwest and California, actively markets whey, and these sales generate approximately 11 percent of the revenue from a modern cheese plant.

Recent R&D has focused on developing industrial films and coatings from either whey-protein isolate (WPI) or whey-protein concentrate, which is 80 percent protein (WPC80). Commercial production is likely by 2005 for several of these new uses (Balagtas et al. 2003). For these technologies, Balagtas et al. (2003) estimated the raw-material cost to potential users for comparison with the cost of existing technologies. Based on these cost comparisons, together with discussions with potential users, the authors evaluated the potential for commercial adoption of each whey application, concluding that near-term adoption is likely for three applications: oxygen-barrier coatings on plastics, gloss coatings on confectioneries, and oxygen-barrier coatings on nuts in candy. (For a more detailed description of the cost estimates and comparisons, refer to Balagtas et al. (2003)) Our discussions with industry also indicated that half of the relevant markets would switch to a whey-protein coating. The additional demand for whey protein associated with adoption of the new applications is likely equivalent to between 1 and 4 percent of the estimated amount of whey protein currently available as a byproduct of cheese manufacture in the U.S.

The Cost of R&D Leading to New Uses of Whey Protein

From 1990 through 2005 (projected date of first commercial adoption of whey coatings), the whey research project will have incurred a cost of $3.8 million (2002 present value), which includes actual outlays through 2002 and an estimate of whey research expenditures for 2003 through 2005 based on recent annual outlays. (We adjusted costs for inflation and used a 4 percent real discount rate to bring annual costs to a 2002 present value. Capitalized figures presented here differ from those in Balagtas et al. (2003) because we used a lower discount rate.) This figure represents the cost of the entire research program and includes the cost of developing new whey applications whose adoption and potential benefits we do not analyze here. Of the total research cost, $2.23 million (approximately 60 percent) will have been paid by research grants from the U.S. dairy industry through CDRF. The California Milk Advisory Board (CMAB) will have contributed $1.68 million (75 percent) and Dairy Management, Inc. (DMI) will have contributed the remainder, \$0.56 million (25 percent).

CMAB and DMI are each funded by a national marketing assessment (check-off) of \$0.15 per one hundred pounds of milk sold. Such assessments affect market prices and quantities and thus are ultimately borne partly by producers and partly by consumers. In the case of the dairy check-off, milk producers and consumers each pay a portion of the check-off investment in the whey research project. The proportion of the check-off borne by producers equals $|\eta|/(\varepsilon + |\eta|)$, where η is the elasticity of demand for U.S. milk and ε is the elasticity of supply of U.S. milk.[1] For example, based on an elasticity of supply of U.S. milk of 1.0 and an elasticity of demand for U.S. milk of −0.35, producers incur approximately 41 percent of the burden of the check-off.[2] Thus, U.S. producers' incidence of the check-off investment in the whey research program is \$0.92 million (41 percent of \$2.23 million), of which \$0.69 million (41 percent of \$1.68 million) is from CMAB through CDRF and \$0.23 million (41 percent of \$0.56 million) is from DMI.

Although there is a single national check-off, individual states and groups of states can retain up to one-third of the check-off revenue generated by local milk sales to fund promotion and research efforts chosen by local producer groups. CMAB, for example, manages one-third of the check-off revenue collected in California. The remaining two-thirds are managed by DMI, which has a national scope. Check-off revenue collected in California accounts for approximately 20 percent of DMI's funding (approximately equal to California's share of U.S. milk production). As stated earlier, approximately 75 percent of CDRF funding is through CMAB and 25 percent is from DMI. Thus, the cost of the whey research program to California producers is \$0.60 million (0.75 × \$0.69 million + 0.2 × (0.25 × \$0.69 million + \$0.23 million)).

In addition to check-off funding, University of California contributed a present value of \$0.11 million in direct research grants and approximately \$1.46 million toward the research program's indirect costs, including researcher and faculty salaries, lab and office space, and research dissemination efforts. To the extent that federal and California tax dollars support the university, federal and California taxpayers pay the university's contribution to the whey research project.

The problem of determining starting and ending points by which to define comparable streams of research costs and benefits is one shared with all economic analyses of R&D (Alston, Norton, and Pardey 1995). We measured research costs as expenditures for this particular research project. We did not consider the costs of basic chemistry and food science knowledge, which served as a point of departure for the research. The project we studied has been a leading innovator in the area of whey-protein-based coatings and also involves basic research that may be applicable to other areas. Development costs incurred by processors were considered when we discussed prospects for adoption with industry representatives. However, we did not have estimates of such costs and did not include them in the analysis. To the extent that such costs affect demand for whey protein, they are distributed among consumers and producers. Our economic analysis considered adoption of only three of the new whey-protein technologies being developed. To calculate the returns to research from those three innovations, we had to determine what portion of the total research cost ($3.8 million) was attributable to them. However, regardless of how costs are attributed to specific applications, the costs are still borne by producers, processors, marketers, consumers, and taxpayers. We modeled and measured the total potential benefits of this research and highlighted the benefits that accrue to U.S. and California milk producers.

The Economic Effects of Increased Whey Demand on U.S. Dairy Markets

Raw milk is used to produce dozens of dairy products. Consequently, the supply of raw milk affects the supply of the end products. Also, demand for the end products determines the demand for raw milk. Supply and demand determine price in each market, but the supplies and demands and thus prices are related across markets by the end products and the raw material, milk, used to make them.

Milk contains fat, protein, other solids, and fluid. Various dairy products use these components in different ratios, and manufacture of one product often generates leftover dairy components for use in other products. Whey, for example, is a byproduct of cheese production. Dairy products also compete with each other for milk components. For example, cheese and whey manufacturers compete for supplies of raw milk with fluid-milk bottlers and butter and powder processors. Demand for raw milk, then, is derived from demand for the processed products. In addition, supply conditions of other complementary and substitute inputs shift the derived demand for milk. To satisfy an increased demand for whey caused by development of new uses, more raw milk must be allocated to cheese and whey production, which affects the amount of milk available for other uses.

In the U.S., forces in the milk market are conditioned by regulatory forces to determine milk prices. Trade barriers, government price supports, direct payments, and the milk-pricing policies contained in milk marketing orders all affect dairy markets and prices. Trade barriers allow milk product prices in the U.S. to remain higher than global market prices. The price support sets minimum government purchase prices for cheese, butter, and nonfat dry milk that in turn set floor prices for raw milk at the farm level. The Milk Income Loss Contract (MILC) program through USDA has recently provided farmers with direct payments that partially offset low milk prices (no payment is made when market prices are relatively high). Federal and California milk marketing orders set minimum prices for milk used in various products. The marketing orders also set differentials that affect the minimum price of milk by end use, but actual minimum milk prices are derived from formulas that use market prices of dairy products. That is, the minimum price that a processor must pay for raw milk is based on market forces that are driven by supply and demand for dairy products. These regulated pricing formulas do not necessarily respond to changes in all dairy-product markets and may respond with some delay, but for our analysis we considered the effects of increased whey demand over a horizon long enough that milk and dairy-product markets adjust. Therefore, the price of raw milk in our analysis reflects market prices of dairy products.

As noted previously, R&D that creates new uses for whey increases demand for whey and affects all other dairy-product markets in addition to the market for raw milk. Our simple simulation model quantified the potential effects of the new whey applications. We represented the key economic relationships as equations in a model of a multimarket dairy industry that uses raw milk as an input in a variety of processed dairy products. Adoption of the new whey technologies simply shifted out whey demand. The model then simulated the percent changes in prices and quantities of milk and dairy products and quantified the benefits to milk producers from the whey research program.

For convenience, Table 15.1 summarizes parameter definitions and base values for the logarithmic differential approximation of the equations sketched here. We considered three dairy products: cheese, whey, and an aggregate of all other dairy products (denoted with the subscript "other"). Demand for these products was written in general form as:

(1) $Q_{whey} = f_{whey}(P_{whey}, \theta)$,

(2) $Q_{cheese} = f_{cheese}(P_{cheese})$, and

(3) $Q_{other} = f_{other}(P_{other})$,

where Q_i is quantity and P_i is the price of product i, f_i represents the demand function for product i, and θ represents the exogenous shift in demand for whey caused by R&D.

Assuming a competitive market for each of the products, the price equals the industry marginal cost in each market. Thus, the equilibrium conditions for cheese and other dairy products are:

(4) $P_{cheese} = c_{cheese}(P_{milk}, P_{input}) - \rho P_{whey}$, and

(5) $P_{other} = c_{other}(P_{milk}, P_{input})$,

where c_i is the marginal cost function for product i. The second term on the right-hand side of Equation (4) is due to the joint production of cheese and whey, and ρ is defined as the quantity of whey produced per unit of cheese output. We wrote this production relationship as:

(6) $Q_{whey} = \rho Q_{cheese}$.

For simplicity, we used a single price of milk, P_{milk}, in all uses.

Table 15.1. Parameters of the Equilibrium Displacement Model

Symbol	Definition of Parameter	Values
θ	Increase in demand for whey as a percentage of current consumption	1–4%
η_{whey}	Price elasticity of demand for whey	–1.0
η_{cheese}	Price elasticity of demand for cheese	–0.5
η_{other}	Price elasticity of demand for other dairy products	–0.5
α_{cheese}	Cost share of raw milk in cheese	0.68
α_{other}	Cost share of raw milk in other dairy products	0.70
γ	Ratio of whey revenue to cheese revenue per unit of cheese output	0.11
φ	Share of all raw milk used in cheese/whey	0.37
ε_{milk}	Elasticity of supply of raw milk	1.0

Assuming milk enters the production function for each product as a fixed proportion input, we could write the relationship between output and milk input as:

(7) $Q_{milk,cheese} = \omega_{cheese} Q_{cheese}$, and

(8) $Q_{milk,other} = \omega_{dairy} Q_{other}$,

where $Q_{milk,i}$ is the milk input required to produce Q_i and ω_i is the milk input requirement per unit of output of product *i*. Since cheese and whey are produced jointly, the quantity of milk used in whey is equal to the quantity of milk used in cheese:

(9) $Q_{milk,whey} = Q_{milk,cheese}$.

The aggregate quantity of milk demanded is the sum of the quantity of milk demanded for cheese/whey and the quantity demanded for all other dairy uses:

(10) $Q_{milk} = Q_{milk,cheese} + Q_{milk,other}$.

The supply of milk closes the model:

(11) $Q_{milk} = g_{milk}(P_{milk})$,

where g_{milk} is the milk industry supply function.

The system of Equations (1) through (11) defines equilibrium in the dairy industry. Comparing equilibria when $\theta = 0$ and $\theta > 0$, we could examine changes in prices and quantities throughout the dairy industry as a result of increased demand for whey. We approximated these changes by totally differentiating the system and converting to elasticity form. For the simulations, the logarithmic differential equations were linear functions relating proportional changes in quantities and prices with parameters represented by the elasticities and shares, as displayed in Table 15.1. The key shifter θ represents a horizontal shift in demand for whey as a proportion of the current quantity demanded. For the time horizon considered, we assumed that prices for all inputs other than milk (e.g., labor, capital) were exogenous to the dairy industry. A full set of simulation equations was reported in Balagtas et al. (2003).

Using producer surplus as a measure of producer well-being, we used the percent changes in milk price and quantity supplied to calculate benefits to dairy farmers. Denoting EQ_{milk} and EP_{milk} as the percent changes in the quantity and price of milk resulting from the shift in demand for whey, we used a linear approximation to the change in producer surplus:

$$(12)\quad \Delta PS = P_{milk,0} Q_{milk,0} EP_{milk} (1 + 0.5\ EQ_{milk}),$$

where $P_{milk,0}$ and $Q_{milk,0}$ are the initial milk price and milk quantity before the shift in whey demand.

The logic of our results is straightforward. Increased demand for whey results in a higher whey price, encouraging increased production of whey. Because whey and cheese are produced jointly and in fixed proportions, cheese production must also increase, which puts downward pressure on cheese prices.[3] Increased demand for milk by cheese and whey manufacturers drives up the price of milk, encouraging greater milk production. At the same time, the higher milk price raises costs for manufacturers of butter, ice cream, and other dairy products, resulting in higher prices and smaller quantities for these products. The magnitudes of the shift in milk demand, changes in the price of milk and quantity of it produced, and increased returns to producers depended on the economic and technological relationships that were embodied in the parameters of our model.

Quantifying the Supply, Demand, and Technological Relationships Used in the Model

Where possible, we assigned values to the parameters of the model based on published agricultural economics literature and public data. Where published estimates and data were lacking, we relied on knowledgeable industry sources and on our own calculations. The magnitude, but not the direction, of the results would vary under different parameter scenarios. We describe the parameter values used for the scenarios reported here in Table 15.1.

We chose U.S. demand elasticities of −0.5 for cheese and −0.5 for all dairy products other than whey and cheese based on estimates published in the agricultural economics literature (see, for example, Johnson, Stonehouse, and Hassan 1992). To our knowledge, no estimate of the elasticity of demand for whey has been published. This parameter, along with the derived supply elasticity of whey, determines the impact of the research-induced shift in the demand for whey on the price of whey and thus on the price of milk. Our estimates of the effects on prices and resulting benefits to producers are approximately proportional to the inverse of the elasticity of demand for whey. We used a long-run demand elasticity of −1.0 to reflect potential substitution from nondairy food and nonfood ingredients. We also briefly discuss results for elasticities of demand for whey of −0.5 and −2.0.

For the relatively long-run time horizon that may apply to the spread of the market effects of new uses, we used a milk supply elasticity of 1.0, which is within the range of estimates found in the agricultural economics literature (see, for example, Chavas and Klemme (1986) and Helmberger and Chen (1994)). Balagtas et al. (2003) used a more inelastic supply that reflected a shorter time horizon for adjustment. For a milk supply elasticity of 0.5, producer benefits would be somewhat larger than those found with a supply elasticity of 1.0.

Other key parameter values were based on 2001 dairy market data from USDA: The cost share of milk in cheese/whey manufacture was 68 percent; the average cost share of milk in the manufacture of all other dairy products was 70 percent; the share of U.S. milk production allocated to cheese/whey production was 37 percent. We also used USDA data to calculate the ratio of whey revenue to cheese revenue for a cheese plant. Based on prices of publicly traded dry whey and forty-pound blocks of cheddar cheese, plus a ratio of 0.56 pounds of dry whey per pound of cheese, whey represents an average of 11 percent of the revenue of a plant that produces both cheese and whey.

The percent increase in demand for whey, θ, was estimated based on the potential market size calculations discussed earlier. We found that new whey applications could increase demand for whey by between 1 and 4 percent. We simulated the effects of a medium adoption rate (3 percent). (Balagtas et al. (2003) provided a range of estimates.)

We used USDA data on milk prices and quantities to calculate the change in producer surplus. The 2001 average all-milk price for the U.S. was $15.05 per one hundred pounds of milk, and U.S. milk production was 165 billion pounds. For the same year, the average all-milk price for California was $13.94, and California milk production was thirty-three billion pounds.

Annual Effects of an R&D-Induced Demand Shift for Whey

Table 15.2 presents the effects in 2002 of a 3 percent shift in demand for whey on prices and quantities of milk and dairy products. We also report in

Table 15.2. Simulation Results: The Annual Effects of a 3 Percent R&D-Induced Increase in Whey Demand

	Percent Change		Percent Change		Percent Change
Prices		**Revenue**		**Quantities**	
Whey	2.86	Whey	0.14	Whey	3.40
Cheese	−0.28	Cheese	0.14	Cheese	−0.18
Other Dairy	0.03	Other Dairy	−0.01	Cheese/Whey Plant[1]	0.21
Raw Milk	0.04	Raw Milk	0.04	Other Dairy	0.02
				Raw Milk	0.80

Gross Annual Benefits, U.S. Milk Producers[2]	$9.96 million
Gross Annual Benefits, California Milk Producers[2]	$1.85 million
Gross Annual Benefits, Total[3]	$42.30 million

Note: These calculations are based on the parameters in Table 15.1.

[1] Based on an average of 11 percent of cheese plant revenue from whey.

[2] Calculated as the change in producer surplus resulting from the shift in demand for whey but excluding the cost of the research program to producers.

[3] Gross annual benefits to producers, processing and marketing firms, and consumers, excluding costs of the research program.

Table 15.2 the annual benefit to U.S. and California milk producers, as well as the total annual benefit to producers, processing and marketing firms, and consumers.

Consider first the direct effects of new uses for whey on the whey and cheese markets. Following a 3 percent increase in demand for whey, the price of whey increases by 2.86 percent. Higher whey prices are accompanied by a reduction in cheese prices of 0.28 percent: as cheese manufacturers receive more for whey, more milk goes to cheese processing, leading to increased cheese production and hence lower cheese prices. However, since whey is a relatively small share (approximately 11 percent) of cheese processors' revenue, the implied reduction in cheese prices is relatively small. The quantities of cheese and whey produced increase by 0.14 percent.

Based on the percent changes in prices and quantities, we calculated the percent change in revenue for cheese and whey manufacturers and for milk producers. Whey revenue increases by 3.4 percent and cheese revenue falls by 0.18 percent. Since whey accounts for only 11 percent of a cheese/whey plant's revenue, revenue for cheese/whey manufacturers increases by only 0.21 percent.

Table 15.3. Present Value of Benefits and Costs of R&D for New Uses of Whey Protein

	Units of Value[1]	Total	U.S. Producers	California Producers
Present Value, Gross Benefits	2002, million dollars	528.73	127.39	23.73
Attributing 100 Percent of Research Costs to Three Marketed Applications				
Present Value, Research Costs[2]	2002, million dollars	3.80	0.92	0.60
Present Value, Net Benefits[3]	2002, million dollars	524.93	126.47	23.13
Benefit-Cost Ratio		139	139	40
Internal Rate of Return (IRR)	percent per annum	36	34	25
Attributing 50 Percent of Research Costs to Three Marketed Applications[3]				
Present Value, Research Costs[2]	2002, million dollars	1.90	0.46	0.30
Present Value, Net Benefits[3]	2002, million dollars	526.82	126.93	23.43
Benefit-Cost Ratio[3]		278	277	79
Internal Rate of Return (IRR)	percent per annum	42	39	29

[1] All present values are in millions of 2002 dollars and based on a real discount rate of 4 percent per annum.

[2] We estimated that the producers' cost of the research program is 41 percent of the total check-off investment. This estimate of the producer incidence of the check-off was based on an elasticity of milk supply of 1.0, an elasticity of milk demand of −0.5, and an elasticity of demand for raw milk of −0.35. We estimated California producers' costs based on the share of the total check-off investment that comes from California milk sales.

[3] Because of rounding, net benefits may not exactly equal gross benefits less costs. Also because of rounding, the benefit-cost ratio with a 50 percent cost attribution may not be two times the benefit-cost ratio with 100 percent cost attribution.

Tracing the results through other markets, we find that raw milk prices and production each rise by 0.04 percent as a result of increased demand from cheese/whey processors. Dairy farm revenue rises by 0.08 percent. As higher milk prices drive up manufacturers' costs, prices of dairy products other than cheese and whey rise by 0.03 percent and production of those products falls by 0.01 percent. These are small percentage impacts, but remember that they apply to a very large industry and the initial spur is a single, relatively small program of research.

Based on the market interactions and changes in prices and quantities previously described, we could determine the implied benefits to producers from increased demand for whey. Increased whey demand results in higher milk prices and increased milk production, making dairy farmers as a group better off. Table 15.2 reports a measure of the annual gross benefits from the whey research program to U.S. dairy farms as a group and to California producers, calculated as the increase in total milk revenue net of marginal production costs (i.e., the change in producer surplus). U.S. producers gain $10 million (2002 dollars) each year because of increased demand for whey. Of that, we calculated that $1.9 million goes to California producers based on their share of U.S. milk production. Note that gross annual benefits were calculated as the change in producer surplus from increased demand for whey, not counting the cost to producers of the whey research program. We consider net benefits in the next section.

Gross annual producer benefits from the shift in demand for whey were approximately proportional to the inverse of the elasticity of demand for whey. Doubling the elasticity of demand for whey from −1.0 to −2.0 reduced the producer gain by approximately half: U.S. producers gained $5 million per year and California producers gained $0.9 million per year. Halving the elasticity of demand for whey from −1.0 to −0.5 approximately doubled the producer gain: U.S. producers gained $19.9 million and California producers gained $3.7 million.

Adoption of the whey technologies also had an impact on dairy processors, marketers, and consumers, and we note these effects briefly. Although milk and dairy consumers incur a portion of the cost of the dairy check-off, consumers have no direct say in how check-off revenue is spent. Traditional consumers of whey suffered an annual loss in economic surplus of approximately 2.9 percent of whey expenditures from higher whey prices. Consumers of cheese are better off, with an annual gain in consumer surplus equal to 0.3 percent of cheese expenditure. Consumers of other dairy products were worse off because of higher prices, but the effects were tiny. For example, the annual loss in consumer surplus in the butter and fluid-milk markets was about 0.3 percent of expenditures on those products.

Aggregating welfare effects in the various linked markets provided an estimate of the gross (annual) benefit to the economy, including producers, processing and marketing firms, and consumers. Our estimate of the total gross benefit of the shift in whey demand, reported in the last line of Table 15.2, was $42.3 million.[4] This measure included benefits to milk producers, benefits and costs to various milk processing subsectors, and benefits and costs to dairy consumers. It did not include the cost of the research program. Further, although we did not explicitly model confectionery and plastics markets, demand for whey-protein coatings from these industries was implicit in our characterization of whey demand. Hence, our estimate of the total benefit included the welfare effects in these markets (Just, Hueth, and Schmitz

1982). As was the case with producer benefits, the total benefit from the shift in demand for whey was approximately proportional to the inverse of the elasticity of demand for whey.

We also simulated the annual effects of a 1 percent and a 4 percent shift in demand for whey but do not report the results here. The effects of those shifts on prices and quantities were linear in the magnitude of the shift in demand for whey; the effects on prices and quantities of a 1 percent increase in demand for whey were one-third the magnitude of those (reported in Table 15.2) for a 3 percent increase in demand.

Capitalized Net Benefits of an R&D-Induced Demand Shift for Whey

The expected gains to dairy farms and total expected benefits reported in Table 15.2 represent annual benefits expressed in 2002 dollars. We calculated total gross benefits over the lifetime of adoption by projecting a pattern of adoption over time and discounting each year's benefits to a specified date. Although future adoption rates were uncertain, the strong interest shown thus far by confectionery and plastic manufacturers suggested that whey-protein coatings will be marketed successfully.

We approximated total capitalized gross benefits by supposing a specific time pattern of adoption. We assumed a time path of adoption such that demand increased by 1 percent for the first five years (2005–2009), by 3 percent for another five years (2010–2014), by 4 percent for ten years (2015–2024), and then back to 1 percent for ten years (2025–2034) before the whey products were fully replaced by newer technologies. Present values of producers' gross benefits and total gross benefits are reported at the top of Table 15.3. Under this adoption pattern, and based on a real annual discount rate of 4 percent, the 2002 present value of gross benefit to producers, processing and marketing firms, and consumers over the period 2005–2034 was $528.73 million.[5] The 2002 present-value gross benefits to all U.S. producers over the same period was $127.39 million, of which $23.73 million accrued to milk producers in California. (Note: capitalized values presented here differ from those in Balagtas et al. (2003) because we used a smaller discount rate.)

The costs of the research project are attributable not only to the three whey uses considered here, which are near commercial adoption, but also to applications in earlier stages of development and those considered less likely to be adopted. Further, some of the research is more basic in nature and has potential for applications not discussed here or even foreseen. To calculate a return to the investment in research resulting in the three applications nearing adoption, we had to apportion research costs to those applications. A conservative approach was to attribute all costs to the three whey applications near adoption, which was equivalent to assuming that these were the only successful products of the research program and ignoring potential benefits from any other applications. In Table 15.3, we report costs under this conservative scenario, as well as under a scenario in which half of the research program's expenditure were attributed to the three applications.

The distribution of costs among producers, consumers, and taxpayers was discussed earlier, and the producers' portion of total research costs is reported in Table 15.3. Recall that total check-off investment in the research program is $2.23 million. We estimated that the producer incidence of the check-off is approximately 41 percent, or $0.92 million, based on an elasticity of U.S. milk supply of 1.0, an elasticity of demand for U.S. milk of −0.5, and an elasticity of demand for raw milk of −0.35 (see note 2). The research cost borne by California producers, $0.60 million, was calculated based on the percentage of the check-off investment that came from check-off revenue collected on California milk sales.

Table 15.3 also reports summary statistics that are commonly used to characterize returns to research and are useful for comparing returns to research on alternative investments. The present value of net benefits is simply the present value of gross benefits less the present value of research costs. The benefit-cost ratio is equal to the present value of gross benefits divided by the present value of costs, or the average return per dollar invested. The (real) internal rate of return (IRR) is the annual interest rate that would just make the net present value of the project zero, such that benefits equal costs in present value terms.

When 100 percent of research expenditures were attributed to the three applications considered here, the present value of total net benefits was $524.93 million, of which $126.47 million accrued to U.S. milk producers and $23.13 million to California milk producers. The total benefit-cost ratio was 139:1, the U.S.-producer benefit-cost ratio was also 139:1, and the California-producer benefit-cost ratio was 40:1. Note that the California-producer benefit-cost ratio was smaller than the U.S.-producer benefit-cost ratio because California producers bear nearly 70 percent of producers' costs while receiving benefits in proportion to their share of U.S. milk production—approximately 20 percent. The IRR for total benefits was 36 percent per annum, the IRR for U.S. producers' investment in the whey research program was 34 percent per annum, and the IRR for California producers' investment was 25 percent per annum. The interpretation is that U.S. producers' investment in the research program is preferable to any alternative investment yielding less than 34 percent per annum. Again, the IRR for California's producers was smaller than that for all U.S. producers because their proportion of producer costs exceeds their proportion of the producer benefits.

Attributing 50 percent of total research expenditures to the three whey applications considered here resulted in a larger present value of net benefits, a larger benefit-cost ratio, and a larger IRR for each group.

Different time patterns of adoption can have a large impact on the present value of net benefits, the benefit-cost ratio, and the IRR. However, given our estimates of annual benefits, returns to investment in whey research were positive and large for a wide range of adoption scenarios. Further, our measures of producer benefits were conservative in that we did not attribute to producers the benefits that accrue to cheese and whey manufacturers, which also accrue to producer-members of dairy cooperatives and dairy-farmer owners of other cheese manufacturing firms (such as Hillmar Cheese) that produce much of the cheese and whey.

Implications for the Dairy Check-off and for Other U.S. Dairy Programs

The foregoing analysis treats milk, cheese, and whey as undifferentiated products. This has become controversial in the context of check-off programs. Plaintiffs in a recent court case (*Cochran v. Anne Veneman and National Dairy Promotion Board* 2004) argued that "traditional" farming practices differentiate their milk from other milk, that their "traditional" milk is a close substitute for other milk, and that they should not be compelled to support promotion for their competitors. It is not clear how this case will be resolved upon appeal or if these differentiation issues apply to check-off funds used for research.

One stated objective of the dairy check-off is reduction of government purchases of dairy products through the dairy price-support program. Our analysis assumed that the dairy price support was not binding over the period for which the adoption applied and that all the relevant prices were market prices. This assumption is consistent with recent experience, in which most manufactured-product price indices have tended to remain above purchase prices. Still, regional and temporal variation in dairy prices have encouraged sales to the government by some manufacturers some of the time, even though government purchases of dairy products have been a modest share of all milk products. Our findings, that likely levels of adoption of new whey applications will result in a small increase in the average price of milk, indicate that the likelihood of government purchases would be reduced and that there would be an associated reduction in government outlays for the dairy price support program. Under low-price market conditions, taxpayers would share in the benefits of the demand shift generated by successful development of new uses.

To be specific, consider the plausible case in which cheese prices are higher than the government purchase price but nonfat dry milk prices are equal to the government purchase price.[6] Additional demand for whey would drive up the price for nonfat dry milk. This would result in reduced government purchases of nonfat dry milk and reduce the government budget cost of that program. Conversely, however, if instead the price of cheese was less than the government purchase price while the prices of other products were higher than government purchase prices, an increase in demand for whey would increase government program costs, and cheese producers would be partially insulated from the negative impacts of the higher price of whey.

The recent MILC program also would affect the share of benefits to producers or taxpayers under low-market-price conditions. MILC began in 2002 and is scheduled to terminate in 2005, in which case it would not interact with adoption of the new whey applications. But efforts are under way to have the program extended. MILC payments make up 45 percent of the loss whenever market prices that determine government-set regulated prices fall below a minimum specified in legislation. Payments to individual farms are limited to 2.4 million pounds of milk per year, or the output of approximately 110 cows, effectively making MILC a fixed payment for farms producing more than 2.4 million pounds, a group that includes almost all milk producers in California (Sumner and Balagtas (in press)). If the program is extended, the higher prices resulting from development and adoption of new uses for whey would reduce the government's expenditures on the payment program and lower the dairy farmers' benefits from MILC.

Summary and Conclusion

New uses for agricultural products promise to add value to raw agricultural products, raise commodity prices, and raise farm producer returns. Producers, consumers, and taxpayers finance the research efforts that create new uses. *Ex ante* economic analysis anticipates the market effects of new uses and is the basis for evaluating investments in R&D before the benefits have been fully realized.

This chapter evaluated R&D likely to generate new uses for whey protein. The magnitude and timing of adoption of the new technologies was gleaned from the researchers and from industry sources (Balagtas et al. 2003). Based on this information, the likely increase in demand for whey was calculated. Our economic simulation analysis traced the effects of increased demand for whey through the markets for dairy products and for milk and quantified the percent changes in prices and quantities in those markets. In this case, check-off funds from federal and California programs covered the bulk of the research costs. We calculated annual net benefits to California and U.S. milk producers and the returns to their investment in the research. We also considered the total, economywide benefits of the research program.

The research leading to the new whey uses, including potential uses not yet ready for adoption, was projected to cost a present value of $3.8 million by the adoption date of 2005. National and California dairy check-off programs will have contributed $2.23 million; of that, approximately 41 percent, or $0.92 million, can be attributed to dairy farms and the rest can be attributed to processors, marketing firms, and consumers. Taxpayers also will have contributed through their support of the University of California, which has paid for some of the research program's costs. To investigate the implications for the check-off program, our analysis focused particularly on returns to investment in R&D by U.S. and California dairy producers.

For a relatively conservative time-pattern of technology adoption, the present value of benefits to U.S. producers (net of the $0.92 million invested by producers in the research program) was approximately $126 million. The benefit-cost ratio for producers was large, and the implied internal rate of return ranged between 34 and 39 percent per annum, depending on what portions of costs were allocated to the three new technologies considered here. Thus, while total benefits to producers were small relative to total dairy farm revenue–U.S. dairy farm cash receipts were $24.7 billion in 2001–benefits were large relative to the cost of the investment.

Nearly 70 percent of the producers' share of the cost of the program was borne by California dairy producers. Yet California dairy producers' share of the producer benefits was less than 20 percent of the total producer benefits, approximately proportionate to their share of U.S. milk production. As a result, the benefit-cost ratio and IRR associated with California's net benefit of $23.7 million were both smaller than the analogous statistics for all producers. The benefit-cost ratio for California producers was between 40:1 and 79:1, and the IRR for California producers ranged from 25 to 29 percent per annum. Thus, although the rate of return for California dairy producers was not as high as that for U.S. producers as a whole, returns for California producers were still high relative to most alternative investments.[7]

This chapter has shown how the use of check-off revenue for demand-enhancing research can provide substantial returns to producers and to society as a whole. The results have potentially important implications as producers consider the allocation of check-off funds between, for example, R&D and promotion.

Notes

The authors are grateful to editor Julian Alston for his insightful comments and suggestions.

1 National supply and demand elasticities are appropriate for measuring the incidence of the dairy check-off because the check-off is determined and implemented nationally. There is no separate check-off for California alone.

2 Assuming fixed proportions of technology in dairy processing and exogenous prices of nonmilk inputs in dairy manufacturing, the elasticity of demand for raw milk derived from the demand for processed dairy products is equal to the elasticity of demand for all dairy products times the cost share of milk in dairy products. In the simulation model that follows, we assumed the elasticity of demand for cheese was −0.5, the elasticity of demand for other dairy products was −0.5, and the elasticity of demand for whey was −1.0. Given the small revenue share represented by whey among all dairy products, an approximate estimate of the elasticity of demand for all (aggregate) dairy products was −0.5. Based on a cost share of approximately 70 percent for raw milk in all dairy products, the elasticity of derived demand for milk was −0.35 (0.7 × 0.5).

3 As noted previously, approximately 30 percent of whey is neither processed nor sold. Unsold whey is produced mainly by smaller, older cheese plants that have not made a capital investment in whey-processing equipment. We expected that most of the increased production of processed whey will come from increased cheese production by cheese plants with whey-processing capacity. In our simulations, we assumed that all increased production of processed whey was associated with increased cheese production and an increased quantity of milk allocated to cheese/whey manufacturing. To the extent that some of the increased demand for whey induces additional whey processing without additional cheese production (and, therefore, without additional milk demand), the effects of the whey research program will be smaller than reported here.

4 We calculated total benefits by simulating the shift in whey demand with elasticity of supply of U.S. milk set equal to 0. With a fixed milk supply, all research benefits accrue to milk producers. Thus, total benefits may be calculated as the change in producer surplus (Equation (12)) when the supply elasticity is 0.

5 Streams of costs and benefits are expressed in constant (2002) dollars. Thus, it was appropriate to use a real discount rate to calculate present values of costs and benefits.

6 This has been the pattern occasionally in recent years and is consistent with industry projections (USDA, *Agricultural Statistics*; Food and Agricultural Policy Research Institute (FAPRI) 2003).

7 We should also note that successful R&D projects conducted in other states have the potential to benefit producers in California. A tacit understanding among states to fund local projects that have important national benefits would affect how one calculates return and could cause the state-specific rate of return to approach the national rate of return.

References

Alston, J.M., and D.E. Beach. "Market Distortions and the Benefits from Research into New Uses for Agricultural Commodities: Ethanol from Corn." *Resource and Energy Economics* 18 (1996):1–29.

Alston, J.M., G.W. Norton, and P.G. Pardey. *Science under Scarcity: Principles and Practice for Agricultural Research Evaluation and Priority Setting*. Ithaca NY: Cornell University Press, 1995.

Balagtas, J.V., F.M. Hutchinson, J.M. Krochta, and D.A. Sumner. "Anticipating Market Effects of New Uses for Whey and Evaluating Returns to R&D." *Journal of Dairy Science* 86 (2003): 1662–1672.

Chavas, J., and R.M. Klemme. "Aggregate Milk Supply Response and Investment Behavior on U.S. Dairy Farms." *American Journal of Agricultural Economics* 68 (1986):55–66.

Coble, K., C.C. Chang, B.A. McCarl, and B.R. Eddleman. "Assessing Economic Implications of New Technology: The Case of Cornstarch-Based Biodegradable Plastics." *Review of Agricultural Economics* 14 (1992):33–43.

Cochran, Joseph S. and Brenda S. v. Anne Veneman, Secretary, U.S. Department of Agriculture and National Dairy Promotion Board. No. 03-2522, U.S. Court of Appeals for the Third Circuit, filed February 24, 2004.

Dairy Management, Inc. *DMI 2002 Annual Report*. Available online at *www.dairycheckoff.com/pdf/DMI_2002_annual_report.pdf*.

Duffield, J., and H. Shapouri. *Biodiesel Development: New Markets for Conventional and Genetically Modified Agricultural Products*. Washington DC: U.S. Department of Agriculture, Economic Research Service Agricultural Economics Report 770, September 1998.

Food and Agricultural Policy Research Institute (FAPRI). *2003 FAPRI U.S. Baseline Briefing Book*. Columbia MO: FAPRI University of Missouri, Columbia, March 2003.

Helmberger, P., and Y. Chen. "Economic Effects of U.S. Dairy Programs." *Journal of Agricultural and Resource Economics* 19 (1994):225–238.

House, R., M. Peters, H. Baumes, and W.T. Disney. *Ethanol and Agriculture: Effects of Increased Production on Crop and Livestock Sectors*. Washington DC: U.S. Department of Agriculture, Economic Research Service Agricultural Economic Report 667, May 1993.

Johnson, S.R., D.P. Stonehouse, and Z.A. Hassan. *Market Demand for Dairy Products*. Ames IA: Iowa State Press, 1992.

Just, R.E., D.L. Hueth, and A. Schmitz. *Applied Welfare Economics and Public Policy*. Englewood Cliffs NJ: Prentice-Hall, Inc., 1982.

Kaiser, H.M., and C.J. Chung. *Impact of Generic Milk Advertising on New York State Markets*. Ithaca NY: Cornell University National Institute for Commodity Promotion and Research Bulletin 99-03, 1999.

Kaiser, H.M., and T.M. Schmit. "Distributional Effects of Generic Dairy Advertising throughout the Marketing Channel." *Agribusiness: An International Journal* 19(3) (2003):289–300.

Lee, H. "Ethanol's Evolving Role in the U.S. Automobile Fuel Market." *Industrial Uses of Agricultural Materials Situation and Outlook Report*. Washington DC: U.S. Department of Agriculture, Economic Research Service, 1993.

Lee, S.Y., K.L. Dangaran, and J.M. Krochta. "Gloss Stability of Whey-Protein/Plasticizer Coating Formulations on Chocolate Surface." *Journal of Food Science* 67 (2002):1121–1125.

Lee, H., J. Glauber, and D.A. Sumner. "Increased Industrial Use of Agricultural Commodities: Policy, Trade, and Ethanol." *Contemporary Economic Policy* 12 (1994):22–32.

Lee, S.Y., T.A. Trezza, J.X. Guinard, and J.M. Krochta. "Whey-Protein-Coated Peanuts Assessed by Sensory Evaluation and Static Headspace Gas Chromatography." *Journal of Food Science* 67 (2002):1212–1218.

Mullen, J.D., J.M. Alston, and M.K. Wohlgenant. "The Impact of Farm and Processing Research on the Australian Wool Industry." *Australian Journal of Agricultural Economics* 33 (1989): 32–47.

Sumner, D.A., and J.V. Balagtas. "Implications for California of the Dairy Provision of the 2002 Farm Bill." *The 2002 Farm Bill and California Agriculture*. R.J. Sexton, ed. Berkeley CA: University of California Division of Agriculture and Natural Resources, Giannini Foundation Research Report, in press.

U.S. Department of Agriculture, National Agricultural Statistics Service. *Agricultural Statistics*. Washington DC, various years.

——. *Dairy Products 2000 Summary*. Washington DC, 2001.

——. *Milk Production, Disposition and Income, 2001 Summary*. Washington DC, 2002.

U.S. Department of Agriculture, Office of Communications. *Agriculture Fact Book 2000*. Washington DC, 2000.

Zhao, X. "Who Bears the Burden and Who Receives the Gain? The Case of the GWRDC R&D Investments in the Australian Grape and Wine Industry." *Agribusiness: An International Journal* 19(3) (2003):367–378.

16

Demand Enhancement through Food-Safety Regulation: A Case Study of the Marketing Order for California Pistachios

Julian M. Alston, Henrich Brunke, Richard S. Gray & Daniel A. Sumner

The U.S. pistachio industry, which is located almost exclusively in California, has experienced phenomenal growth over the past thirty years. The value of U.S. production of pistachios grew from nearly zero in 1976 to $333 million by 2002. The U.S. share of the rapidly growing global pistachio industry has also increased. Outside California, the main pistachio growing region is the Middle East, including Turkey and particularly Iran.

In recent years, a group of California pistachio growers led an initiative to establish a federal marketing order that would mandate quality standards and establish an inspection program to assure consistency in the quality of California pistachios. The main provisions of the marketing order set standards and require testing for quality and for aflatoxin, a cancer-causing mold found in many nuts and grains. Proponents argued that the marketing order would increase consumer confidence and reduce the chance of an aflatoxin event in the pistachio market, thereby stimulating demand and enhancing consumer benefits and producer returns. Hearings were held in 2002. In early 2004, the proposal was supported by more than 90 percent of the growers who voted, and the marketing order is scheduled to take effect in February 2005.

This chapter begins with some brief background on the California pistachio industry and food-safety issues that led to development of the marketing order. We then discuss the economic rationale for the industry's collective action and the process and requirements for establishing a new marketing order. These aspects have much in common with other collective-action programs discussed in other chapters of this book, but there are some differences, and the fact that this is a new program provides an opportunity to examine the issues and costs that arise when a program is introduced—aspects that are often (sometimes appropriately) ignored when discussing the economics of collective-action programs of long standing.

The mandated quality-assurance and food-safety program for pistachios shares many elements with mandated generic commodity promotion programs. Both are funded by check-offs. Those assessments are borne by both producers and consumers in proportions that depend on quantities of the good produced and consumed and on elasticities of supply and demand. In both cases the primary purpose is to enhance producer returns by stimulating demand.

There are, however, important differences. In particular, commodity promotion programs are supposed to enhance demand every year so that their benefits are experienced for the most part during the time period when the cost of the check-off is incurred. Hence, the criterion for a program's effectiveness is whether its year-by-year benefits more than offset corresponding year-by-year costs. The pistachio marketing order, on the other hand, involves a cost borne every year for a program that aims to reduce the odds and severity of a major negative shock to demand posed by an aflatoxin outbreak in some future year. The criterion here involves comparing year-by-year costs with the expected benefits from changes in the odds of an outbreak and in its severity as an episodic, major negative shock to demand. Consequently, evaluating the marketing order for California pistachios is best accomplished by a comparison of the expected values of costs and benefits over many years. Our analysis of the pistachio marketing order, therefore, involves a stochastic dynamic simulation rather than the more typical comparative-static analysis.

The California Pistachio Industry

World production of pistachios has expanded rapidly during the past twenty years, and U.S. production has increased as a share of that growing total.[1] Iran is still the world's largest producer, but the United States is established as the second largest pistachio producer in the world (followed by Syria and Turkey) and is now also the second largest exporter after Iran (California Pistachio Commission 2004).[2]

Almost all U.S. pistachios are produced in California.[3] California's production has increased more than 200-fold since 1976, when the first commercial crop of 1.5 million pounds was harvested. California produced a record crop of 302 million pounds in 2002, up from the previous record of approximately 242 million pounds in 2000.[4] Table 16.1 presents time-series data on the industry. The longer-term trends show steadily growing acreage, yields, quantities, and value of production and corresponding downward trends in price with important fluctuations around those trends. California also has steadily expanded its exports as a share of world trade and of production.

Total California pistachio acreage grew from 34,726 acres in 1980 to 106,000 acres in 2002. Normally, it takes a pistachio tree seven to eight years to mature before it produces an economically significant crop and twelve to fifteen years to reach full potential. Bearing acreage in 2002 was estimated by the California Agricultural Statistics Service (CASS) to be 83,000 acres, up 222 percent from 25,773 bearing acres in 1980. The growth in area and

Table 16.1. Data on California Pistachio Area, Production, Yields, and Value, 1980–2002

Year	Bearing Acres	Nonbearing Acres	Production in Million Pounds	Yield in Pounds per Acre	Value in Million Dollars	Average Return in Dollars per Pound
1980	25,773	8,989	27.2	1,055	55.8	2.05
1985	32,332	18,739	27.3	838	36.6	1.37
1990	53,700	11,100	117.3	2,375	129.6	1.02
1991	55,700	13,300	76.4	1,465	100.7	1.25
1992	56,500	13,900	146.5	2,592	150.9	1.03
1993	57,000	15,700	150.9	2,648	161.5	1.07
1994	57,507	16,633	128.3	2,232	118.1	0.92
1995	60,300	13,400	147.7	2,449	160.9	1.09
1996	64,300	17,100	104.3	1,622	121.0	1.16
1997	65,373	17,062	179.5	2,746	202.9	1.13
1998	68,000	19,300	187.5	2,757	193.1	1.03
1999	71,000	21,000	122.4	1,724	162.8	1.33
2000	74,578	21,730	241.6	3,239	239.2	1.01
2001	78,000	23,500	160.3	2,055	166.7	1.01
2002	83,000	23,000	302.4	3,644	335.7	1.11

Source: CASS 2002.

production has been steady for the past twenty-two years and is expected to continue. Nonbearing acreage reached 23,000 acres in 2002 (CASS 2002). The value of the crop varies with the quantity produced. The long-term trend is increasing quantities and increasing value of the crop but falling returns per pound, reflecting the fact that supply has been growing faster than demand. The trend for the past twenty-two years in price per pound (even in nominal terms) has been gradually downward–the price has fallen from the high in 1980 of $2.05 a pound to $1.11 per pound in 2002.

In 2002, California had approximately 650 pistachio producers. There is a single pistachio-producer cooperative and nineteen handlers who process pistachios. About 70 percent of California pistachio producers produce less than 100,000 pounds per year; 21 percent produce more than 100,000 and less than 500,000 pounds; and about 9 percent produce more than 500,000 pounds. Based on an average grower price of $1.10 per pound, about 91 percent of California's pistachio producers generate less than $550,000 in annual revenue and 9 percent generate more than $550,000. About 85 percent of California pistachio handlers handle less than ten million pounds per year and about 15 percent handle more than ten million pounds per year. The largest handler processes about 50 percent of the industry's total production.

The California Pistachio Commission (CPC) was organized under the supervision of the California Department of Food and Agriculture (CDFA). It is distinct from the committee that will be responsible for administering the terms and provisions of the marketing order. CPC began operation in 1981 to provide support through public relations, government relations, marketing, and production research and is funded by an assessment of $0.035 per pound of pistachios produced in California. CPC also receives funding under the U.S. Department of Agriculture's (USDA's) Market Access Promotion Program–an average of about $800,000 per year from 1997/98 through 2001/02 (USDA, Foreign Agricultural Service 2003). In 2002/03, CPC spent almost $9 million, more than $6 million of which went to promotion (Carman and Alston in Chapter 2 of this volume). Details on CPC's history and activities can be found at its website, *www.pistachios.org*.

Food-Safety Issues and Aflatoxin in Pistachios

Aflatoxin is the main issue behind the creation of the marketing order for California pistachios. An event of aflatoxin poisoning in pistachios or some other product–even simply concern about the possibility of such an event–can adversely affect demand.[5]

Many produce-related food scares have occurred in recent years. For the period from 1990 to 1999, the Center for Science in the Public Interest (CSPI) lists fifty-five cases in the United States alone. A recent produce-related food scare involved cantaloupes and *Salmonella* in 2000, 2001, and 2002. *Salmonella* had infected several people in the United States and Canada and, in response to potential health risks, certain cantaloupe brands were recalled nationwide (U.S. Department of Health and Human Services, Food and Drug Administration 2002). In 1996, the California strawberry industry lost an estimated 5 percent of total revenue due to a *Cyclospora* scare (see Chapter 14; details can be found in Calvin 2003). An earlier, well-known event demonstrating the public's sensitivity to food safety was the 1989 Alar scare associated with apples. A television broadcast reported that Alar was used in apple production and that it was "the most cancer-causing substance in the food supply." Apple demand dropped dramatically overnight, and apple growers suffered losses estimated at hundreds of millions of dollars (details can be found in van Ravenswaay and Hoehn 1991).[6]

The main event that directly related to pistachios occurred in Europe. Iranian pistachio imports were banned by the European Union (EU) in September 1997 because the country's shipments exceeded allowed levels of aflatoxins (*The Economist* 1997). The ban lasted nearly three months and was lifted in December 1997 (European Commission, Food and Veterinary Office 1998). However, demand for pistachios was affected for a longer period. The United Nation's Food and Agricultural Organization (FAO) (2003) presents data showing that aggregate imports into the EU, including those from the United States, the main alternative source of pistachios, dropped from 102,698 metric tons in 1997 to 59,619 metric tons in 1998 (about 40 percent for the year).

In 1997, when imports were curtailed by the ban, Germany imported 47,494 metric tons of pistachios worth $175.3 million (United Nations, FAO 2003). The average for the preceding five years (1993–1997) was 43,459 metric tons per year (United Nations, FAO 2003). In 1998, Germany imported only 18,937 metric tons, just 40 percent of the quantity imported the previous year. German imports during the next two years were also well below 1997 quantities. Imports were 27,059 metric tons in 1999 and only 25,090 metric tons in 2000. The value of the imports fell to $78.9 million in 2000, or just 45 percent of the value in 1997. This drastic and protracted reduction in imports after the ban was lifted points to a decrease in consumption, perhaps resulting from negative publicity regarding the ban in the media. In 1999, *Oeko-Test*, a German consumer report, sampled pistachios from German supermarkets and found that aflatoxin levels in eight of eleven samples were higher than allowed and that the highest levels were in California pistachios (Hermes 1999).

In the years since 1997, shipments of pistachios have exceeded maximum aflatoxin levels on several occasions and in several countries, making news worldwide. In 2000 in Germany alone, several articles were published in national (*Der Spiegel* 2000, *Süddeutsche Zeitung* 2000) and regional newspapers following discoveries of high aflatoxin levels in pistachio ice cream. Aflatoxin problems have continued to appear worldwide. For example, recalls based on high aflatoxin levels in pistachios occurred recently in Australia and in February 2000 in Japan. Retailers were concerned about pistachios from Iran and California. Also in 2000, testing in France found high levels of aflatoxins.[7]

The central argument for establishing the marketing order was that collective action or government intervention was required to raise and assure the actual and perceived quality and safety of California pistachios given the potential for free-riders, an argument that we develop in detail later. The marketing order states that no handler shall ship for domestic human consumption pistachios that exceed an aflatoxin level of fifteen parts per billion (ppb).[8] An aflatoxin inspection certificate must cover all domestic pistachio shipments. The marketing order further outlines procedures for aflatoxin testing that are necessary before an aflatoxin inspection certificate can be obtained. The aim is to be able to trace every certified lot of an individual handler from testing through shipment. Certification by an accredited laboratory would assure that every lot of California pistachios shipped to the domestic market meets requirements and does not exceed proscribed aflatoxin levels.

Rationale for Collective Action in Pistachio Markets

Mandated collective-action programs like CPC and the federal marketing order for California pistachios use the coercive powers of the state or federal government to oblige individual producers to participate and contribute assessments. The programs are voluntary in the sense that their establishment requires the support of a sufficiently large majority of producers, but they do not require unanimous support and consumers, who bear a portion of the costs, do not have voting rights. Unlike truly voluntary collective-action programs, such as cooperatives or clubs, once established, these programs are mandatory for all producers of the commodity in the defined area.

The conventional economic justification in principle for use of the government's taxing and regulatory powers in this fashion to create mandated collective-action programs is that there are collective goods within the industry–research, promotion, grade standards, packing regulations, public relations, and the like–that would be undersupplied otherwise.[9] In practice, whether the pistachio marketing order will yield net benefits to producers, consumers, the state, and the nation will depend on the nature and extent of the "public-good" or "external" costs and benefits associated with its minimum quality standards, mandatory testing for aflatoxin, and other provisions and on the private cost of implementing the program.

The various regulations under the marketing order offer different types of public-good characteristics, of which some are more easily justified than others. Standardized grades and packaging have public-good characteristics in that they reduce transaction costs (e.g., see Freebairn 1967, 1973). An argument for quality regulation can be made when quality is not easily distinguishable prior to purchase and the market can be spoiled as a result of distortions in incentives for producers and marketers to provide and communicate accurate information about quality (e.g., Akerlof 1970). The public-good element is that consumers who experience a particular quality of pistachios from one supplier carry that experience over to their relationship with other suppliers. Especially in the case of a food-quality issue, bad experiences associated with any pistachio producer are likely to have a large, long lasting impact on the whole industry, but individual producers will not take the industrywide consequences of their actions entirely into account.

Regulations regarding visual standards–freedom from blemishes or minimum size regulations, for example–are generally not as easy to justify on the grounds of public goods since they relate to aspects of quality that are not hidden from consumers. Such regulations may provide *de facto* supply control by diverting some of the volume of the product to nonfood uses altogether or *de facto* price discrimination by diverting a larger proportion of the crop to the processing market, which has a more elastic demand response.[10] One rationale for minimum quality standards for pistachios, however, is that aflatoxin is more often found in small and damaged nuts and eliminating those nuts from the market is consequently an indirect way of reducing the risk of aflatoxin.

Maximum aflatoxin standards and inspection and certification have a food-safety role as well as a collective-good element for the industry because aflatoxin is a serious, and in some cases deadly, poison. However, the standards of the marketing order are in addition to and tighter than those already established by the U.S. government for food safety (but not as tight as standards applied elsewhere, such as in Europe).

An industrywide food-safety issue could arise as a result of evidence of death or illness associated with consumption of pistachios containing aflatoxin with consequences for demand for pistachios in general rather than simply for firms directly responsible for incidents. As a result, consumers may choose not to purchase products, retailers may choose not to stock a product that might be subject to recall or lawsuits, or governments may choose not to allow products to be sold because of heightened concerns over food safety.

The same type of market problem could arise without a case of food poisoning. It could result from an aflatoxin event where levels of aflatoxin are detected that exceed the twenty parts per billion allowed by the Food and Drug Administration. Negative perceptions could also result from adverse attention to aflatoxin even if no excess amounts were discovered. For instance, there may be adverse effects on the pistachio market from the perception of a threat based on adverse publicity associated with excessive levels of aflatoxin in other products in the United States or anywhere else in the world.

Perceptions of a food-quality problem are not specific to individual suppliers; they affect the industry in a collective way. Therefore, the private incentive to assure high-quality nuts that are perceived as safe does not reflect the full industrywide or public benefit of these actions. In such cases, voluntary actions motivated by private incentives provide less safety and quality assurance than is in the interest of the industry (and, perhaps, the general consuming public). In this case, each farm and firm benefits from a strong reputation for pistachios in general but the individual farm or firm cannot single-handedly assure such a reputation. Consequently, individual farms and firms have a private incentive to keep their direct costs low and invest less in safety testing and quality assurance than would be optimal from the view of the whole market. This is a free-rider problem in which individuals cannot be precluded from sharing in the benefits even when they fail to contribute, and one individual benefiting from the better collective reputation does not preclude benefits to others.[11]

Two characteristics of the pistachio market make public-good concerns particularly important in the context of food-safety assurances and quality standards. First, as with many fresh fruits and nuts, pistachios have little established brand identification. Thus, a customer who has a poor experience with a purchase of pistachios or who hears negative news about their safety is unlikely to associate it with a specific brand or supplier. Unlike branded, packaged consumer items, any negative news about a largely unbranded item affects the industry at large. Second, many pistachio purchasers consume the product infrequently, purchase relatively small quantities, and know relatively little about pistachios. One would therefore expect a strong reaction from the industry to an aflatoxin event in pistachios compared with a similar event for a more familiar food, especially in the context of food-safety concerns. The wholesale trade would be even more sensitive to an event if a recall were necessary. Consequently, the pistachio industry has strong reasons in principle for acting collectively to assure industrywide compliance with quality and food-safety standards. But this is only an in-principle case. Whether collective action of this type can provide net benefits to the industry also depends on how effective the program is in reducing either the likelihood of a food scare or its severity and on the costs of the program.

The Form of Intervention and Distribution of Costs and Benefits

We have discussed potential public-good arguments that apply to food safety and to pistachio inspection and testing. However, we should consider the alternative approaches to food safety and the potential public good associated with food standards.

First, despite the public-good aspects noted previously, individual consumers, marketers, processors, and farms have strong incentives to avoid harm from aflatoxins. Even if information is costly for individual consumers, under current market conditions, reputable marketing firms have an incentive to establish monitoring programs and to deal with suppliers they trust. Suppliers also have incentives to maintain their reputations for reliably safe products. The market-based approach to food-safety concerns is for marketers and suppliers to establish testing and monitoring programs of their own. This would facilitate their ability to assure consumers of the safety of products purchased from their outlets. Indeed, competition over traceability and reliable food safety may yield higher food standards than those established by the government. A market-based approach would allow more producer flexibility and more choice on the part of retail buyers. The fact that this approach has not been implemented for pistachios could simply mean that the anticipated private benefits are less than the costs.

At the other end of the spectrum, the government sets mandatory food-safety standards for many products, including pistachios. These standards can be enforced with mandatory and continuous government inspection like that which occurs in meat-packing plants. In the case of pistachios, current government standards for maximum aflatoxin levels are less restrictive than the limits proposed under the marketing order. Furthermore, the government standards are not enforced with any mandatory inspection or other active program. If aflatoxin were considered a major health threat, the standards could be tightened via regulation and an active mandatory inspection procedure could be implemented without an industry-sponsored marketing order. Evidently, interest in such programs has not been strong enough to stimulate legislation or new regulation.

Under the middle approach of an industry-sponsored program requiring active consent by a majority of firms, a majority of producers must expect a net benefit, but there may still be differential benefits and net losses by some. For example, branding is becoming more prevalent in the industry and product differentiation may become more difficult when the marketing order creates generic standards. If this were important, we would expect firms that are working to establish brand identities to resist the marketing order. In fact, however, large producers and processors that are most actively pursuing branding were leaders of the committee proposing the marketing order.

Distributional differences may arise from differences among firms in the costs incurred in meeting the higher standards or in testing. This issue was investigated during development of the marketing order and no evidence of differential costs was found. The argument is that higher safety standards could disproportionately impose additional costs on operations with less fixed investment in an industry since such firms have more elastic supply functions (especially for downward shocks) and less to lose from food-safety events that cause a downshift in demand. This argument seems relatively weak when applied to the pistachio industry, where trees require many years to reach maturity and most of the acreage is relatively young. Furthermore, much of the industry's processing capacity is relatively new. In the hearing process, which we discuss next, there was no active resistance to the marketing order from any segment of the industry.

Development of the Marketing Order for California Pistachios

USDA's Agricultural Marketing Service (AMS) specifies how an industry should proceed to establish a federal marketing order. First, after identifying mutual marketing problems, a steering committee of key industry members submits a preliminary proposal to USDA. This proposal should indicate the degree of industry support and the problems the program would address. Second, a request for a hearing on the proposal must be submitted that includes a proposed hearing site and approximate date. The hearing is presided over by a USDA administrative law judge, and a verbatim record is compiled of the testimony of opponents, proponents, and others. Third, the proponent group bears the burden of proof and must therefore present evidence in support of the need for the program. In addition to evidence of its benefits to the industry and consumers, a positive ruling generally requires evidence that the proposed marketing order will not disadvantage small businesses (producers, processors, or marketers) and generally has widespread support. Fourth, based on the evidence from the hearing, USDA issues a recommended decision and then, after allowing time for comment, a positive final decision by USDA allows the proposed order to be put to a grower referendum. Finally, at least two-thirds of the growers voting by number or by volume must approve the proposal before the U.S. Secretary of Agriculture can issue the marketing order.

The pistachio industry spent several years preparing to submit its proposal and succeeded in making its case to USDA. In July 2002, hearings were held in Fresno, California, on the proposal to establish a federal marketing order for pistachios grown in California. The stated objective of the program was to enhance grower returns through delivery of higher-quality pistachios to consumers.[12] Hitherto, industry quality-control practices have been limited to voluntary testing for aflatoxin and other quality requirements under the California Pistachio Marketing Agreement, which was entered into by a number of pistachio handlers under the authority of CDFA. The agreement has been limited to issues relating to blending of artificially and naturally opened pistachios and with bleaching pistachios. Under this agreement, aflatoxin testing and sampling guidelines were used only for exports to specified countries.

The proposed federal marketing order sets standards for the quality of pistachios produced and handled in California by establishing a maximum aflatoxin tolerance level, maximum limits for defects, a minimum size requirement, and mandatory inspection and certification. These standards will apply solely to pistachios from California that are marketed in the United States. An eleven-member committee consisting of eight producers, two handlers, and one public member will administer the program, which will be financed by assessments on handlers of pistachios grown in the production area.

The two-day hearing in Fresno in 2002 featured written submissions and oral testimony from growers, processors, and other industry members. The proponent committee retained experienced legal counsel to put together a case for the positive net benefits from the marketing order and to satisfy all of the legal specifications required under USDA regulations. The proponents' case included testimony by food-safety experts so that the health aspects of aflatoxins and the usefulness of the testing regime could be established in the hearing record and so the hearing judge and USDA officials could ask questions. Economic testimony and responses to questions by USDA were used to establish the argument that the proposed order would benefit producers and consumers and not unfairly burden small businesses. Overall, the process required hundreds of hours from proponent committee members and is likely to have cost hundreds of thousands of dollars in legal fees and related costs.

After the hearing, USDA took more than a year to study the record and prepare its positive preliminary recommendation. Based on evidence contained in the hearing record, USDA found that the proposed order met the requirements that it would have a net benefit for the industry and also benefit small firms. After the hearing, the proposed order was recommended by AMS and released for comment in 2003 and a grower vote in 2004. Results of the vote were announced on March 1, 2004. AMS reported that valid ballots representing 338 California pistachio producers were cast. Of those voting in the referendum, 90.8 percent favored establishing the order (these voters represented 90 percent of the total volume of production). Overall, the process took more than three years from beginning active preparation to approval. The order is scheduled to take effect in February of 2005.

A Model of the Industry and the Impacts of the Marketing Order

We developed a multiperiod, stochastic, dynamic simulation model and used it to simulate markets for California pistachios and project production, prices, and allocation of pistachios for fifty years, beginning with the year 2000. Yields vary over time as a reflection of alternate bearing and random influences. Aflatoxin events also occur at random. In our model, both the probability of an event and the severity of the demand response to a given event are lower in the case in which the marketing order is in place. For each "draw" of a time series of future yields, we simulated the outcomes for economic variables in the industry with and without the marketing order. By considering one hundred draws of future time paths of yields, we were able to estimate the effects of the marketing order on variables of interest, represented by both average values and the range of values for the variables (or other measures of variability).

We specified equations representing domestic and export demands for pistachios, including storage demand, using estimates of elasticities and data on market shares, quantities, and prices. The marketing order applies solely to the domestic market. It imposes regulations that entail costs of compliance with requirements for aflatoxin testing and meeting quality standards that are borne in the first instance by processors plus other relatively minor costs that are financed by an assessment on processors. The potential increases in demand, on average, relative to a scenario without a marketing order include the effects of (1) a reduced probability of a negative shock to demand associated with an aflatoxin event and a reduction in the size of the negative shock associated with a given event, and (2) an increase in demand in every year owing to greater consumer and buyer confidence in the product associated with USDA testing and certification. Later in this section we discuss these two demand elements, but first we turn to the costs of compliance they are meant to offset.

Cost of Testing and other Compliance Issues

The quantitative economic analysis required information on the costs of aflatoxin testing and compliance with other quality standards that are regulated under the marketing order. The initial incidence of this additional cost would be on the processing sector. Depending on the current level of testing and other characteristics, processing firms face different costs of complying with the standards. A telephone conference with the seven major processors in California (Pistachio Processor Group, Fresno CA, July 2002) provided data on the costs that various processors in the industry would face under the marketing order.

In his testimony at the hearings on July 25, 2002, Daniel Sumner described in detail how to estimate these costs for three different types of processors depending on their current testing practices, the size of their processing operation, labor requirements and wage rates for inspectors, lot sizes, and various other factors. The resulting estimate of the direct per-unit cost of compliance was a weighted average (across the different types of processors) of $0.00525 per pound on the two-thirds of production to which the proposed marketing order rules apply. This figure seems to represent a consensus of views in the industry.[13] The weighted cost of compliance applied across all California pistachio production was $0.0035 per pound. These figures were based on an assumption that few undersized pistachios are currently sold in the standard market such that the implementation of the other features of the marketing order would not have significant implications for the total quantity available to the domestic market.[14]

Effects on the Probability of an Aflatoxin Event and its Consequences

Direct evidence does not exist on the probability of an aflatoxin event that would cause a major negative shock to demand or on the likely severity of such a shock. We assumed that increased aflatoxin testing reduces the probability of such an event and may reduce the severity of a shock as well, but, again, we have no direct quantitative evidence on relevant magnitudes. To calibrate the potential effects of a pistachio food scare, we used information from other produce-related food scares in the United States along with information from an event involving pistachios in Germany, and we conducted sensitivity analyses in which we varied the relevant parameters.

Some direct evidence is available on the market response to an aflatoxin event in pistachios but not for an event in the United States. As described previously, the EU banned pistachio imports for three months in the last quarter of 1997 because of aflatoxins. After the ban was lifted, German imports were substantially reduced over the next three years—by an estimated 40 to 50 percent. The pistachio market is relatively small, and other nuts and snack foods are likely to be close substitutes for pistachios. This may be why losses for pistachios in the EU market were larger than observed after some other food scares in the United States, but the differences also may reflect some differences between the United States and other countries—in terms of institutions and consumer behavior—that affect the response.

Taking a conservative approach, we assumed that an aflatoxin event in the domestic market for U.S. pistachios in year t would cause a 30 percent reduction in demand in the year of the event (i.e., $\delta_t = 0.3$, where δ stands for the proportion by which demand decreases in year t). The evidence from Germany suggested that the negative demand effects from a single aflatoxin event would continue to affect demand for several years. In the model, the negative demand shock decayed at a rate of 30 percent per year (i.e., $\delta_{t+n} = 0.7^n \delta_t$).

Aflatoxin events do not happen every year, but the market always faces some probability of a food scare. The marketing order cannot eliminate the chance of a food scare associated with aflatoxin in pistachios, but it has provisions that make such an event less likely. The benefit from additional testing is a reduction of the probability of an aflatoxin event or food scare. For the base case of no mandatory testing, we used an annual probability of 4 percent for an outbreak that affects demand as previously specified. We assumed that, with mandatory testing, the chance of an aflatoxin outbreak would fall to 2 percent and that an event, when it occurred, would have a smaller effect on demand. We assumed an initial downward shock of 15 percent under the marketing order and 30 percent without the order.

Effects on Consumer and Buyer Confidence

Demand for pistachios could increase as a result of official USDA certification ensuring a good quality product. Many agricultural products take advantage of USDA grading and other services, and buyers for major food outlets and many consumers in the United States are well aware of USDA standards. In general, USDA standard-setting is thought to convey a positive benefit in a market as reflected by use of this claim in product promotion, labels, and displays. We were not aware, however, of empirical evidence of the magnitude of the impact of such certification. We used a small increase in demand to reflect greater buyer confidence in pistachios due solely to USDA's participation in the standards process.

In addition, but similarly, demand for pistachios increases with mandatory minimum quality standards and buyers' more favorable perception of safety. It is anticipated that the standards in the order will reflect well on the product as a whole and shift out demand for all pistachios because buyers will anticipate a smaller probability of acquiring low-quality shipments. This demand effect involves two aspects. The first is the general notion that buyers are willing to pay more for nuts of a higher quality. Second, a minimum quality standard assures buyers that they have a smaller chance of a low-quality shipment. This effect relies on more information being available to buyers that all pistachios from the marketing order area meet minimum standards. One of the provisions under the marketing order prohibits the sale of inferior pistachios. Although these represent a tiny fraction of total production, removing them from the market altogether will result in an increase in the general quality of pistachios, albeit a small one. To reflect both of these elements in the simulations, we allowed for a small increase in demand in every year, relative to the base case, in response to introduction of the marketing order–an increase in U.S. consumers' willingness-to-pay for pistachios equal to 1 cent per pound (about 1 percent of recent prices).

Benefit-Cost Analysis

To estimate the impact of the marketing order, we computed and compared a pair of fifty-year simulations (i.e., one with and one without the marketing order) using baseline values for the parameters as shown in Table 16.2. For each year of the fifty-year simulation, the model determined a market-clearing price, bearing acres, acres planted, yield, production, domestic quantity demanded, export quantity demanded, ending stocks, revenue, and consumer surplus. To capture the effects of random yield variability and aflatoxin-related demand shocks, the fifty years of simulated equilibrium values were calculated for a set of one hundred equally likely futures that differed in terms of values for randomly generated yields and aflatoxin shocks. Hence, in a given scenario, each simulated variable of interest has a fifty-year time path that is affected by the marketing order with a random distribution in each period. It is important to keep this time path and the random nature of the variables in mind as we report summary statistics and average impacts.[15]

Under the marketing order, as shown in Table 16.2, the annual probability of an aflatoxin event was reduced from 4 percent to 2 percent, and the demand impacts of such an event were assumed to be half as large (i.e., an initial drop in demand of 15 percent versus 30 percent applied to both the domestic market and relevant export markets). In addition to this benefit, the marketing order was assumed to increase domestic consumers' willingness-to-pay for

Table 16.2. Key Parameters for the Simulation Model

Parameter	Baseline Value
Underlying Market Conditions	
Elasticity of Domestic Demand	-1.00
Elasticity of Export Demand	-3.30
Elasticity of Demand for Stocks	-2.00
Long-Run Annual Growth Rate of Demand (percent)	3.60
Elasticity of New-Plantings Response to Profitability	1.00
Impact Parameters without a Marketing Order	
Probability of an Aflatoxin Event (percent per year)	4.00
Initial Impacts of an Event – Domestic Demand (percentage reduction)	30.00
Initial Impacts of an Event – Foreign Demand (percentage reduction)	6.45
Impact Parameters with a Marketing Order	
Probability of an Aflatoxin Event (percent per year)	2.00
Initial Impacts of an Event – Domestic Demand (percentage reduction)	15.00
Initial Impacts of an Event – Foreign Demand (percentage reduction)	3.23
Compliance Costs (cents per pound)	0.525
Domestic Demand Enhancement from Certification (cents per pound)	1.00

Table 16.3. Simulation Results: Benefit-Cost Analysis of the Pistachio Marketing Order

Consequences of the Marketing Order	Baseline	High Impact	Low Impact
Induced Changes in Average of Annual Values, 2000–2050			
Price of California Pistachios (real cents per pound)	0.501	0.726	0.371
Bearing Area of California Pistachios (acres)	1,866	2,716	1,279
Production of California Pistachios (million pounds)	12.55	18.31	8.63
Domestic Consumption of California Pistachios (million pounds)	11.54	16.91	8.15
Exports of California Pistachios (million pounds)	1.01	1.40	0.51
Stocks of California Pistachios (million pounds)	-0.62	-1.05	-0.47
Present Values for Net Benefits in Year 2003 in Millions of Dollars			
Changes in U.S. Consumer Surplus (*CS*)	165.4	246.7	109.8
Changes in California Producer Surplus (*PS*)	68.9	103.7	49.6
National Benefits ($NS = CS + PS$)	234.3	350.4	159.4
Net Changes in Foreign Surplus (*FS*)	-25.0	-36.5	-19.2
Global Net Benefits ($GS = NS + FS$)	209.3	313.9	140.2
Present Values for Marketing Order Costs in Year 2003 in Millions of Dollars			
Cost of Compliance (*CC*)	36.7	34.9	38.4
Benefit-Cost Ratios			
Global Benefit-Cost Ratio ($1 + [GS / CC]$)	6.7	10.0	4.7
National Benefit-Cost Ratio ($1 + [NS / 1.1\ CC]$)	6.9	10.2	4.8
Grower Benefit-Cost Ratio ($1 + [PS / 0.15\ CC]$)	13.5	20.8	9.6

pistachios by 1 cent per pound. The cost of compliance with the marketing order, 0.525 cents per pound consumed domestically, was reflected as a reduction in the price to growers for domestic sales.

The impacts of the marketing order are reported in the first column of Table 16.3. To summarize the effects of the marketing order over the fifty-year simulation, we report average effects over the fifty years for some variables; for others, we report the net present value in 2003 of the effects over the fifty years. The marketing order modestly increases the average price received by growers (by 0.5 cents per pound, or 0.6 percent), along with average bearing acres (by 1,870 acres or 1.3 percent) and production (by 12.6 million pounds per year or 1.5 percent). These increases in production are associated generally with increases in domestic consumption (by 11.5 million pounds per year or 2.8 percent) and in exports (by 1 million pounds per year or 0.2 percent) and decreases in stocks (by 0.6 million pounds per year or 0.3 percent). These averages mask the fact that, as noted previously, the effects on some of these variables change over time because of trends (the production response to the order increases with time whereas the domestic demand response begins immediately) and from year to year (through the interaction of order-induced changes in bearing acres and variable yields). This is true in particular for the effects of the marketing order on exports—the small average effects reflect negative impacts in some years, especially initially, and positive impacts in others, especially the later years.

The dynamics of the impact of the marketing order on revenue per bearing acre, bearing acres, and domestic consumer surplus are particularly interesting. The marketing order increases grower price and revenue per acre by increasing consumer confidence and reducing the odds and the impact of an aflatoxin event. The impact on revenue is greatest in the first few years after introduction of the order because supply is not affected during this period of time. The increase in revenue per acre eventually causes an increase in the time path of bearing acres. The increase in bearing acres results in increased production, driving down prices and revenue per acre and dissipating the benefits for producers. Consumers gain initially from improved food safety, and that benefit is then augmented by subsequent reductions in price resulting from increases in production.

The net benefits from the marketing order—reflecting the consequences of both the assessment and the regulations and of the demand and supply responses to them—are expressed as present values (in 2003) of changes in economic surplus accruing to different groups. These net benefits included \$68.9 million to domestic producers and \$165.4 million to domestic consumers, yielding a total national net benefit of \$234.2 million.[16] From a global perspective, U.S. net benefits were slightly offset by net losses in foreigner surplus (the "consumer surplus" measured off the demand for U.S. exports) worth \$25 million, leaving global net benefits with a present value in 2003 equal to \$209.2 million.[17]

We also estimated the total cost of the marketing order (in terms of expenditures incurred by processors in complying), which had a present value in 2003 of \$36.7 million. The initial incidence of this cost falls on processors, but the incidence is redistributed over time through supply and demand responses. To evaluate the final incidence, we ran a simulation with just the assessment (modeled as a reduction in domestic buyers' willingness-to-pay of 0.525 cents per pound). In present-value terms, we found that 15 percent of the cost was borne by growers and 85 percent by domestic and foreign consumers combined. Domestic consumers paid 95 percent of the total cost (this is more than the 85 percent attributed to domestic and foreign consumers, combined, because foreign consumers are net beneficiaries of a tax on domestic consumers). Hence the incidence of the global cost of $36.7 million was $39.7 million on the United States, including $5.5 million on U.S. producers. By dividing each measure of net benefits by the corresponding measure of the incidence of the costs, we obtained a ratio of net benefits to costs (i.e., $(B - C) / C$), to which we added 1 to compute conventional benefit-cost ratios (B / C) for domestic producers (13.5), the United States (6.9), and the world (6.7).

To examine the general sensitivity of the results to modeling assumptions, we devised a "high-impact" scenario and a "low-impact" scenario, and the summary results for the simulations under these scenarios are reported in the second and third columns of Table 16.3. For the high-impact scenario, we altered most of the parameters of the model by 10 percent in the direction that would increase the impact of the marketing order; for the low-impact scenario, we altered the parameters by 10 percent in the opposite direction. These scenarios reveal how the results are affected by a modest but consistent upward or downward bias in parameter values. As shown at the bottom of the table, the combined effect of the parameter changes creates more than 10 percent variation in the estimated impacts of the marketing order. Compared with a benefit-cost ratio for producers of 13.5 in the base scenario, the ratio is 20.8 (50 percent higher) in the high-impact scenario and 9.6 (30 percent lower) in the low-impact scenario. Similarly, for the United States as a whole, compared with a benefit-cost ratio of 6.9 in the base scenario, the ratio is 10.2 (50 percent higher) in the high-impact scenario and 3.8 (45 percent lower) in the low-impact scenario. Nevertheless, the benefit-cost ratios are all significantly greater than zero even in the low-impact scenario, indicating that the order entails substantial net benefits for both producers and the nation as a whole.

Conclusion

An aflatoxin event could impose serious costs on the California pistachio industry. The marketing order is intended to reduce the odds of an event, to mitigate the consequences if an event should occur, and to provide some quality assurance to buyers to offset the negative consequences of concerns over the potential for a food scare affecting pistachios. We modeled the market for California pistachios to provide an *ex ante* assessment of the benefits and costs and other consequences of the marketing order looking forward fifty years from its anticipated introduction in 2004. Our approach used a stochastic, dynamic simulation of the industry under scenarios with and without the marketing order to compare the stream of simulated outcomes and the consequences for measures of economic welfare for producers in the industry, for consumers, for the nation as a whole, and globally.

Assessing the implications of the marketing order required incorporating into the simulation a number of parameters representing the odds of an aflatoxin event, its consequences for demand, and the extent to which a marketing order could reduce those magnitudes. Many of the parameters are hard to estimate because relevant historical data for pistachios are not available. As well as simulating the consequences implied by "best-guess" values for key parameters, we undertook sensitivity analysis. Across the full range of parameters used in the analysis, the benefit-cost analysis was always favorable to the marketing order: the measured benefits to producers, the nation, and the world always well exceeded the corresponding measure of costs, typically by many times. The benefit-cost ratios were generally greater than 5 and often greater than 10, providing substantial leeway to accommodate potential errors in assumptions and still have favorable findings. In present-value terms, the benefits to producers were estimated at $68.9 million. Two-thirds of the benefits, $165.4 million, would accrue to domestic consumers. These are significant values and are large relative to the cost of compliance with the program, which is a very small amount—about half of one percent of the current value of domestic sales.

Notes

1 Statistical information in this section was supplied by the California Pistachio Commission unless otherwise noted.

2 Iran's exports peaked in 1996, when it exported 308 million pounds of pistachios, but fell to 127 million pounds in 1997, when Iranian pistachios were banned by the European Union because of aflatoxins. Iranian exports returned gradually to near pre-ban levels in the following few years.

3 In 2000, Arizona had 2,700 acres and produced four million pounds of pistachios, just 1.5 percent of national production in that year (Arizona Agricultural Statistical Service 2003) and an amount that is too small to significantly affect the national market for pistachios. New Mexico had 391 acres of pistachios in 1999, less than half a percent of total acreage (New Mexico Agricultural Statistics 2000).

4 Pistachios exhibit alternate bearing with low yields tending to follow high yields. Thus, 2001 was a relatively low-yield, low-production year. The yield cycle is an important factor in quantity produced, price received, total value of the crop, and gross revenue per bearing acre (U.S. Department of Agriculture, Risk Management Agency 2003).

5 Aflatoxicosis is poisoning that results from ingestion of aflatoxins in contaminated food or feed. The most pronounced contamination has been encountered in tree nuts, peanuts, and other oilseeds, including corn and cotton seed. Aflatoxicosis in humans rarely has been reported; however, cases are not always recognized. In affluent countries, aflatoxin contamination rarely occurs in foods at levels that cause acute aflatoxicosis in humans, but there have been important aflatoxin events in pistachios. (Information here is taken from *http://vm.cfsan.fda.gov/~mow/chap41.html.*)

6 Additional studies of the impact on demand of food-safety events and information on demand can be found in Smith, van Ravenswaay, and Thompson (1988), Brown and Schrader (1990), Richards and Patterson (1999), and Piggott and Marsh (2004).

7 Information here was supplied by Karen Reinecke, president of the California Pistachio Commission, in 2002 in preparation for the federal marketing order hearing.

8 This is a tighter standard than the current maximum allowed by the U.S. Food and Drug Administration of twenty parts per billion.

9 This is the standard public-good argument for government intervention. The services in question are "public goods" in the sense that they are nonrival (benefit by one consumer does not reduce availability for others) and not price excludable (if the service is provided at all, it is very costly to preclude benefits by others even if they refuse to pay), but these public-good benefits are confined to the producers and consumers of a particular commodity and are associated with consumption or production of the commodity. The collective goods could be provided using the general revenues of the relevant state or national government, but it is likely to be fairer and more efficient to finance their provision using a tax on the commodity with which the collective goods are associated.

10 Alston et al. (1995) analyzed the impacts of the allocated reserve policy applied by the Almond Board of California under a federal marketing order, which is an example of this type of supply control that can be mimicked by using quality regulations to divert some fraction of production from the market. Chalfant and Sexton (2002) analyzed an interesting example of *de facto* price discrimination associated with grade standards in the California prune industry.

11 Naturally, some of the free-rider problems could be avoided if pistachios were differentiated, say by brands. Then individual firms could provide private testing and quality assurance and capture at least some of the benefits. But branding and product differentiation is costly and not predominant in the industry. Moreover, the existence of brands does not eliminate all (or even most) of the externalities that can be associated with food scares. We thank John Crespi for causing us to elaborate this point.

12 Details here are based on the proposed rule published in the *Federal Register* (2003).

13 The *Federal Register* (2003, p. 46017) reports "The average cost of compliance, as identified by several witnesses and reiterated in Dr. Sumner's analysis, is approximately one half cent per pound of domestic pistachio production, or $0.00525 per pound."

14 In any event, under the marketing order, undersized pistachios could still be diverted to the export market so the consequences of this element of the regulations for prices, quantities, and values would be negligible.

15 Further details on the simulation model and alternative parameterizations may be found in the more comprehensive full report on the study (Gray et al. 2004).

16 On 78,000 bearing acres in 2001, the producer benefit is worth $2,120 per acre, but the benefits would not be confined to these acres.

17 The positive effect on export quantity seems to contradict the higher average price and the reduction in foreign "consumer" surplus associated with the order. The effect on foreign "consumer" surplus is complicated. First, there are some benefits to foreigners from the order because in the baseline case there is a spillover effect of an aflatoxin event from U.S. demand to foreign demand and the order-induced reduction in probability and severity of an aflatoxin event applies to export markets as well as to domestic ones. These benefits are offset at least somewhat by the larger domestic demand responses that drive up prices, especially in the early years; in later years, those effects in turn are offset at least somewhat by the consequences of the U.S. supply response to the policies. The benefits to foreigners are greater in the earlier years and, given discounting, the net present value is negative even though the average effect on quantity of exports, undiscounted, is slightly positive.

References

Akerlof, G.A. "Market for 'Lemons': Qualitative Uncertainty and the Market Mechanism." *Quarterly Journal of Economics* 84(August) (1970):488–500.

Alston, J.M., H.F. Carman, J.A. Chalfant, J.M. Crespi, R.J. Sexton, and R.J. Venner. *The California Prune Board's Promotion Program: An Evaluation*. Berkeley CA: University of California Division of Agriculture and Natural Resources, Giannini Foundation Research Report 344, March 1998.

Alston, J.M., H.F. Carman, J.E. Christian, J.H. Dorfman, J.R. Murua, and R.J. Sexton. *Optimal Reserve and Export Policies for the California Almond Industry: Theory, Econometrics and Simulations*. Berkeley CA: University of California Division of Agriculture and Natural Resources, Giannini Foundation Monograph 42, February 1995.

Alston, J.M, G.W. Norton, and P.G. Pardey. *Science under Scarcity: Principles and Practice for Agricultural Research Evaluation and Priority Setting*. Ithaca NY: Cornell University Press, 1995.

Arizona Agricultural Statistics Service. "Publications." Phoenix AZ, 2003. Available online May 2003 at *www.nass.usda.gov/az/rlsetoc.htm*.

Brown, D.J., and L.F. Schrader. "Cholesterol Information and Shell Egg Consumption." *American Journal of Agricultural Economics* 72 (1990):548–555.

California Agricultural Statistical Service. *Pistachio Acreage and Forecast Reports*. Sacramento CA, 2002. Available online April 2002 at *www.nass.usda.gov/ca*.

——. *Pistachio Acreage Report, 2000*. Sacramento CA, 2001. Available online March 2002 at *www.nass.usda.gov/ca/rpts/acreage/pistachio/008pstac.htm*.

California Pistachio Commission. "Proposed United States Standards for Grades of Pistachio Nuts." Sacramento CA, 2003a. Available online March 2003 at *www.pistachios.org*.

——. "History." Sacramento CA, 2003b. Available online December 2003 at *www.pistachios.org*.

——. "Statistics." Sacramento CA, 2004. Available online April 2004 at *www.pistachios.org/data/statistics.xls*.

Calvin, L. "Produce, Food Safety, and International Trade: Response to U.S. Foodborne Illness Outbreaks Associated with Imported Produce" *International Trade and Food Safety, Economic Theory and Case Studies*. J.C. Buzby, ed. Washington DC: U.S. Department of Agriculture, Economic Research Service Agricultural Economic Report 828, 2003.

Center for Science in the Public Interest (CSPI). "Produce-Related Outbreaks, 1990." Available online July 2002 at *www.cspinet.org/new/prodhark.html.*

Chalfant, J.A., and R.J. Sexton. "Marketing Orders, Grading Errors, and Price Discrimination." *American Journal of Agricultural Economics* 84(1) (2002):53–66.

Der Spiegel. "Giftiges Pistazien Eis (Poisonous pistachio ice cream)." Issue 26, June 24, 2000.

European Commission, Food and Veterinary Office. "Summary of Inspection of Iran's Food and Veterinary Office, 1998." Available online May 2003 at *http://europa.eu.int/comm/food/fs/inspections/fnaoi/reports/contaminants/iran/fnaoi_rep_iran_1492-1998_de.html.*

Federal Register. Vol. 68, No. 149, 45990-46033, August 4, 2003, Part III, Department of Agriculture, Agricultural Marketing Service, 7 CFR Part 983. Details available at *www.ams.usda.gov/fv/moformal/fv029831sd.pdf.*

Freebairn, J.W. "Grading as a Market Innovation." *Review of Marketing and Agricultural Economics* 35(3) (1967):147–162.

——. "The Value of Information Provided by a Uniform Grading Scheme." *Australian Journal of Agricultural Economics* 35(3) (1973):127–139.

Gray, R.S., D.A. Sumner, J.M. Alston, and H. Brunke. "Economic Consequences of Mandated Grading and Food Safety Assurance: *Ex Ante* Analysis of the Federal Marketing Order for California Pistachios." Draft monograph, Department of Agricultural and Resource Economics, University of California, Davis, March 2004.

Hermes, Peter. "Iranisches Roulett (Iranian Roulette)." *Oeko-Test* (monthly German Consumer Report), November 1999.

New Mexico Agricultural Statistics Service. "NM Publications." Las Cruces NM, 2003. Available online May 2003 at *www.nass.usda.gov/nm.*

Piggott, N.E., and T.L. Marsh. "Does Food Safety Information Impact U.S. Meat Demand?" *American Journal of Agricultural Economics* 86(1) (2004):154–174

Richards, T.J., and P.M. Patterson. "The Economic Value of Public Relations Expenditure: Food Safety and the Strawberry Case." *Journal of Agricultural and Resource Economics* 24(2) (1999):440–462.

Smith, M.E., E.O. van Ravenswaay, and S.R. Thompson. "Sales Loss Determination in Food Contamination Incidents: An Application to Milk Bans in Hawaii." *American Journal of Agricultural Economics* 70 (1988):513–520

Süddeutsche Zeitung "Verschimmeltes Pistazieneis (Moldy Pistachio Ice Cream)." June 26, 2000.

The Economist. "Nuts to the EU." September 20, 1997, p. 52.

United Nations, Food and Agriculture Organization. Rome Italy, 2003. "Statistics." Available online June 2003 at *www.fao.org/waicent/portal/statistics_en.asp.*

U.S. Department of Agriculture, Agricultural Marketing Service. *Steps in Developing a Fruit and Vegetable Marketing Order or Marketing Agreement Program.* Washington DC, 2003. Available online July 2003 at *www.ams.usda.gov/fv/moabsteps.html.*

——. "Pistachios Grown in California; Secretary's Decision and Referendum Order on Proposed Marketing Agreement and Order Number 983." Docket Numbers AO-F&V-983-2; FV02-983-01. Washington DC, 2003. Available online January 2004 at *www.ams.usda.gov/fv/moformal/fv029831sd.pdf.*

U.S. Department of Agriculture, Economic Research Service. *Fruit and Tree Nut Situation and Outlook Yearbook.* Washington DC, various issues. Available online October 2003 at *www.ers.usda.gov/Briefing/FruitAndTreeNuts/marketoutlook.htm.*

U.S. Department of Agriculture, Foreign Agricultural Service. "Market Access Program Allocations, Fiscal Year 2002." Washington DC, 2002. Available online October 2003 at *www.fas.usda.gov/excredits/quarterly/2002/2002-sum/03-01%20TABLE-MAP.pdf.*

U.S. Department of Agriculture, Risk Management Agency. "An Economic Assessment of Pistachios: Executive Summary." Washington DC, 2003. Available online June 2003 at *www.rma.usda.gov/pilots/feasible/txt/pistchio.txt.*

U.S. Department of Health and Human Services, Food and Drug Administration. "Press Release: Nationwide Recall of Susie Brand Cantaloupes due to Potential Health Risk." Washington DC, 2002. Available online May 2003 at *www.fda.gov/oc/po/firmrecalls/kunik05_02.html.*

U.S. Food and Drug Administration, Center for Food Safety and Applied Nutrition. *Bad Bug Book.* Washington DC, 2004. Available online October 2004 at *http://vm.cfsan.fda.gove/~mow/chap41.html.*

van Ravenswaay, E.O., and J.P. Hoehn. "The Impact of Health Risk Information on Food Demand: A Case Study of Alar and Apples." *Economics of Food Safety.* J.A. Caswell, ed. New York NY: Elsevier, 1991.

Part IV

Conclusion

17

The Economics of Commodity Promotion Programs: Summary and Synthesis

Julian M. Alston, John M. Crespi, Harry M. Kaiser & Richard J. Sexton

As noted in Chapter 1, California's mandated commodity promotion programs are important and controversial. Economists have a role to play in providing economic analysis to support better-informed decisions by industry participants and policymakers about the design and operation of these programs. This fact has been recognized both by government, which requires regular reviews and benefit-cost assessments for the programs, and by the courts, which have used economic assessments as evidence about the programs. These demands have stimulated both private and public investments in economic evaluations, and a considerable investment has been made in benefit-cost assessments of programs in California and nationally.

This book presents the results of a number of benefit-cost studies of California commodity programs, makes these results available and accessible to a broad audience, and puts them into context. We document thirteen case studies, all conducted recently and involving quantitative analysis of the benefits and costs of mandated commodity programs in California.

During the course of our preparing this book, litigation over commodity promotion programs has continued, and decisions have been taken at various levels in the judicial system. Some key cases are still in process, including one before the U.S. Supreme Court regarding the beef program, demonstrating the topical nature of the issue and the timeliness of this work.

Coverage and Character of the Case Studies

A primary basis for choosing the cases studies to be included as chapters was that a benefit-cost study had already been done and results were available for publication. This meant that the coverage of the book might not be fully representative of the sixty-two active programs in California (documented in Chapter 2). Studies have tended to focus on programs that have been subject to litigation and that involve sizable budgets. Nevertheless, the two programs in California with the largest budgets–the California Fluid Milk Processor Board and the California Milk Advisory Board–have not been the subject of benefit-cost analysis in the public domain and so are not included.

These factors notwithstanding, the coverage of the case studies is broad in terms of both types of programs and commodity coverage–they encompass almonds, avocados, dried plums (prunes), eggs, raisins, strawberries, table grapes, walnuts, and milk and dairy products.[1] Many of the studies evaluated the impacts of generic advertising or other forms of commodity promotion on domestic or export markets. In addition, we include a variety of other cases involving less direct approaches to enhancing demand for farm products–nutrition education, research on new uses for farm products, mandated inspection and quality assurance, and public relations to counter a food scare. The cases included in the book provide examples of benefit-cost analysis of many of the different kinds of activities undertaken by commodity programs, but the coverage is still incomplete since some of the activities undertaken by marketing programs are intrinsically difficult to quantify and have not been evaluated by economists.

The case studies include some common methodological elements, but there are also important differences that have implications for interpretation of results and comparability across studies. All the studies employed (at least conceptually) a model of supply and demand for an agricultural commodity with competitive market clearing as outlined in Chapter 1. The studies focused on the benefits to agricultural producers (often including first handlers of farm products as well as farmers) as opposed to society more broadly (including consumers and others as well as producers) from enhanced demand generated by collective promotion and other activities funded by commodity assessments.

The studies all combined estimates of parameters that measured the size of the shifts in demand induced by the programs with other parameters that represented the supply and demand for the commodity to infer benefits to producers from the shift. The approaches used to estimate these parameters differed among studies, particularly between *ex ante* (forward-looking) and *ex post* (backward-looking) studies.

Most of our case studies are *ex post* and used econometric methods to estimate at least some of the parameters, but most of the studies also used assumed or "guesstimated" values for parameters that were based on some combination of economic theory, knowledge of the industry and its markets, and reviews of estimates in the literature. The studies of commodity promotion (or advertising) programs are all *ex post* analyses. They used econometric methods to estimate parameters representing how demand responds to promotion (and sometimes price) and typically used aggregate (market-level) historical time-series data. A few studies had access to cross-sectional data that they used to buttress a primary analysis that was based on time-series analysis. By definition, historical data could not be used to obtain estimates of parameters in the *ex ante* analyses of new-uses research (Chapter 15) and of quality testing and certification for California pistachios (Chapter 16). Parameters for those studies were developed in other ways based on different kinds of data.

Given that check-off assessment rates are typically well less than 1 percent of the value of sales, promotion programs can be profitable even if they have only very small effects on demand, and it is difficult to identify and measure small effects precisely using econometric analysis of aggregate time-series data when many factors that can affect demand are changing together. The precision of the estimates depends on the accuracy of the structure of the model and on the precision of the estimates of the parameters. The studies profiled here varied in the extent to which they paid attention to the roles of modeling assumptions and other choices in the analysis that could influence their estimates of the demand response to the programs and the benefits to producers. Still, all of the studies that used econometric methods found statistically significant program-induced market impacts.

The concepts of benefits and benefit-cost ratios differed across studies, and these differences have important implications for the extent to which their measures are comparable. Some reported an *average* benefit-cost ratio (ABCR) in which the total benefits from the program are divided by the total costs of the program. In an *ex post* analysis of an existing program, computing this ratio correctly requires a simulation of market outcomes in the absence of the program while accounting for the impacts of both the program and the collection of assessments to fund it. In contrast, some studies reported a *marginal* benefit-cost ratio (MBCR), which measures the additional benefits resulting from a small increment to the program (e.g., the additional benefits resulting from an increase of one dollar in spending on the program).[2]

A further complication arises when we consider the potential for the costs of the assessment to be passed on to consumers through the operation of supply and demand–shifting the incidence of the assessment. It is not the total expenditure on programs (which is the *initial incidence* on producers of the assessment collected from them) that matters for producer profits; rather, it is the *final incidence* of the assessment on them (the fraction that is not passed on to consumers). Some studies measured the ABCR or MBCR using the change in producer surplus as the relevant concept of benefits and total expenditures on programs as the relevant concept of costs. Others used the producer incidence of the assessment required to finance the expenditure rather than the total expenditure as the relevant concept of costs.

These different concepts of benefit-cost ratios–marginal versus average and using the producers' share of total costs rather than total costs of the program–can imply quite substantial differences in the measures. This point is illustrated in Chapter 4, in which the authors report estimates of benefit-cost ratios for promotion of California table grapes using multiple concepts. As can be seen in Table 4.2, the values for MBCRs are typically about half the size of the values for ABCRs (this reflects the square root functional form used, which imposes a particular degree of diminishing marginal returns). It is important to note that MBCRs and ABCRs based on the producer incidence of the assessment are not sensitive to the assumed value for the price elasticity of supply (because the benefits and costs change roughly in proportion to one another when the supply elasticity changes), but ratios based on total program costs are highly sensitive to supply elasticity (reducing the supply elasticity from 5.0 to 1.0 causes the benefit-cost ratios to increase roughly by a factor of five).

The natural concept for the *ex post* assessment of programs for policy purposes is an ABCR (rather than an MBCR) in which the denominator measures the producer incidence of the assessment to finance the specific expenditure (rather than total expenditure). A rationale that has been offered for the use of total expenditure, rather than the producer incidence, in the denominator is that the resulting estimate is conservatively low.[3] However, an objective of "conservatism" could be applied in a more informative way by taking a pessimistic interpretation of an unbiased estimate rather than by choosing an approach that is inherently biased to an extent that is not apparent in the estimate alone. Further, an approach that relies on total expenditure means the estimate is sensitive to the elasticity of supply, a parameter that is notoriously difficult to estimate and always difficult to interpret; and it is a parameter that is almost always assumed as part of the study.

The marginal concept (MBCR) is less naturally relevant than the ABCR to the question of whether a program pays for itself. The MBCR is the appropriate concept for informing decisions about marginal changes to programs, such as whether the current expenditure is optimal or whether it should be decreased or increased. It is common to assume diminishing marginal returns to program expenditures, but this assumption is not always built into models. In some cases, then, MBCRs will coincide with ABCRs (if we have constant returns to scale of promotion throughout) while in other cases MBCRs will be less than ABCRs (if we have diminishing returns throughout). On this basis, one might feel comfortable using the MBCR as a conservatively low estimate of the ABCR. However, if fixed costs are important, a model with diminishing marginal returns is not sufficient to assure that the MBCR is less than the ABCR (though we are not aware of any empirical work that has included a substantial fixed-cost element).[4] On the other hand, approximation formulas–based simply on elasticities of demand with respect to price and promotion and promotion intensity–can be used to estimate the MBCR with considerably less effort than is required to estimate the ABCR.[5] Hence it may still be worth reporting MBCRs rather than ABCRs in a context where the ABCR is the relevant concept if the alternative is not to have any measure of the benefit-cost ratio.

Main Findings

To summarize the main findings, we first present a brief synopsis of each chapter. Then we present an overall synthesis of the results in the case studies and their interpretations, making use of tabular summaries of the key features of each study and its results.

Overview of Mandated Marketing Programs in California

Mandated marketing programs covered almost 55 percent of California agricultural commodities in 2002/03 based on value, including 79 percent of California production of animal products, 73 percent of fruit and nut production, and 43 percent of vegetable production. These programs include state and federal marketing orders as well as commodity commissions and councils, and they operate under broad enabling legislation and, in some cases, individual statutes.

In Chapter 2, Carman and Alston present a brief history of these programs; a discussion of the legislative framework and processes for introducing, establishing, and maintaining the programs; and an overview of the different types of activities undertaken by the programs–including advertising and promotion, research, minimum quality standards and mandated inspection, volume control, and standardization of containers or packs. They then provide details on current programs in California.

California currently has sixty-two active marketing programs, including twelve federal marketing orders, twenty-seven state marketing orders and agreements, twenty commissions, and three councils. Budgeted expenditures for the programs–estimated at $208 million in 2003/04, or about 1.2 percent of the $16.8 billion value of the commodities covered–have increased significantly over the past twenty years through the combined effects of several factors: expansion of the number of programs and commodities covered, increases in the value of the commodities, and modest increases in the average intensity of the programs (program expenditures per dollar of value of the commodity).

Carman and Alston also present details on the Buy California Marketing Agreement, which was introduced in 2001 to promote consumption of California agricultural products in California. This activity is funded by a combination of federal and state government funds and contributions from commodity organizations, mostly from mandated marketing programs.

Litigation over Promotion Programs

The U.S. Supreme Court has held that advertising is a type of speech and that freedom of speech includes the right to refrain from speaking. Because commodity promotion programs compel growers to jointly support the program and pay for an advertisement, such programs must tread gingerly through the Constitution's First Amendment guarantees of freedom of speech and freedom of association. As this book goes to print, the U.S. Supreme Court is preparing to hear its third commodity promotion case. The ruling on that case, which concerns beef promotion, is expected to provide some clarification to two earlier rulings by the Supreme Court on generic advertising; it will no doubt be key to the future of commodity promotion programs, at least as they currently exist. But how did we get here?

Chapter 3 by Crespi provides a historical analysis of the programs and the litigation surrounding them. Crespi argues that it is a mistake to see these programs as, in the words of Supreme Court Justice Antonin Scalia, "time-warped" out of the Great Depression. Although marketing orders did begin with Depression-era legislation, that legislation evolved after years of experiments with other forms of collective farm marketing programs. Further, Crespi shows that today's litigation arose when it did in the early 1980s only because of Supreme Court rulings on earlier First Amendment cases that had nothing to do with agriculture. When the Court made its rulings in the *Abood* case in 1977 and the *Central Hudson* case in 1980, it became inevitable that generic advertising programs would be implicated. Unfortunately, the Court's rulings on generic advertising in 1997 (*Wileman*) and 1999 (*United Foods*) did little to clarify which of the programs pass constitutional muster. Unless the Supreme Court can provide further guidance in the upcoming beef promotion case, Crespi foresees more litigation in the future.

Promotion of California Table Grapes

The California Table Grape Commission was established in 1968. In recent years, California table grapes have, in round figures, had a farm value approaching $1 billion per year and expenditures on promotion have been around $10 million per year, about two-thirds of which has been directed at the domestic market.

In Chapter 4, Alston et al. report the results of an evaluation of promotion in the "domestic" market (combining the United States and Canada) based on an econometric model of per capita demand using annual data over the period 1968–1993. The promotion variable represented the sum of expenditures on advertising, merchandising, and public relations, and this variable entered the models in square root form. Econometric estimates are reported for several alternative models, followed by a benefit-cost analysis using the "preferred" model. (The original analysis by Alston et al. (1997) on which this chapter is based reported results from extensive diagnostic testing and model selection procedures, as well as Monte Carlo simulations, and presented confidence intervals as well as means for the benefit-cost ratios.)

Benefit-cost ratios were computed using a range of assumed values for the elasticity of supply and using discount rates of 0 and 3 percent per annum. Using the midpoint of the range of supply elasticities (an elasticity of 1.0) and a discount rate of 3 percent, the ABCR was 144.9 and the MBCR was 80.1 when the measure of cost was the final incidence on producers. When the measure of cost was total expenditure on promotion, the ABCR was 38.8 and the MBCR was 21.5. With any of these measures, the benefits are very great relative to costs, and the estimates are also quite precise. The results of additional econometric work using monthly data on domestic markets and models of export demand reinforce the results from the annual aggregate model, allowing for further confidence in the results.

Egg Promotion

Between 1984 and 2002, the California Egg Commission (CEC) developed and implemented advertising and promotion campaigns aimed at increasing egg consumption and enhancing returns for California egg producers.[6] CEC had an annual budget of about $4 million with approximately 90 percent of that budget directed toward advertising, public relations, promotion, and education and research. California produces more eggs than any other state in the nation–more than twenty million laying hens produce more than four billion eggs with a total farm value of $282 million per year.

In Chapter 5, Schmit, Reberte, and Kaiser describe an econometric model of supply and demand in the California egg industry that they used to evaluate the impact of generic egg advertising in the state on producer prices and returns over a ten-year period. Econometric estimation indicated that the generic advertising efforts had a positive impact on producer prices and net profits. Specifically, they estimated a long-run price flexibility of demand with respect to advertising of 0.13–i.e., a 1 percent increase in advertising results in a 0.13 percent increase in the producers' egg price for a given quantity.

The model was used to simulate the market with existing advertising expenditures and with expenditures 1 percent higher than actual. A 1 percent change in advertising expenditures resulted in an average 0.13 percent increase in producer prices (similar to the flexibility since supply is highly inelastic), which implied that the MBCR was 6.9.[7] In other words, each additional dollar spent on advertising generated $6.90 in producer profits. The resulting impact on net returns was evaluated over a 90 percent confidence interval of the advertising parameter estimates, indicating returns exceeded costs to producers over the entire interval. The estimated returns to generic egg advertising in California indicate that advertising efforts in the state have been quite profitable.

Dried Plum Promotion

The California Dried Plum Board (CDPB), established in 1952, operates under a state marketing order. The board levies assessments to fund production research and generic promotion of California dried plums (previously called prunes). California is the world leader in production of dried plums, accounting for 99 percent of U.S. production and 70 percent of the world's supply. In Chapter 6, Alston et al. report a study of the economic impacts of industry-financed market development and promotion activities for California dried plums, specifically asking whether CDPB's marketing expenditures directed to the domestic market have affected the demand for and sales of California dried plums and whether net revenues to producers from the programs increased enough to offset program costs.

The authors constructed and estimated an econometric model of demand for California dried plums using monthly data from September 1992 through July 1996. Estimation results revealed dried-plum demand to be inelastic with a price elasticity of −0.3 to −0.5, depending on the model specification. The elasticity of demand with respect to CDPB promotion was estimated to be about 0.05, and the elasticity of demand with respect to promotion by the leading marketer, Sunsweet Growers, was about 0.16.

To translate these demand shifts into measures of economic return, the authors constructed an equilibrium simulation model of the industry by combining the preferred demand model with a domestic-market residual supply function that was calibrated to replicate observed industry outcomes. Alternative supply elasticities of 0, 0.5, 1.0, 2.0, and 5.0 were used to reflect a range of lengths of run. Results from the simulation analysis suggest that, over the four-year period analyzed, investments by dried-plum growers in CDPB-funded promotion yielded growers marginal returns of at least $2.65 for every dollar of assessment incidence borne by them. MBCRs of this magnitude indicate that the industry could have profitably invested even more in promotion during this period.

Avocado Promotion

The California Avocado Commission (CAC) is a state marketing order that seeks, in part, to increase demand for California avocados. In 2003, the California avocado crop had a farm value of $373 million, and CAC spent almost $10 million on advertising and promotion in 2003/04. Determining the effectiveness of CAC's avocado promotion is tricky given that avocado production is highly variable. In Chapter 7, Carman and Craft describe an economic model of the California avocado market that incorporates this supply variation, as well as producers' planting responses to demand changes. The authors developed and used this model to determine the impact of marketing expenditures on supply and demand for California avocados and, ultimately, to determine whether the benefits resulting from CAC's marketing expenditures outweighed the cost of the assessments.

Based on their preferred econometric estimates, Carman and Craft found that increases in the price of avocados were associated with both decreases in production and increases in advertising expenditures by CAC. The estimated price-flexibility of demand with respect to quantity was −1.3, which implies that a 1 percent decrease in production would result in a 1.3 percent increase in the price of avocados. The estimated price-flexibility of demand with respect to CAC advertising expenditures was 0.13, which implies that a 10 percent increase in advertising expenditures would result in a 1.3 percent increase in the price of avocados.

Carman and Craft looked at both short-run and long-run benefits to avocado producers and found that short-run average benefits were about $5 to $6 per dollar expended by CAC. However, over time, producers responded by planting more acreage. When Carman and Craft accounted for production responses and examined the returns over the long run, the benefits were approximately $1.7 of revenue for every dollar spent by CAC. The long-run marginal return was lower, approximately 1.4, suggesting that CAC's expenditures were nearing the industry's profit-maximizing ratio (recall that a MBCR of 1 implies the profit-maximizing level of investment in promotion).

Almond Promotion

The controversial nature of mandatory commodity marketing programs and the litigation these programs have faced since their inception, particularly recently, provided some of the impetus for this book. As discussed in Chapter 3, some programs have been terminated while others have survived court challenges. Some insight into the economic effects of the litigation can be gleaned from the analysis of the promotion program by the Almond Board of California (ABC) in Chapter 8 by Crespi and Sexton. During the crop years 1994/95 to 1996/97, ABC's advertising expenditures were curtailed as a result of a lower-court decision. They resumed in 1997, when the order was overturned following the Supreme Court's *Wileman* decision. This suspension provided a natural experiment that enabled Crespi and Sexton to assess the economic impact on almond producers.

Beginning with an econometric model of U.S. demand for almonds, Crespi and Sexton first established that advertising by both ABC and Blue Diamond Almonds (the largest almond processor) had positively affected demand for almonds in the United States. In particular, they found that a 10 percent increase in promotion expenditures was associated with an increase in almond demand of approximately 1.3 percent. Next, using the results of previous studies, Crespi and Sexton determined a range of probable supply elasticities facing the U.S. almond industry and used those elasticities in a simulation model of the U.S. almond market that took into account the effect of promotion on export markets as well. Crespi and Sexton generated MBCRs for the industry in the range of 3 to 10, depending on the supply elasticity used, which indicate that more spending on almond advertising would have been profitable. Next, the authors used their model to perform a hypothetical analysis of the market had the suspension never occurred. They concluded that the suspension of ABC advertising from 1994/95 to 1996/97 resulted in accrued losses to producers (i.e., foregone profits) ranging from $84 million to $231 million over that period.

Walnut Promotion

The Walnut Marketing Board (WMB), which was established in 1948 as a federal marketing order representing walnut growers and handlers, promotes use of walnuts through publicity, product promotions, research, and education programs. Expenditures on domestic promotions have run about $1.5 million annually in recent years. The farm value of walnuts in 2002 was $305 million.

In Chapter 9, Kaiser reports the results of an analysis of the domestic marketing efforts of WMB using time-series data from 1980 through 1999 and an econometric model of per capita demand. A key finding was that WMB's marketing programs had a positive and statistically significant impact on per capita walnut demand since 1980. The estimated marketing elasticity of 0.005 means that a 1 percent increase in expenditures on walnut marketing resulted in an increase in per capita domestic demand for walnuts of 0.005 percent. While this elasticity is not large relative to the other demand determinants, it is statistically significantly different from zero and positive.

The estimated demand equation was used to simulate market conditions with and without WMB efforts. The simulation results indicated the WMB's programs had a major impact on annual walnut utilization in the United States, increasing total walnut utilization by 124,665 tons between 1984 and 1999, or 7,333 tons per year. WMB also had a positive impact on the walnut grower's price that ranged from just under 3 cents per pound in the case of an inelastic supply response to just under 0.5 cents per pound in the case of a very elastic supply response. ABCRs computed for WMB's marketing activities exceeded 1 for every elasticity of supply response considered in the simulation and ranged from 9.72 for the most inelastic supply response considered to 1.65 for the most elastic supply response. Given the wide range of supply responses evaluated, the evidence indicates that WMB's marketing programs have been profitable for the domestic walnut industry.

Export Promotion of California Raisins

One objective of the Raisin Administrative Committee (RAC) is to conduct export promotion programs in various countries with the principal aim of increasing exports of California raisins. RAC is part of a federal raisin marketing order that has been operating in California since 1949. Most of RAC's export promotion has targeted Japan and the United Kingdom. In the late 1990s, expenditures on export promotion of raisins totaled 274 million yen ($2.5 million based on current exchange rates) in Japan and 832,402 pounds ($1.5 million) in the United Kingdom. In recent years, the farm value of California raisins has ranged from $500 million to $750 million.

In Chapter 10, Kaiser, Liu, and Consignado report results of their evaluation of the economic effectiveness of raisin export promotion in Japan and the United Kingdom using an econometric model of per capita import demand in the two countries and annual data from 1965 through 1998. Based on the estimated elasticities of demand with respect to promotion (0.029 in Japan and 0.133 in the United Kingdom), the statistical evidence supports the notion that the programs have had a positive impact on demand for California raisins in both countries. Furthermore, the simulation results indicate that the magnitudes of the impacts were not trivial; they accounted for an increase in sales of 6,107 metric tons per year in Japan and 18,116 metric tons per year in the United Kingdom.

The resulting ABCRs for markets in Japan and the United Kingdom were 5.1 and 15.3, respectively, and the overall ABCR for the two markets combined was 7.3.[8] A 90 percent confidence interval indicates that the lower-bound estimates for the ABCR in Japan and the United Kingdom were 4.6 and 14.6, respectively. These confidence intervals give credence to the finding that the revenue-expansion benefit of the raisin industry's export promotion programs has been considerably greater than the cost of the programs. Finally, a separate, marginal simulation analysis indicated that, while RAC's export promotion spending level in Japan has been optimal, the industry should explore investing more money in its export promotion activities in the United Kingdom.

Demand Enhancing Strategies for Strawberries

The California Strawberry Commission (CSC), initiated in 1955, has an annual budget of $6.8 million (2002) used for promotion and research activities. Strawberries are an important crop in California with annual values of processed and fresh production exceeding $1.1 billion in 2002, placing strawberries in the top ten commodities in California in terms of gross value of production. In Chapter 11, Carter, Chalfant, and Goodhue investigate the impacts of a consumer-oriented advertising campaign by CSC that focused on the health benefits of strawberries.

The authors estimated a price-dependent model of weekly demand for strawberries using weekly time-series data from 1990 through 2003. The price of California strawberries was modeled as a function of fresh volume, lagged fresh volume, lagged price, processing volume, substitute fruit volumes, time of season, year, and the percentage of markets advertising strawberries. This model is somewhat unique among those in this book in that advertising was measured as the percentage of markets in the United States featuring advertising for a particular week, and the advertising was sponsored not only by CSC but also by retailers nationwide.

The authors found that the advertising variable was statistically significant with a price flexibility of 0.16–a 1 percent increase in the share of national markets that are "on ad" or advertising strawberries leads to a 0.16 percent increase in the California strawberry price. If we assume that a 1 percent increase in advertising expenditure has an effect on sales equivalent to that of a 1 percent increase in the share of the national market on ad, we can compute an approximate measure of the MBCR. This approximate MBCR is equal to the estimated price flexibility with respect to advertising divided by the advertising intensity. Given CSC expenditures on advertising of about $4 million annually and a gross value of sales of about $1.1 billion in 2003, the advertising intensity was about 0.36 percent. The implied MBCR is 44. This relatively high MBCR indicates that the industry could have profitably invested in even more advertising.

National Promotion Programs

National promotion programs for dairy, beef, pork, and cotton have accounted for the lion's share of the total funds spent on agricultural product promotion in the United States. In 2001, these programs generated expenditures in excess of $500 million and accounted for some 83 percent of the total funds available from national check-offs. California has been a significant contributor to all four programs and the major contributor to the dairy program. Given the magnitude of funds expended under these programs and the legislative mandate that their effectiveness be evaluated at least once every five years, the economic impacts of these programs have been the subject of considerable research.

In Chapter 12, Kinnucan and Zheng provide a synopsis and critique of mandated evaluations for dairy, beef, pork, and cotton and a comparison of results from these studies to other evaluations in the scholarly literature for the same commodities. Kinnucan and Zheng note considerable variation in econometric approaches among the studies and also a considerable diversity of methods used to compute benefit-cost ratios. They identify the core issues in each case and critique the various approaches with the goal of indicating areas for attention and possible improvement in future evaluations.

On balance, the mandated evaluation studies suggest that promotion has been highly effective in shifting demand and yielding high MBCRs and ABCRs. However, Kinnucan and Zheng caution that these benefit-cost estimates must be treated with circumspection owing to outstanding issues regarding model specification and estimation and their inconsistency in some cases with results from other studies. Overall, the authors conclude, there is reason to believe that the private benefits generated by these programs have exceeded costs. What is less certain, in their view, is the extent to which the programs have been socially beneficial, not only from the perspective of consumers who help finance the programs through shifting of tax incidence but also from the perspective of the entire farming community because promotions for one commodity may spill over and adversely affect demand for related commodities.

Nutrition Education Programs

The Dairy Council of California (DCC) has been operating since 1945. In Chapter 13, Alston, Chalfant, and James report the results of their evaluation of an industry-funded nutrition education program, *Exercise Your Options* (*EYO*), that is provided to junior-high-school children in California. The *EYO* program was conducted by DCC and funded by a per-unit assessment on milk.

The authors used food records completed by students before and after the program, along with food records from a control group, to estimate the effects of learning about nutrition on consumption patterns. They found evidence of a statistically significant change toward a healthier diet, including an increase in dairy consumption. They estimated the benefits and costs to the dairy industry associated with the implied increase in demand for milk, taking into account the effects of milk-price policy. Their intermediate estimate of the value of benefits was $89,317 over the first thirty days and $62,040 over the subsequent thirty days, assuming 100 percent of the increase was in the form of fluid milk. The total implied benefit of $151,357 is slightly greater than the cost of the program, $150,126.

The authors also reported a range of benefit-cost ratios by allowing benefits to persist for up to twelve months with different rates of decay of program impacts and allowing for different assumptions about the distribution of the demand increase between fluid milk and other products. Benefit-cost ratios ranged from 0.13 to 1.74, increasing both with months of *EYO* persistence and with the fluid-milk share. The authors did not report a most-likely value but rather indicated the ranges of combined assumptions over which the ABCR was greater than 1.0. They note that the measure of benefits to producers (and thus the ABCR) was a lower-bound estimate of the social benefits since it did not account for the implications for reducing social costs of nutrition-related illnesses such as osteoporosis, coronary heart disease, and cancer.

Benefits from Public Relations Expenditures: The Case of Strawberries

The California Strawberry Commission (CSC) is a California marketing program established in 1955 to conduct strawberry production research and undertake generic promotion of California strawberries. The industry faced significant scares about the safety of consuming strawberries in 1996 and 1997 resulting from disease outbreaks that were attributed, incorrectly in one case, to eating strawberries. In Chapter 14, Richards and Patterson examine specifically the impacts of these health scares on demand for California strawberries and grower profits and the effectiveness of CSC's efforts to counter negative media reports associated with these disease outbreaks—what Richards and Patterson term "spin control."

Richards and Patterson developed a dynamic monthly demand system for fresh fruits (strawberries, apples, bananas, grapes, and oranges) that included quantitative measures of both positive and negative media reports associated with consuming strawberries. Conceptual underpinning for the econometric model was provided by attribution theory, which presents an explanation for the way decision-makers respond to signals provided by others. Signals are regarded as credible representations of the true intentions of the provider if their cause can be attributed to interests of the provider that go beyond self-interest and reflect a wider concern for other individuals or society as a whole. Based on this theory, Richards and Patterson hypothesized that negative information has a stronger impact on demand than offsetting information provided by CSC. This hypothesis is supported by the econometric results, in which bad news had almost twice the impact on strawberry demand as the same quantity of good news.

Estimated impacts on producer profits from the 1996 and 1997 health scares depend on assumptions about the price elasticity of supply, and Richards and Patterson provide simulations for supply elasticities ranging from 0.01 (almost perfectly inelastic) to 5.00 (very elastic). For a midrange supply elasticity of 1.0, they estimated that a 10 percent shock in unfavorable media exposure would reduce producer returns by $218.6 million over the long run, but use of defensive media reports ("spin") could reduce those losses to $105.2 million, giving the defensive media efforts a value of $113.4 million. Since CSC is estimated to have spent about $0.25 million countering the 1997 health scare, the analysis suggests that this money was very well spent. (The implied ABCR, comparing total benefits and total expenditure on the activity, was 454:1.) Given the inherent tendency for bad news to be more potent than offsetting good news, Richards and Patterson suggest that CSC would be wise to invest resources in an industrywide food-safety program designed to prevent health scares from occurring.

New-Uses Research

The California Dairy Research Foundation (CDRF) provides funds for research and development. CDRF receives approximately three-quarters of its budget from national check-off revenue generated in California and the remaining one-quarter of its budget from national check-off revenue generated in all states.[9]

In Chapter 15, Balagtas and Sumner report the results of their evaluation of the economic returns to an ongoing research program at the University of California that has been funded jointly by CDRF and the university. The program is developing new uses for whey protein–a byproduct of cheese manufacture–in the form of films and coatings for both food and nonfood applications.

Over the period 1990 to 2005, the research was projected to have cost a present value of $3.8 million. National and California dairy check-off programs will have contributed $2.23 million, of which approximately 41 percent, or $0.92 million, can be attributed to dairy farms and the rest to processors, marketing firms, and consumers. To estimate the benefits, the authors inferred an estimate of the increase in demand for milk (implied by estimates of the likely adoption of the new whey-based products) that they incorporated in a simulation model of the dairy industry. For a relatively conservative time pattern of technology adoption, the estimated present value of benefits to U.S. producers (net of the $0.92 million invested by producers in the research program) was approximately $126 million. Thus, while total benefits to producers were small relative to total dairy farm revenue–U.S. dairy farm cash receipts were $24.7 billion in 2001–benefits were large relative to the cost of the investment. Nearly 70 percent of the producers' cost of the program was borne by California dairy producers, but California producers received less than 20 percent of total producer benefits, approximately proportional to their share of U.S. milk production. Nevertheless, the ABCR for California producers was high, between 40:1 and 79:1, and the corresponding internal rate of return for California producers ranged from 25 to 29 percent per annum.

Quality Testing and Certification

The value of U.S. production of pistachios grew from about zero in 1976 to $333 million in 2002, mostly in California. A new federal marketing order was approved in 2004 to mandate quality standards and an inspection program to assure consistency in the quality of California pistachios. The main provisions set standards and require testing for quality and for aflatoxin, a cancer-causing mold found in many nuts and grains. Proponents argued that the marketing order would increase consumer confidence and reduce the chance of an aflatoxin event in the pistachio market, thereby stimulating demand and enhancing consumer benefits and producer returns.

In Chapter 16, Alston et al. report an *ex ante* evaluation of the benefits and costs of the marketing order conducted prior to its approval. The authors evaluated impacts of the marketing order looking forward fifty years from its likely introduction in 2004; they used a stochastic, dynamic simulation of the industry under scenarios with and without the proposed marketing order in place to compare the stream of simulated outcomes and the consequences for measures of economic welfare of producers in the industry, consumers, the nation as a whole, and the world. Assessing the implications of the marketing order required incorporating into the simulation a number of parameters representing the odds of an aflatoxin event, its consequences for demand, and the extent to which a marketing order would reduce those magnitudes. Using the baseline parameter values, the benefit to California producers was estimated at $68.9 million in present value terms. Two-thirds of the national benefit, $165.4 million, would accrue to domestic consumers. These are significant values and are large relative to the $36.7 million cost of compliance with the program (about 0.5 percent of the current value of domestic sales). The corresponding ABCR for California growers is 13.5, which takes into account the final incidence of the compliance cost.

Summary and Synthesis of Findings from the Case Studies

The main findings from the case studies reported here are represented in summary terms in Tables 17.1 and 17.2. Table 17.1 covers the eight case studies of California-specific commodity promotion programs detailed in Chapters 4 through 11; Table 17.2 covers the four case studies of California-specific nonpromotion programs detailed in Chapters 13 through 16. The tables have some elements in common, but differ in detail.

Table 17.1 reports a number of characteristics of each of the case studies in Chapters 4 through 11. All of these studies were *ex post* evaluations based on econometric estimation of demand models, including variables to represent expenditure on promotion (or elements of promotion such as advertising). Reported characteristics include details on: (1) the specifics of the organization, the commodity, and the program being evaluated (in terms of its size and nature); (2) the specifics of the econometric analysis in terms of the time period of the econometric estimation, whether the model was quantity- or price-dependent, and measures of the responsiveness of demand to price and promotion; (3) the characteristics of the benefit-cost analysis in terms of the time period of the market simulation, whether the benefit-cost ratio referred to an average or marginal concept, and whether the costs referred to total costs or the producer incidence of costs; and, finally, (4) resulting benefit-cost ratios.

The programs included in these case studies varied in size (in recent years, they spent between $2.4 million and $14 million annually), partly reflecting differences in the sizes of the industries and the markets to which they applied. The program intensities (expenditures on the program relative to the size of the industry in terms of gross value of production) were typically close to 1 percent and generally in the range of 0.5 percent to 2 percent.

All of the studies found statistically significant demand responses to price and promotion, and the measures of demand response to promotion were generally large (i.e., elasticities in the range of 0.05 to 0.16) relative to the intensities of promotion, so that a favorable benefit-cost ratio was implied. The benefit-cost ratios ranged from 0.8 to 144.9, but some of this range reflected different concepts. ABCRs were generally higher than their counterpart MBCRs, and the ratios based on producer incidence of costs were higher than those based on total expenditures. If the appropriate measure of profitability is an ABCR based on producer incidence, then the evidence from the studies is more favorable than would be indicated by an ABCR based on total expenditures or an MBCR; hence, some of the ratios in Table 17.1 tend to understate the profitability of the investments. The range is much narrower if we set aside the results from Alston et al. on California table grapes, which yielded very high benefit-cost ratios. Among the other studies, the range was from 0.8 to 80.1 across all the different concepts of benefit-cost ratios. Since a benefit-cost ratio greater than 1.0 is sufficient for profitability, the evidence consistently indicates that the programs have been highly profitable. These benefit-cost ratios are generally higher than those from the evaluations of national commodity programs for dairy, beef, pork, and cotton, which are summarized in Chapter 12.

The four case studies summarized in Table 17.2 refer to evaluations of California programs investing in activities other than promotion with a view to enhancing demand for the relevant farm product. The details reported in Table 17.2 are similar to those in Table 17.1 but with some differences, since none of the case studies in Table 17.2 entailed direct econometric estimation of demand response to the program expenditure being evaluated and they involved a variety of different types of data and modeling approaches. The benefit-cost ratios in Table 17.2 are all ABCRs, and they all express benefits relative to total expenditure on the program activities being evaluated rather than producer incidence of assessments (in part because the activities in question typically use only a fraction of total income from assessments in the relevant industry). The range is enormous, but generally the evidence indicates very large benefits relative to program expenditures on research on new uses for whey ($ABCR_T$ = 40–79), mandated inspections and quality standards for pistachios ($ABCR_T$ = 13.5), and "spin control" for strawberries in the event of a food scare ($ABCR_T$ = 454). The exception is the dairy council's nutrition education program, for which the range was smaller ($ABCR_T$ = 0.13–1.74) and close to 1.0.

Finally, to provide some context for the results in Table 17.1 and 17.2, corresponding results from the literature for other U.S. commodity promotion programs are summarized in Table 17.3. We included only studies that reported a benefit-cost ratio for commodity promotion within the United States or Canada, which required us to exclude a large number of studies. We included studies of which we were aware and could easily access copies but did not conduct a formal literature search. Hence, the table provides a representative

Table 17.1. Eight Case Studies of California Commodity Promotion Programs

Characteristics	California Table Grape Commission (Chapter 4)	California Egg Commission (Chapter 5)	California Dried Plum Board (Chapter 6)	California Avocado Commission (Chapter 7)
Type of Organization	State commodity commission.	State commodity commission.	State marketing order.	State commodity commission.
Commodity Value	$1,805 million in 2002.	$282 million in 2003.	$72 million in 2000.	$373 million in 2003.
Study	Alston et al.	Schmit & Kaiser	Alston et al.	Carman & Craft
Activities Evaluated	Domestic Promotion, U.S. and Canada.	Advertising.	Domestic Promotion.	Promotion in U.S. market.
Prog. Expenditure 2003/04[1]	$12.1 million.	Program ended 2002.	$5.2 million.	$9.8 million.
Spending on Activities Evaluated	$5.4 million in 2002.	$4 million in 1995.	$5.2 million in 2003/04.	$7 million in 1994.
Period of Econometric Estimation	1968–1993	1985:1–1995:2	9/1992–7/1996	1961–1994
P- or Q-Dependent Demand	Q-dependent	P-dependent	Q-dependent	P-dependent
Elasticity (E) or Flexibility (F) of Demand Response to Price (P) or Promotion (A)	$E_P = -0.5$ $E_A = 0.16$	$F_P = -1.7$ $F_A = 0.13$	$E_P = -0.4$ $E_A = 0.05$	$F_P = -1.3$ $F_A = 0.13$
Period of Simulation	1968–1993	1985:1–1995:2	9/1992–7/1996	1961–1994
MBCR or ABCR	MBCR, ABCR	MBCR	MBCR	MBCR, ABCR
Total Cost (T) or Incidence (I)	T, I	T	T, I	T, I
Estimated Benefit-Cost Ratio[2]	$ABCR_I =$ 44.9 $MBCR_I =$ 38.8 $ABCR_T =$ 80.1 $MBCR_T =$ 21.5	$MBCR_T = 6.9$	$MBCR_I =$ 2.7 $MBCR_T = 0.8$	$ABCR_{T(SR)} =$ 5.0 $ABCR_{T(LR)} =$ 1.7 $ABCR_{I(LR)} =$ 2.2 $MBCR_{T(LR)} =$ 1.4

1 Values are taken from Carman and Alston (Chapter 2, this volume), Table 2.3.

2 $ABCR_I$ ($MBCR_T$) refers to the average (marginal) benefit-cost ratio with costs measured using producer incidence (total expenditure).

Table 17.1 (continued)

Characteristics	Almond Board of California (Chapter 8)	Walnut Marketing Board (Chapter 9)	Raisin Administrative Committee (Chapter 10)	California Strawberry Commission (Chapter 11)
Type of Organization	Federal marketing order.	Federal marketing order.	Federal marketing order.	State commodity commission.
Commodity Value	$1,419 million in 2003.	$305 million in 2002.	$435 million in 2002.	$1,100 million in 2003.
Study	Crespi & Sexton	Kaiser	Kaiser, Liu & Consignado	Carter et al.
Activities Evaluated	Promotion in U.S. market.	Domestic Promotion, PR, Education Research.	Export Promotion.	Advertising.
Prog. Expenditure 2003/04[1]	$14 million	$2.9 million	$4.6 million	$4.84 million
Spending on Activities Evaluated	$5.7 million in 1997.	$2.9 million in 2003/04.	$1.8 million in 1998.	~$4 million in 2003.
Period of Econometric Estimation	1961–1997	1980–1999	1965–1998	1990–2002, weekly
P- or Q-Dependent Demand	Q-dependent	Q-dependent	Q-dependent	P-dependent
Elasticity (*E*) or Flexibility (*F*) of Demand Response to Price (*P*) or Promotion (*A*)	$E_P = -0.7$ $E_A = 0.13$	$E_P = -0.08$ $E_A = 0.005$	Japan $E_P = -1.12$ UK $E_P = -1.28$ Japan $E_A = 0.029$ UK $E_A = 0.133$	$F_P = -0.65$ $F_A = 0.16$
Period of Simulation	1990–1997	1984–1999	Japan 1965–1998 UK 1985–1998	2003
MBCR or ABCR	MBCR	ABCR	MBCR, ABCR	MBCR
Total Cost (T) or Incidence (I)	T	T	T	I
Estimated Benefit-Cost Ratio[2]	$MBCR_T = 6.2$	$ABCR_T =$ 1.65–9.72	$ABCR_T =$ Japan 5.1 UK 15.3 Total 7.3	$MBCR_I = 44.0$[3]

[3] $MBCR_I$ for CSC advertising was computed by the editors using the approximation $MBCR_I = F_A(V / A)$, the flexibility of demand with respect to advertising multiplied by the value of sales and divided by the advertising expenditure.

Table 17.2. Four Case Studies of Nonpromotion Programs for California Commodities

Characteristics	Dairy Council of California (Chapter 13)	California Strawberry Commission (Chapter 14)	California Dairy Research Foundation (Chapter 15)	Pistachio Marketing Board (Chapter 16)
Type of Organization	Council.	State commodity commission.	Foundation funded by check-off programs.	Federal marketing order.
Commodity Value	\$4,032 million in 2003.	\$1,100 million in 2003.	\$4,032 million in 2003.	\$333 million in 2002.
Study	Alston, Chalfant & James	Richards & Patterson	Balagtas & Sumner	Alston et al.
Activities	Nutrition education, the *EYO* program.	Public relations related to food scare affecting strawberries.	New-uses research for whey.	Mandated inspection and quality assurance for pistachios.
Total Program Expenditure in 2003/04[1]	\$4.59 million	\$4.84 million	NA	NA
Expenditure on Activities Evaluated	\$0.15 million in 1996	\$0.25 million	NA	NA
Methods Used	Monthly simulation over one year based on analysis of survey data.	Dynamic monthly demand system for fresh fruits, including measure of + and − media.	Simulation using dairy market model and estimates of whey product adoption.	Simulation using stochastic, dynamic model of market for California pistachios.
Period of Evaluation	1996	1996–1997	1990–2005	2004–2054
MBCR or ABCR	ABCR	ABCR	ABCR	ABCR
Total Cost (T) or Incidence (I)	T	T	T	I
Estimated Benefit-Cost Ratio	$ABCR_T$ Low = 0.13 High = 1.74	$ABCR_T = 454$	$ABCR_T$ Low = 40 High = 79	$ABCR = 13.5$

[1] Taken from Carman and Alston (Chapter 2, this volume), Table 2.3.

selection rather than a comprehensive report of the literature. This table includes just the benefit-cost ratios from the studies, divided into marginal versus average measures and then arranged first chronologically and then alphabetically within years. "Median" refers to the reported benefit-cost ratio for studies that reported just one estimate and to the midpoint of the range for studies that reported multiple estimates (reflecting different parameter values, modeling assumptions, types of expenditures, or differences in other features of the analysis). Looking at the median values in Table 17.3, MBCRs range from 1.0 to 44.5 and ABCRs range from 1.0 to 29.1. The bulk of the estimates are in single digits but well over 1.0, indicating that promotion programs have been profitable and, in the case of marginal measures, that they could have been profitably expanded. Even when we include the high and low values from Table 17.3, rather than just the midpoints, the overall range is from 1.0 to 30.9, which is much narrower than the range for the California case studies shown in Tables 17.1 and 17.2 but encompasses the range of California case studies of promotion except for the case of table grapes.

Conclusion

Mandated marketing programs have been important in California since the middle of the twentieth century but have been particularly so during the past twenty-five years. These programs have undertaken a variety of activities, but much of the growth in recent years, and the lion's share of the expenditures across all of the programs combined, has been in generic promotion. It is not surprising, then, that the focus of litigation and of the majority of the economic studies has been on the promotion programs.

The studies reported here used a variety of approaches reflecting various factors, including the state of the art at the time of the study, the purpose for which the study was undertaken, and the resources available for the work. All of the studies reported in detail in this book estimated a benefit-cost ratio (or the ingredients for one to be computed)–either the MBCR or the ABCR or both. An important dimension of differences among the studies is in the concepts of benefits and costs that they measured and used to compute the ratios. Some studies used a measure of changes in producer gross revenue as a measure of benefits, whereas others used a measure of producer surplus (or profit). Producer surplus is more appropriate generally; gross revenue is a good proxy if costs are approximately constant, as is reasonable to assume for perennial crops in the short run. On the cost side, the main difference was between studies that estimated the producer incidence of assessments used to

Table 17.3. Estimated Benefit-Cost Ratios for Advertising and Promotion for Various Commodities

Commodity and Program	Study[1]	Benefit-Cost Ratio[2] Low	Median	High
Studies Reporting Marginal Benefit-Cost Ratios			*MBCR*	
Australian Wool - Promotion in U.S. market	Dewbre et al. (1987)	-	2.0	-
Almonds - Export promotion (Taiwan, Japan, Hong Kong)	Halliburton & Henneberry (1995)	3.7	4.8	5.9
California Tree Fruit - Promotion	Freed et al. (1998)	0.7	2.4	4.1
Dairy - U.S. advertising	Kaiser (1997)	1.8	3.3	6.7
Oregon Hazelnuts - Promotion	Gopinath (2000)	-	26.0	-
Idaho-Oregon Onions - Promotion	Gopinath & Cornelius (2000)	1.0	1.0	1.1
U.S. Eggs - Advertising	Kaiser & Schmit (2000)	1.8	4.3	6.7
Pecans - Export promotion (Asia and EU)	Onunkwo & Epperson (2000)	6.5	6.7	6.8
Vidalia Onions - Promotion	Costa et al (2001)	-	22.5	-
Cotton - U.S. promotion	Crawford et al. (2001)	-	44.5	-
Pears - U.S. promotion	Erikson et al. (2001)	3.4	10.4	18.4
Studies Reporting Average Benefit-Cost Ratios			*ABCR*	
Canadian Butter - Advertising	Goddard & Amuah (1989)	-	1.0	-
Washington Apples - Advertising	Ward & Forker (1991)	-	7.0	-
California Walnuts - Export promotion	Weiss et al. (1996)	-	6.0	-
Cotton - U.S. promotion	Nichols et al. (1997)	3.2	3.4	3.5
U.S. Red Meat - Export promotion, Pacific Rim excl. Japan	Le, Kaiser & Tomek (1998)	-	15.6	-
U.S. Soybeans - Export promotion, production research	Williams et al. (1998)	-	8.3	-
U.S. Beef - Advertising	Ward (1998)	4.9	5.8	6.7
U.S. Pork - Advertising	Davis et al. (2001)	4.8	15.5	26.2
Hawaiian Papayas - Promotion	Ferguson et al. (2001)	10.1	21.1	31.2
Florida Tomatoes - Promotion	Van Sickle & Evans (2001)	27.2	29.1	30.9
Florida Orange Juice - Advertising	Capps et al. (2003)	2.9	4.5	6.1

[1] Studies are listed separately according to whether they reported marginal versus average benefit-cost ratios and then first in chronological and then in alphabetical order within years.
[2] Some studies reported only one benefit-cost ratio while others reported a range. Here we include the high and low values when a range was reported as well as the midpoint of the range. For studies reporting only one value, we listed the value under Median as well.

finance the program (i.e., they allowed for some of the costs of assessments to be passed on to consumers) and studies that used the total cost of the relevant activities as the measure of costs (and we note that some studies used the total cost of all program expenditures on all activities as a proxy for expenditures on specific activities).

The ideal measure of the benefit-cost ratio for evaluating whether a program is profitable for growers is not controversial. It is the ABCR, in which "benefit" refers to the additional producer surplus (or profit) given by the program activities being evaluated relative to producer surplus without such activities, and "cost" refers to the additional expense to producers from the assessment used to finance the activities in question (i.e., the producer incidence of that element of the assessment). Some of the studies reported here used that ideal measure. When studies used alternatives, almost always the implication was to reduce the ratio. Sometimes authors have justified this practice by saying that the estimate was therefore "conservative," but this practice also entails a systematic bias in some of the evidence in the literature and in how it is read and understood, since the details of the interpretation are rarely made clear or emphasized.[10] The practice also adds to the difficulty of comparing results among studies. The resulting systematic downward bias may have reduced the range of estimates in the literature in general, as well as reduced the ratios reported in some of our case studies. Nevertheless, the evidence in our case studies is very strongly supportive of the view that promotion programs have been profitable for producers. The benefit-cost ratios are generally considerably greater than 1.0 and generally in the range of published estimates for commodity promotion programs; nonpromotion programs were even more profitable.

Even when studies use the same concepts for benefits and costs, there can be problems of comparability. Results can differ among studies because market conditions or assumptions vary such that the true benefit-cost ratios differ or because the studies used different procedures and made different errors. It is important to recognize that errors are inescapable. We can never "know" the true values for benefit-cost ratios or the parameters on which estimates of the ratios should be based. At best, perhaps, we can seek to obtain measures that are statistically unbiased and reasonably precise. The studies reported here differed in how they approached the issues of bias and precision. Some reported just a single estimate of a benefit-cost ratio; some reported a range of benefit-cost ratios reflecting different parameter values, and some went as far as using Monte Carlo experiments to generate probability distributions of benefit-cost ratios; some reported different benefit-cost ratios for different types of expenditures, for different lengths of run of evaluation, or for different concepts of benefits and costs; some computed confidence intervals for upper and lower bounds on the benefit-cost ratios; others reported estimates using a variety of techniques or data sources.

The studies have tended to rely on statistical analysis of aggregate time-series data. This kind of approach has the drawback that the analysis is inherently dated—the evidence is explicitly about the past—which limits its direct applicability to questions about the future. In addition, it is challenging econometrically to isolate and precisely measure the effects of commodity-program expenditures versus other factors using typically available types of data. Stronger reinforcement for the findings might be obtained by using different kinds of data, such as cross-sectional data (e.g., for particular states, cities, or other submarkets) or data from controlled experiments rather than from further work with similar aggregate time series; but such data are not often available.

These caveats notwithstanding, the overwhelming conclusion from the studies reported and reviewed in this book is that mandated commodity marketing programs have been very profitable for California's agricultural producers. In every case, the evidence suggests that one can be reasonably confident that the benefits have well exceeded the costs and that it would have been profitable for producers to have increased expenditures on the programs.

In spite of this evidence, disputes over the future of the programs continue. Some producers and handlers oppose the programs either on ideological grounds or because they believe that they do not benefit from them. The ideological and legal aspects are beyond the scope of economic analysis. Economists can contribute to the debate by providing information about the economic consequences of the programs. The information provided by our work can be enhanced by investing in making estimates of aggregate benefit-cost ratios more meaningful (by measuring appropriate concepts and reporting them clearly and consistently), more precise (by obtaining more and better data and by using the best methods), and more credible (by increasing professional scrutiny and criticism of the estimates). These comments relate primarily to measures of overall program impact. Some issues may depend not just on whether programs pay but for whom they pay. In addition to more and better information about the aggregate impacts of the programs, it may be appropriate to invest more in measuring distribution of impacts—but this is even more difficult to do and results are more likely to be controversial.

Notes

1 As noted, benefit-cost studies of major milk-advertising campaigns are not available. Chapter 13 evaluates a nutrition education program conducted by the Dairy Council of California and Chapter 15 evaluates new-uses research funded by dairy check-offs.

2 In addition to benefit-cost ratios, the egg study (Chapter 5) reported an internal rate of return (IRR) to advertising and the whey study (Chapter 15) reported an IRR to new-uses research. The procedure is similar to calculating an MBCR or ABCR; however, here the streams of benefits and costs over time are discounted to solve for the discount rate for which the benefit-cost ratio would be 1.0. A program is considered profitable if its IRR exceeds the opportunity cost of the funds invested.

3 Kinnucan and Zheng (Chapter 12) point out that some studies have used total assessments rather than total program expenditure and charged all overhead, administration, and all other programs to the particular element being evaluated.

4 Kinnucan and Zheng (Chapter 12) made this point.

5 Alston et al. (1998) derived the approximation formula and applied it to California dried plums. The formula is presented in Chapter 6, which summarizes that study.

6 In 2002, faced with legal challenges, a referendum was held and 89 percent of those firms voting chose to permanently suspend operations.

7 This benefit-cost ratio is based on a marginal increase in total expenditures rather than the producer incidence of an additional assessment used to finance that marginal increase in expenditure.

8 These benefit-cost ratios were all based on total expenditures rather than on the producer incidence of the assessment used to finance that expenditure.

9 The California Milk Advisory Board (CMAB) and Dairy Management, Inc. (DMI) are each funded by a national marketing assessment, a check-off, of $0.15 per one hundred pounds of milk sold.

10 Such "conservatism" might be less popular if the estimates were smaller, at which point a downward bias in the ratio could lead to a ratio less than 1.0. In their meta-analysis of the literature on returns to agricultural R&D, Alston et al. (2000) discussed the tendency for studies to be biased toward obtaining estimates that are "acceptable" to the audience, and dimensions of acceptability include whether the finding is favorable (a benefit-cost ratio significantly greater than 1.0) and credible (not too much greater than 1.0).

References

Alston, J.M., C. Chan-Kang, M.C. Marra, P.G. Pardey, and T.J. Wyatt. *A Meta-Analysis of the Rates of Return to Agricultural R&D: Ex Pede Herculem*. Washington DC: International Food Policy Research Institute Research Report 113, 2000.

Capps, O., Jr., D.A. Bessler, and G.W. Williams. "Evaluating the Economic Impacts Associated with the Advertising Effort by the Florida Department of Citrus: Final Report." Unpublished, Forecasting and Business Analytics, L.L.C., College Station TX, 2003.

Costa, E.F., J.E. Epperson, C.L. Huang, J.C. McKissick, B. Thomas, and K. Wolf. *Impacts of Advertising and Promotions on the Demand for Scanned Purchases of Vidalia Onions*. Center for Agribusiness and Economic Development CR-01-42, University of Georgia, 2001.

Crawford, W., B.C. Murray, R.H. Beach, W.J. White, C. Viator, N. Piggott, and M. Wohlgenant. *An Economic Analysis of the Cotton Research and Promotion Programs*. Research Triangle Institute Project 8024, Final Report, September 2001.

Crespi, J.M., and R.J. Sexton. "Have Expenditures to Advertise Almonds Been Effective?" *NICPRE Bulletin* 3 (2000):1–4.

Davis, G.C., O. Capps, Jr., D.A. Bessler, J.H. Leigh, J.P. Nichols, and E. Goddard. *An Economic Evaluation of the Pork Checkoff Program*. College Station TX: Texas A&M University, Department of Agricultural Economics Technical Report 01-1, January 2001.

Dewbre, J., B. Richardson, and S. Beare. "Returns from Wool Promotion in the United States." Australian Government Publishing Service Occasional Paper 100, Canberra Australia, 1987.

Erikson, G., R.C. Mittelhammer, and R.T. Schotzko. "An Evaluation of the Effectiveness of Pear Advertising and Promotion: Final Report." Unpublished, Washington State University, 2001.

Ferguson, C.A., S.T. Nakamoto, and N. Sawada. "An Economic Analysis of the Hawaii Papaya Administrative Committee's Marketing Program." Honolulu HI: College of Tropical Agriculture and Human Resources, University of Hawaii at Manoa, 2001.

Freed/Landry & Associates. "The Impact of Various Factors on Grower Price and Estimates of Value Contributed by the Marketing Program of California Tree Fruit Agreement." Mimeo. April 1998, 30 pages.

Goddard, E.W., and A.K. Amuah. "Demand for Canadian Fats and Oils: A Case Study of Advertising Effectiveness." *American Journal of Agricultural Economics* 71 (1989):741–749.

Gopinath, M. "Evaluation of Promotion Programs of the Hazelnut Marketing Board." Mimeo. Department of Agricultural and Resource Economics, Oregon State University, June 2000.

Gopinath, M., and J. Cornelius. "Evaluation of Promotion Programs of the Idaho-Eastern Oregon Onion Committee." Mimeo. Department of Agricultural and Resource Economics, Oregon State University, June 2000.

Halliburton, K., and S.R. Henneberry. "The Effectiveness of U.S. Nonprice Promotion of Almonds in the Pacific Rim." *Journal of Agricultural and Resource Economics* 20 (1995):108–121.

Kaiser, H.M. "Impact of National Dairy Advertising on Dairy Markets, 1984–95." *Journal of Agricultural and Applied Economics* 29 (1997):303–314.

——. "The Domestic Impacts of the Walnut Marketing Board's Marketing Activities." *NICPRE Quarterly* 1 (2002):1–4.

Kaiser, H.M., and T.M. Schmit. "An Economic Evaluation of Generic Egg Advertising by the American Egg Board." Unpublished, Department of Applied Economics and Management, Cornell University, 2000.

Le, C.T., H.M. Kaiser, and W.G. Tomek. "Export Promotion and Import Demand for U.S. Red Meat in Selected Pacific Rim Countries." *Agribusiness: An International Journal* 14 (1998): 95–106.

Nichols, J.P., O. Capps, Jr., G.C. Davis, and D.A. Bessler. "Evaluating Returns to the Cotton Checkoff Program." *NICPRE Quarterly* 2 (1997):1–4.

Onunkwo, I.M., and J.E. Epperson. "Export Demand for U.S. Pecans: Impacts of U.S. Export Promotion Programs." *Agribusiness: An International Journal* 16 (2000):253–266.

Van Sickle, J.J., and E. Evans. "The Florida Tomato Committee's Education and Promotion Program: An Evaluation." Mimeo. Food and Resource Economics Department, University of Florida, December 2001.

Ward, R.W. "Evaluating the Beef Promotion Checkoff." *NICPRE Quarterly* 4 (1998):1–4.

Ward, R.W., and O.D. Forker. "Washington Apple Advertising: An Economic Model of Its Impact." Gainseville FL: Washington Apple Commission WAC 91:1, University of Florida, Gainesville, 1991.

Weiss, K.R., R.D. Green, and A.M. Havenner. "Walnuts in Japan: A Case Study of Generic Promotion under the USDA's Market Promotion Program." *Agricultural Commodity Promotion Policies and Programs in the Global Agri-Food System, Proceedings of NEC-63 Conference.* Ithaca NY: National Institute for Commodity Promotion Research and Evaluation (NICPRE), Cornell University, 1996.

Williams, G.W., C.R. Shumway, H.A. Love, and J.B. Ward. "Effectiveness of the Soybean Checkoff Program." Department of Agricultural Economics, Texas Agricultural Market Research Center Report, Texas A&M University, May 1998.

Contributors

Julian M. Alston is a professor in the Department of Agricultural and Resource Economics at the University of California, Davis.

Joseph V. Balagtas is an assistant professor in the Department of Agricultural Economics at Purdue University.

Henrich Brunke is a postgraduate research economist at the University of California Agricultural Issues Center.

Hoy F. Carman is a professor in the Department of Agricultural and Resource Economics at the University of California, Davis.

Colin A. Carter is a professor in the Department of Agricultural and Resource Economics at the University of California, Davis.

James A. Chalfant is a professor and chair of the Department of Agricultural and Resource Economics at the University of California, Davis.

Jason E. Christian was a research associate in the Department of Agricultural Economics at the University of California, Davis.

Ted Consignado is an economist for the California Raisin Administrative Committee.

R. Kim Craft is an associate professor of managerial economics in the School of Business at Southern Utah University.

John M. Crespi is an assistant professor in the Department of Agricultural Economics at Kansas State University.

Rachael E. Goodhue is an associate professor in the Department of Agricultural and Resource Economics at the University of California, Davis.

Richard S. Gray is a professor and head of the Department of Agricultural Economics at the University of Saskatchewan.

Jennifer S. James is an assistant professor in the Department of Agricultural Economics and Rural Sociology at Pennsylvania State University.

Harry M. Kaiser is a professor in the Department of Applied Economics and Management at Cornell University.

Henry W. Kinnucan is a professor in the Department of Agricultural Economics and Rural Sociology at Auburn University.

Donald J. Liu is an associate professor in the Department of Applied Economics at the University of Minnesota.

Erika Meng is an economist in the Impacts Targeting and Assessment Program at the International Maize and Wheat Improvement Center (CIMMYT).

Paul M. Patterson is an associate professor in the Morrison School of Agribusiness and Resource Management at Arizona State University.

Nicholas E. Piggott is an associate professor in the Department of Agricultural Economics and Resource Economics at North Carolina State University.

J. Carlos Reberte is an economist with American Express, Inc.

Timothy J. Richards is the power associate professor in the Morrison School of Agribusiness and Resource Management, Arizona State University.

Todd M. Schmit is a research associate in the Department of Applied Economics and Management at Cornell University.

Richard J. Sexton is a professor in the Department of Agricultural and Resource Economics at the University of California, Davis.

Daniel A. Sumner is the Frank H. Buck, Jr. professor in the Department of Agricultural and Resource Economics at the University of California, Davis, and director of the University of California Agricultural Issues Center.

Yuqing Zheng is a graduate research assistant in the Department of Agricultural Economics and Rural Sociology at Auburn University.

Index

S

T

U